Lecture Notes in Computer Science   9270

Commenced Publication in 1973
Founding and Former Series Editors:
Gerhard Goos, Juris Hartmanis, and Jan van Leeuwen

More information about this series at http://www.springer.com/series/7407

Andreas Maletti (Ed.)

Algebraic Informatics

6th International Conference, CAI 2015
Stuttgart, Germany, September 1–4, 2015
Proceedings

 Springer

Editor
Andreas Maletti
Institut für Maschinelle Sprachverarbeitung
University of Stuttgart
Stuttgart
Germany

ISSN 0302-9743 ISSN 1611-3349 (electronic)
Lecture Notes in Computer Science
ISBN 978-3-319-23020-7 ISBN 978-3-319-23021-4 (eBook)
DOI 10.1007/978-3-319-23021-4

Library of Congress Control Number: 2015947117

LNCS Sublibrary: SL1 – Theoretical Computer Science and General Issues

Springer Cham Heidelberg New York Dordrecht London

Printed on acid-free paper

Springer International Publishing AG Switzerland is part of Springer Science+Business Media
(www.springer.com)

Preface

This volume contains the papers presented at the 6th International Conference on Algebraic Informatics (CAI 2015), which was held September 1–4, 2015, in Stuttgart, Germany.

The conference covers a wide range of algebraic topics relevant to computer science ranging from weighted logics and automata to cryptography. This volume contains the 15 regular papers that the Program Committee decided to accept together with the contributions from four keynote speakers. In total, 25 submissions were reviewed by at least three Program Committee members, i.e., 60% of the submissions. The Program Committee consisted of 18 leading international researchers covering the wide spectrum of topics represented by the conference.

CAI 2015 was the sixth conference in the series, which was started by Symeon Bozapalidis (Greece) in 2005. Besides CAI 2005, the second and third conferences (2007 and 2009) were held in Thessaloniki (Greece) under the auspices of Symeon Bozapalidis and George Rahonis. Franz Winkler organized the fourth conference in Linz (Austria) and CAI 2013 was held on Porquerolles Island (France) organized by Traian Muntean, Dimitrios Poulakis, and Robert Rolland. The sixth entry in the series was graciously awarded to Stuttgart by the international Steering Committee.

First and foremost, I would like to thank the Steering Committee for awarding CAI 2015 to Stuttgart. The Program Committee assisted by the external reviewers worked hard to assess the quality of the submitted papers. My wholehearted thanks to them, and I fully appreciate their commitment to the conference. In addition, I want to express my deepest gratitude to the keynote speakers Volker Diekert, Jarkko Kari, Werner Kuich, and Mehryar Mohri for sharing their insights and presenting their contributions at CAI 2015. The local organization team consisting of Fabienne Braune, Sybille Laderer, Daniel Quernheim, and Nina Seemann worked tirelessly to create an enjoyable meeting experience, and I could not have organized the conference without them. I applaud their efforts. EasyChair, true to their name, made the paper collection, reviewing, and publication process as easy as possible and the team at Springer made sure that you can hold these proceedings in your hands today. Last, but not least, I want to thank the German Research Foundation grant MA 4959/1-1, which provided financial assistance to the conference.

June 2015

Andreas Maletti

Organization

Program Committee

Symeon Bozapalidis	Aristotle University of Thessaloniki, Greece
Fabienne Braune	University of Stuttgart, Germany
Bruno Courcelle	University of Bordeaux, France
Frank Drewes	Umeå University, Sweden
Manfred Droste	Universität Leipzig, Germany
Tero Harju	University of Turku, Finland
Gregory Kucherov	University of Paris-Est, Marne-la-Vallée, France
Sybille Laderer	University of Stuttgart, Germany
Andreas Maletti	University of Stuttgart, Germany
Traian Muntean	Aix-Marseille Université, France
Alexander Okhotin	University of Turku, Finland
Friedrich Otto	Universität Kassel, Germany
Jean Eric Pin	CNRS and Université Paris 7, France
Daniel Quernheim	University of Stuttgart, Germany
George Rahonis	Aristotle University of Thessaloniki, Greece
Robert Rolland	Aix-Marseille Université, France
Kai Salomaa	Queen's University, Kingston, ON, Canada
Nina Seemann	University of Stuttgart, Germany
Heiko Vogler	Technische Universität Dresden, Germany
Mikhail Volkov	Ural Federal University, Yekaterinburg, Russia
Franz Winkler	J. Kepler Universität, Linz, Austria
Zoltán Ésik	University of Szeged, Hungary

Additional Reviewers

Ananichev, Dmitry	Naehrig, Michael
Augot, Daniel	Noll, Thomas
Carayol, Arnaud	Paperman, Charles
Cho, Dajung	Poulakis, Dimitrios
Choudhury, Salimur	Rück, Hans-Georg
Heller, Pavel	Schneider, Martin
Ionica, Sorina	Schröder, Lutz
Iván, Szabolcs	Staton, Sam
Kuske, Dietrich	Strüngmann, Lutz
Lugiez, Denis	Tarannikov, Yuriy
Mandrali, Eleni	Van Der Merwe, Brink
Messerschmidt, Hartmut	Verma, Rakesh

Contents

Learning Weighted Automata

Borja Balle[1] and Mehryar Mohri[2,3](✉)

[1] School of Computer Science, McGill University, Montréal, Canada
[2] Courant Institute of Mathematical Sciences, New York, NY, USA
`mohri@cs.nyu.edu`
[3] Google Research, New York, NY, USA

1 Introduction

Weighted finite automata (WFA) are finite automata whose transitions and states are augmented with some weights, elements of a semiring. A WFA induces a function over strings. The value it assigns to an input string is the semiring sum of the weights of all paths labeled with that string, where the weight of a path is obtained by taking the semiring product of the weights of its constituent transitions, as well as those of its origin and destination states. The mathematical theory behind WFAs, that of *rational power series*, has been extensively studied in the past [26,52,40,16] and has been more recently the topic of a dedicated handbook [23]. WFAs are widely used in modern applications, perhaps most prominently in image processing and speech recognition where the terminology of *weighted automata* seems to have been first introduced and made popular [34,43,49,41,44], in several other speech processing applications such as speech synthesis [55,1], in phonological and morphological rule compilation [35,36,47], in parsing [45], bioinformatics [25,2], sequence modeling and prediction [22], formal verification [3], in optical character recognition [18], and in many other areas.

These applications, as well as a number of theoretical questions, have strongly motivated the problem of *learning WFAs*, that is that of finding a WFA closely estimating a semiring-valued target function, using for training a finite sample of strings labeled with their target values. This problem has a rich history since its simpler instances date back to the origins of computer science. We will therefore discuss only briefly some of the key results of the literature.

A special instance of this problem is that of learning (unweighted) finite automata, which coincide with WFAs defined over the Boolean semiring. A series of negative results are known for this problem when the target itself is a finite automaton and when the complexity criterion used is the size of the automaton learned. In particular, the problem of finding a consistent deterministic finite automaton (DFA) of minimum size was shown to be NP-hard by Gold [29]. This result was later extended by Angluin [4]. Pitt and Warmuth [50] further strengthened these results by showing that even an approximation within a polynomial function of the size of the smallest consistent automaton is NP-hard. Their hardness results apply also to the case where prediction is made using non-deterministic finite automata (NFA) (see also [21]). Kearns and Valiant [37] presented for the same problem hardness results of a different nature relying on

© Springer International Publishing Switzerland 2015
A. Maletti (Ed.): CAI 2015, LNCS 9270, pp. 1–21, 2015.
DOI: 10.1007/978-3-319-23021-4_1

cryptographic assumptions. Their results imply that no polynomial-time algorithm can learn consistent NFAs polynomial in the size of the smallest DFA from a finite sample of accepted and rejected strings if any of the generally accepted cryptographic assumptions holds, for example if RSA public key cryptosystem is secure.

These results imply the computational intractability of the general problem of *passively* learning finite automata for several learning models, including the mistake bound model of Haussler et al. [31] or the PAC-learning model of Valiant [56]. In contrast, an *active* model of learning automata was introduced by Angluin [4,5], where the learner can make membership and equivalence queries. For this model, it was shown that finite automata can be learned in time polynomial in the size of the minimal automaton and that of the longest counterexample [4] (see also [38] and [46]).

Fewer results have been reported in the literature for the general case of learning WFAs over a non-Boolean semiring. Bergadano et al. [15] extended the positive result of [4] in the scenario where membership and equivalence queries can be made, to the problem of learning WFAs defined over any field. Using the relationship between the size of a minimal weighted automaton over a field and the rank of the corresponding Hankel matrix, the learnability of many other concepts classes such as disjoint DNF can be shown [13]. In the passive setting, the problem of learning a *probabilistic* WFA using a finite sample drawn according to the same distribution has been the subject of a series of publications in recent years using a *spectral method,* starting with the work of Hsu et al. [32] for learning hidden Markov models (HMMs). The main technique used in these publications consists of a singular value decomposition (SVD) of a Hankel matrix. Balle and Mohri [11] further showed that spectral methods combined with a constrained matrix completion algorithm can be used to learn arbitrary WFAs (not necessarily probabilistic) from finite samples drawn according to a distribution unrelated to the target WFA.

This paper surveys a number of key theoretical results and algorithms for learning WFAs. In Section 2, we introduce the main definitions and notation used throughout the paper. The notion of Hankel matrix turns out to play a key role in the definition of several learning algorithms for WFAs. In Section 3, we discuss several important properties of Hankel matrices and their use in the reconstruction of WFAs. In Section 4, we use these results to describe three algorithms for learning WFAs, as well as their theoretical guarantees.

2   Definitions and Properties

In this section, we briefly introduce some basic notions and notation related to semirings and weighted automata needed for the discussion in the following sections.

2.1   Semirings

A weighted finite automaton (WFA) $\mathcal{A}$ is a finite automaton whose transitions and states carry some weights. For various operations to be well defined, the weights must belong to a *semiring*, that is a ring that may lack negation. More formally, $(\mathbb{S}, \oplus, \otimes, \overline{0}, \overline{1})$ is a semiring if $(\mathbb{S}, \oplus, \overline{0})$ is a commutative monoid with identity element $\overline{0}$, $(\mathbb{S}, \otimes, \overline{1})$ is a monoid with identity element $\overline{1}$, $\otimes$ distributes over $\oplus$, and $\overline{0}$ is an annihilator for $\otimes$, that is $a \otimes \overline{0} = \overline{0} \otimes a = \overline{0}$ for all $a \in \mathbb{S}$.

As an example, $(\mathbb{R}_+ \cup \{+\infty\}, +, \times, 0, 1)$ is a semiring called the *probability semiring*. The semiring isomorphic to the probability semiring via the negative log is the system $(\mathbb{R} \cup \{-\infty, +\infty\}, \oplus_{\log}, +, +\infty, 0)$, where $\oplus_{\log}$ is defined by $x \oplus_{\log} y = -\log(e^{-x} + e^{-y})$; it is called the log *semiring*. The semiring derived from the log semiring via the Viterbi approximation is the system $(\mathbb{R} \cup \{-\infty, +\infty\}, \min, +, +\infty, 0)$ and is called the *tropical semiring*. It is the familiar semiring of shortest-paths algorithms.

A semiring is said to be *commutative* when the multiplicative operation $\otimes$ is commutative. It is said to be *idempotent* if $x \oplus x = x$ for all $x \in \mathbb{S}$. The Boolean semiring and the tropical semiring are idempotent.

2.2   Weighted Automata

Given an alphabet Σ, we will denote by $|x|$ the length of a string $x \in \Sigma^*$ and by ϵ the *empty string* for which $|\epsilon| = 0$.

The second operation of a semiring is used to compute the weight of a path by taking the $\otimes$-product of the weights of its constituent transitions. The first operation is used to compute the weight of any string x, by taking the $\oplus$-sum of the weights of all paths labeled with x.

For a WFA $\mathcal{A}$ defined over a semiring $(\mathbb{S}, \oplus, \otimes, \overline{0}, \overline{1})$, we denote by $Q_{\mathcal{A}}$ its finite set of states and by $E_{\mathcal{A}}$ its finite set of transitions, which are elements of $Q_{\mathcal{A}} \times \Sigma \times \mathbb{S} \times Q_{\mathcal{A}}$.[1] We will also denote by $\alpha_{\mathcal{A}} \in \mathbb{S}^{Q_{\mathcal{A}}}$ the vector of initial weights, by $\beta_{\mathcal{A}} \in \mathbb{S}^{Q_{\mathcal{A}}}$ the vector of final weights, and by $w_{\mathcal{A}}[e] \in \mathbb{S}$ the weight of a transition $e \in E_{\mathcal{A}}$. More generally, we denote by $w_{\mathcal{A}}[\pi]$ the weight of a path $\pi = e_1 \cdots e_n$ of $\mathcal{A}$ which is defined by the $\otimes$-product of the transitions weights: $w_{\mathcal{A}}[\pi] = w_{\mathcal{A}}[e_1] \otimes \cdots \otimes w_{\mathcal{A}}[e_n]$. For any path π, we also denote by $\text{orig}[\pi]$ its origin state and by $\text{dest}[\pi]$ its destination state.

It is sometimes convenient to define the set of *initial states* $I_{\mathcal{A}} = \{q \in Q_{\mathcal{A}} : \alpha_{\mathcal{A}}[q] \neq \overline{0}\}$ and similarly the set of *final states* $F_{\mathcal{A}} = \{q \in Q_{\mathcal{A}} : \beta_{\mathcal{A}}[q] \neq \overline{0}\}$. A path from $I_{\mathcal{A}}$ to $F_{\mathcal{A}}$ is then said to be an *accepting path*.

A WFA $\mathcal{A}$ over an alphabet Σ defines a function mapping the set of strings Σ^* to $\mathbb{S}$ that is abusively also denoted by $\mathcal{A}$ and defined as follows:

$$\forall x \in \Sigma^*, \quad \mathcal{A}(x) = \bigoplus_{\pi \in P_{\mathcal{A}}(x)} \left(\alpha_{\mathcal{A}}[\text{orig}[\pi]] \otimes w_{\mathcal{A}}[\pi] \otimes \beta_{\mathcal{A}}[\text{dest}[\pi]] \right),$$

---

[1] All of our results can be straightforwardly extended to the case where $E_{\mathcal{A}}$ is a multiset, thereby allowing multiple transitions between the same two states with the same labels and weights.

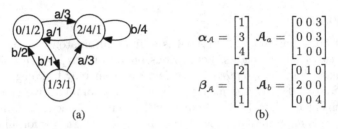

Fig. 1. (a) Example of WFA $\mathcal{A}$. Within each circle, the first number indicates the state number, the second after the slash separator the initial weight and the third the final weight. In particular, $\mathcal{A}(ab) = 1 \times 3 \times 4 \times 1 + 3 \times 3 \times 4 \times 1 + 4 \times 1 \times 1 \times 1$. (b) Corresponding initial vector $\alpha_{\mathcal{A}}$, final vector $\beta_{\mathcal{A}}$, and transition matrices $\mathcal{A}_a$ and $\mathcal{A}_b$.

where $P_{\mathcal{A}}(x)$ denotes the (finite) set of paths in $\mathcal{A}$ labeled with x. By convention, $\mathcal{A}(x) = \bar{0}$ when $P(x) = \emptyset$.

For any $a \in \Sigma$, let $\mathcal{A}_a \in \mathbb{S}^{Q_{\mathcal{A}} \times Q_{\mathcal{A}}}$ be the matrix $[\mathcal{A}_a]_{pq} = \oplus_{e \in P_{\mathcal{A}}(p,a,q)} w_{\mathcal{A}}[e]$, where $P_{\mathcal{A}}(p, a, q)$ is the set of transitions labeled with a from p to q. Then, (2.2) can be equivalently written as follows in terms of matrices with entries in $\mathbb{S}$:

$$\forall x = x_1 \cdots x_k \in \Sigma^*, \quad \mathcal{A}(x) = \alpha_{\mathcal{A}}^{\top} \mathcal{A}_{x_1} \cdots \mathcal{A}_{x_k} \beta_{\mathcal{A}}.$$

This is similar to the linear representation of recognizable formal power series [52,40,16]. Figure 1 illustrates these definitions with a specific example of WFA. The *size of a WFA* is denoted by $|\mathcal{A}|$ and defined as the sum of the number of states and the number of transitions of $\mathcal{A}$: $|\mathcal{A}| = |Q_{\mathcal{A}}| + |E_{\mathcal{A}}|$. In the absence of any ambiguity, we will drop all $\mathcal{A}$-subscripts in the definitions just presented.

3   Hankel Matrices and WFA Reconstruction Algorithms

A key algebraic tool used in the design of the learning algorithms we will present is the notion of *Hankel matrix*. Thus, in this section, we present an extensive analysis of Hankel matrices and their properties. We will show how sufficiently informative finite sub-blocks of the Hankel matrix of a WFA can be used to reconstruct a WFA.

From here on, we will assume that the semiring $\mathbb{S}$ is in fact a *field*. This enables us to define the rank of a matrix with entries in $\mathbb{S}$ and devise effective algorithms for solving linear systems with unknowns and coefficients in $\mathbb{S}$. We note, however, that some of the results stated in this section can be extended to rings.

3.1   Definitions

Let $\mathbf{H} \in \mathbb{S}^{\Sigma^* \times \Sigma^*}$ be an infinite matrix with rows and columns indexed by strings in Σ^*. We denote by $\mathbf{H}(u, v)$ its entry with row index $u \in \Sigma^*$ and column index $v \in \Sigma^*$. The following definitions are essential for the rest of the paper.

Definition 1 (Hankel matrix). *We will say that* $\mathbf{H}$ *is a* Hankel matrix *if* $\mathbf{H}(u,v) = \mathbf{H}(u',v')$ *for all* $u, u, v, v' \in \Sigma^*$ *such that* $uv = u'v'$ *and will denote by* rank($\mathbf{H}$) *the rank of* $\mathbf{H}$.

Definition 2 (Hankel matrix of a function). *The* Hankel matrix $\mathbf{H}_f$ *of a function* $f \colon \Sigma^* \to \mathbb{S}$ *(or formal series over* $\mathbb{S}$*) is the matrix defined by* $\mathbf{H}_f(u,v) = f(uv)$, *for all* $u, v \in \Sigma^*$. *Conversely, any Hankel matrix* $\mathbf{H}$ *defines a function* $f \colon \Sigma^* \to \mathbb{S}$ *by setting* $f(u) = \mathbf{H}(u, \epsilon)$ *for all* $u \in \Sigma^*$ *and thus* $\mathbf{H} = \mathbf{H}_f$.

3.2 Hankel Matrices of Rational Functions

A function $f \colon \Sigma^* \to \mathbb{S}$ is said to be *rational* when it can be represented by a WFA $\mathcal{A}$, that is when $f(x) = \mathcal{A}(x)$ for all $x \in \Sigma^*$ [52,40,16]. The following theorem of Fliess [28] (see also [20]) provides an important characterization of rational functions in terms of the finiteness of rank($\mathbf{H}_f$).

Theorem 1 (Fliess [28]). *Let* $\mathbb{S}$ *be a field. Then, the rank of the Hankel matrix* $\mathbf{H}_f$ *associated to a function* $f \colon \Sigma^* \to \mathbb{S}$ *is finite if and only if* f *is rational. In that case, there exists a WFA* $\mathcal{A}$ *representing* f *with* rank($\mathbf{H}_f$) *states and no WFA representing* f *admits fewer states.*

Thus, when rank($\mathbf{H}_f$) $< +\infty$, a WFA representing f with rank($\mathbf{H}_f$) states ($|Q_\mathcal{A}| = $ rank($\mathbf{H}_\mathcal{A}$)) is *minimal*. Note that this minimality is defined only in terms of the number of states, unlike the notion of minimal deterministic WFA [41,42]. In fact, such minimal WFAs often have a large number of transitions.

Proof. Suppose first that there exists a WFA $\mathcal{A}$ representing f. Then, for any $u, v \in \Sigma^*$, we can write

$$f(uv) = \mathcal{A}(uv) = (\boldsymbol{\alpha}_\mathcal{A}^\top \mathbf{A}_u)(\mathbf{A}_v \boldsymbol{\beta}_\mathcal{A}) \ . \tag{1}$$

Observe that $\boldsymbol{\alpha}_\mathcal{A}^\top \mathbf{A}_u$ is a row vector in $\mathbb{S}^{1 \times Q_\mathcal{A}}$ and $\mathbf{A}_v \boldsymbol{\beta}_\mathcal{A}$ a column vector in $\mathbb{S}^{Q_\mathcal{A} \times 1}$. Let $\mathbf{P}$ be the matrix in $\mathbb{S}^{\Sigma^* \times Q_\mathcal{A}}$ defined by $\mathbf{P}_\mathcal{A}(u, \cdot) = \boldsymbol{\alpha}_\mathcal{A}^\top \mathbf{A}_u$ for all $u \in \Sigma^*$ and $\mathbf{S}_\mathcal{A} \in \mathbb{S}^{\Sigma^* \times Q_\mathcal{A}}$ the matrix defined by $\mathbf{S}_\mathcal{A}(v, \cdot) = (\mathbf{A}_v \boldsymbol{\beta}_\mathcal{A})^\top$ for all $v \in \Sigma^*$. Then, in view of (1), for all $u, v \in \Sigma^*$,

$$f(uv) = (\boldsymbol{\alpha}_\mathcal{A}^\top \mathbf{A}_u)(\mathbf{A}_v \boldsymbol{\beta}_\mathcal{A}) = (\mathbf{P}_\mathcal{A} \mathbf{S}_\mathcal{A}^\top)(u, v) \ .$$

This proves that $\mathbf{H}_f = \mathbf{P}_\mathcal{A} \mathbf{S}_\mathcal{A}^\top$. Since $\mathbf{P}_\mathcal{A}$ and $\mathbf{S}_\mathcal{A}$ are in $\mathbb{S}^{\Sigma^* \times Q_\mathcal{A}}$, the rank of $\mathbf{H}_f$ is upper bounded by $|Q_\mathcal{A}|$, the number of states of $\mathcal{A}$, and is therefore finite.

Assume now that rank($\mathbf{H}_f$) $= n < +\infty$. For any $v \in \Sigma^*$, we denote by $\mathbf{H}_f(\cdot, v)$ the column of $\mathbf{H}_f$ indexed by v. Let $(\mathbf{H}_f(\cdot, v_1), \ldots, \mathbf{H}_f(\cdot, v_n))$ be a basis for all columns. Then, there exist $\beta_1, \ldots, \beta_n \in \mathbb{S}$ such that the column $\mathbf{H}_f(\cdot, \epsilon)$ can be expressed as $\mathbf{H}_f(\cdot, \epsilon) = \sum_{i=1}^n \beta_i \mathbf{H}_f(\cdot, v_i)$. Since for all $w \in \Sigma^*$, $f(w) = \mathbf{H}(\epsilon, w) = \mathbf{H}(w, \epsilon) = \sum_{i=1}^n \beta_i \mathbf{H}_f(w, v_i)$, this implies that $f = \sum_{i=1}^n \beta_i \mathbf{H}_f(\cdot, v_i)$. Now, for all $i \in [1, n]$ and $a \in \Sigma$, the column $\mathbf{H}_f(\cdot, av_i)$ can also be expressed in terms of the basis: there exist (γ_{ji}^a) such that $\mathbf{H}_f(\cdot, av_i) = \sum_{j=1}^n \gamma_{ji}^a \mathbf{H}_f(\cdot, v_j)$. Let $\mathbf{A}_a$ be the matrix defined by $(\mathbf{A}_a)_{ji} = (\gamma_{ji}^a)$. Then, we

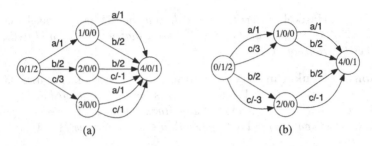

(a)   (b)

Fig. 2. Illustration of standardization. (a) WFA $\mathcal{A}$ defined over the field $(\mathbb{R}, +, \times, 0, 1)$. (b) WFA $\mathcal{B}$ obtained by standardization of $\mathcal{A}$. In this instance, the first stage of standardization leaves the WFA unchanged. In the second stage, state 3 is eliminated since it is a linear combination of states 1 and 2 in the following sense: let f_3 be the function defined by setting state 3 to be the only initial state with initial weight 1, and similarly with states 2 and 3, then, $f_3 = f_1 - f_2$.

can show by induction on the length of w that for all $w = a_1 \cdots a_k \in \Sigma^*$, $\mathbf{H}_f(\cdot, wv_i) = \sum_{j=1}^n (\mathbf{A}_w)_{ji} \mathbf{H}_f(\cdot, v_j)$, where $\mathbf{A}_w = \mathbf{A}_{a_1} \cdots \mathbf{A}_{a_k}$. Indeed, if the equality holds for w_1 and w_2, then for $w = w_1 w_2$ and for all $u \in \Sigma^*$ we have $\mathbf{H}_f(u, wv_i) = \mathbf{H}_f(uw_1, w_2 v_i)$ and:

$$
\mathbf{H}_f(uw_1, w_2 v_i) = \sum_{j=1}^n (\mathbf{A}_{w_2})_{ji} \mathbf{H}_f(uw_1, v_j) = \sum_{j=1}^n (\mathbf{A}_{w_2})_{ji} \mathbf{H}_f(u, w_1 v_j)
$$

$$
= \sum_{j=1}^n (\mathbf{A}_{w_2})_{ji} \sum_{k=1}^n (\mathbf{A}_{w_1})_{kj} \mathbf{H}_f(u, v_k) = \sum_{k=1}^n (\mathbf{A}_{w_1} \mathbf{A}_{w_2})_{ki} \mathbf{H}_f(u, v_k) \ .
$$

Thus, for any $w = a_1 \cdots a_k \in \Sigma^*$,

$$
f(w) = \sum_{i=1}^n \beta_i \mathbf{H}_f(\epsilon, wv_i) = \sum_{i=1}^n \beta_i \sum_{j=1}^n (\mathbf{A}_w)_{ji} \mathbf{H}_f(\epsilon, v_j) = \boldsymbol{\alpha}^\top \mathbf{A}_{a_1} \cdots \mathbf{A}_{a_k} \boldsymbol{\beta} \ ,
$$

where $\alpha_j = \mathbf{H}_f(\epsilon, v_j)$ and $\beta_j = \beta_j$ for all $j \in [1, n]$. This proves that f can be represented by a WFA with $n = \text{rank}(\mathbf{H}_f)$ states. $\square$

3.3 Standardization of WFAs

Theorem 1 proves the existence of a minimal WFA for the representation of a rational function f. In this section, we briefly describe an algorithm for computing a minimal WFA $\mathcal{B}$ from an input WFA $\mathcal{A}$ representing f. The first algorithm for this problem is due to Schützenberger [54] (see also [53]) and is known as a *standardization* of the representation of the linear representation of a rational power series. A more efficient version of this algorithm was later given by Cardon and Crochemore [19]. Here, we give a brief description of that algorithm.

The algorithm consists of first finding a basis $(\mathbf{v}_1, \ldots, \mathbf{v}_m)$ of row vectors in $\mathbb{S}^{1 \times Q_{\mathcal{A}}}$ for the vector space generated by $\{\boldsymbol{\alpha}_{\mathcal{A}}^\top \mathbf{A}_w : w \in \Sigma^*\}$ such that for any $j \in [1,m]$ and $a \in \Sigma$, $\mathbf{v}_j \mathcal{A}_a$ is in $\mathrm{span}(\mathbf{v}_1, \ldots, \mathbf{v}_m)$. The basis can be constructed incrementally by starting with $v_1 = \boldsymbol{\alpha}_\mathcal{A}^\top$ and by augmenting the current sequence of vectors $(\mathbf{v}_1, \ldots, \mathbf{v}_t)$ as follows. For any $j \in [1,t]$ and $a \in \Sigma$, the vector $\mathbf{w}$ is chosen in $\mathrm{span}(\mathbf{v}_1, \ldots, \mathbf{v}_m, \mathbf{v}_j \mathcal{A}_a)$ such that if $\mathbf{v}_j \mathcal{A}_a$ is linearly dependent of $(\mathbf{v}_1, \ldots, \mathbf{v}_t)$, then $\mathbf{w} = 0$; otherwise, such that $(\mathbf{v}_1, \ldots, \mathbf{v}_t, \mathbf{w})$ is triangular modulo the order of components and v_{t+1} is set to $\mathbf{w}$. Additionally, the components of $\mathbf{v}_j \mathcal{A}_a$ are computed with respect to $(\mathbf{v}_1, \ldots, \mathbf{v}_t, \mathbf{w})$ when $\mathbf{w} \neq 0$, with respect to $(\mathbf{v}_1, \ldots, \mathbf{v}_t)$ otherwise. Testing the dependency of $\mathbf{v}_j \mathcal{A}_a$ with respect to $(\mathbf{v}_1, \ldots, \mathbf{v}_t)$ and determining $\mathbf{w}$ such that $(\mathbf{v}_1, \ldots, \mathbf{v}_t, \mathbf{w})$ be triangular in the independent case can be done as in Gaussian elimination. This helps define a WFA $\mathcal{B}'$ equivalent to $\mathcal{A}$ and whose number of states is $\dim(\mathrm{span}(\{\boldsymbol{\alpha}_\mathcal{A}^\top \mathcal{A}_w : w \in \Sigma^*\})) = \dim(\mathrm{span}(\{\boldsymbol{\alpha}_{\mathcal{B}'}^\top \mathcal{B}'_w : w \in \Sigma^*\}))$. The components of $\mathbf{v}_j \mathcal{A}_a$ computed by the algorithm help define the transitions of $\mathcal{B}'$. The time complexity of the algorithm is in $O(|\Sigma||Q_\mathcal{A}|^3)$ semiring operations since at each iteration, the complexity of determining $\mathbf{w}$ is in $O(|\Sigma||Q_\mathcal{A}|^2)$.

The second stage of the algorithm is symmetric. It consists of starting with $\mathcal{B}'$ and constructing a WFA $\mathcal{B}$ whose number of states is $\dim(\mathrm{span}(\{\mathcal{B}'_w \boldsymbol{\beta}_{\mathcal{B}'} : w \in \Sigma^*\})) = \dim(\mathrm{span}(\{\mathcal{B}_w \boldsymbol{\beta}_\mathcal{B} : w \in \Sigma^*\}))$. The second stage therefore coincides with the first stage if we first reverse the WFA $\mathcal{B}'$ and permute $\boldsymbol{\alpha}_{\mathcal{B}'}$ and $\boldsymbol{\beta}_{\mathcal{B}'}$. Since $|Q_\mathcal{A}\mathcal{B}'| \leq |Q_\mathcal{A}|$, the overall time complexity of the algorithm is in $O(|\Sigma||Q_\mathcal{A}|^3)$.

The two consecutive stages guarantee that the resulting WFA $\mathcal{B}$ is minimal.

3.4 Hankel Masks and Bases

A Hankel basis for an infinite Hankel matrix with finite rank essentially identifies a finite sub-block of that matrix which contains as much information as the infinite matrix itself. The existence of such bases is paramount for the design of learning algorithms for WFAs. Here, we will prove the existence of Hankel bases, provide bounds on their sizes, and briefly discuss the problem of finding one in practice. We start by giving several definitions.

Definition 3 (Hankel Mask). *Let $\mathcal{P}, \mathcal{S} \subseteq \Sigma^*$ be two subsets of the set of all strings. Then, the pair $\mathcal{B} = (\mathcal{P}, \mathcal{S})$ is called a Hankel mask. The elements of $\mathcal{P}$ are called the prefixes and those of $\mathcal{S}$ the suffixes of the mask.*

Definition 4 (Hankel sub-block). *Let $\mathbf{H} \in \mathbb{S}^{\Sigma^* \times \Sigma^*}$ be a Hankel matrix. Given a Hankel mask $\mathcal{B} = (\mathcal{P}, \mathcal{S})$, we write $\mathbf{H}_\mathcal{B} \in \mathbb{S}^{\mathcal{P} \times \mathcal{S}}$ to denote the Hankel sub-block of $\mathbf{H}$ with rows indexed by elements of $\mathcal{P}$ and suffixes indexed by those of $\mathcal{S}$. Thus, for all $u \in \mathcal{P}$ and $v \in \mathcal{S}$ we have $\mathbf{H}_\mathcal{B}(u, v) = \mathbf{H}(u, v)$.*

Observe that $\mathbf{H}_\mathcal{B}$ inherits from $\mathbf{H}$ the Hankel property. Furthermore, since $\mathbf{H}_\mathcal{B}$ is a sub-block of $\mathbf{H}$, we always have $\mathrm{rank}(\mathbf{H}_\mathcal{B}) \leq \mathrm{rank}(\mathbf{H})$. This motivates our next definition.

Definition 5 (Hankel basis). *We say that the Hankel mask $\mathcal{B} = (\mathcal{P}, \mathcal{S})$ is a Hankel basis for $\mathbf{H}$ if $\mathrm{rank}(\mathbf{H}_\mathcal{B}) = \mathrm{rank}(\mathbf{H})$.*

Since the rank of a matrix is upper bounded by its dimensions, all Hankel masks satisfy $\mathrm{rank}(\mathbf{H}_{\mathcal{B}}) \leq \min\{|\mathcal{P}|, |\mathcal{S}|\}$. The next result is an immediate consequence of the definition of the rank of a Hankel matrix indicating that this bound is attainable.

Proposition 1. *Let $\mathbf{H}$ be a Hankel matrix with $\mathrm{rank}(\mathbf{H}) = n$. Then there exists a Hankel basis $\mathcal{B} = (\mathcal{P}, \mathcal{S})$ for $\mathbf{H}$ with $|\mathcal{P}| = |\mathcal{S}| = n$.*

Definition 6 (Prefix-closed and suffix-closed sets). *A subset $\mathcal{W} \subseteq \Sigma^*$ is prefix-closed if $w = uv \in \mathcal{W}$ implies $u \in \mathcal{W}$. Similarly, $\mathcal{W} \subseteq \Sigma^*$ is suffix-closed if $w = uv \in \mathcal{W}$ implies $v \in \mathcal{W}$. Note that if $\mathcal{W}$ is either prefix-closed or suffix-closed, then it must contain ϵ.*

The standardization procedure for WFA described in Section 3.3 provides further information about the structure of minimal Hankel bases.

Proposition 2. *Let $\mathbf{H}$ be a Hankel matrix with $\mathrm{rank}(\mathbf{H}) = n$. Then, there exists a Hankel basis $\mathcal{B} = (\mathcal{P}, \mathcal{S})$ for $\mathbf{H}$ with $|\mathcal{P}| = |\mathcal{S}| = n$, where $\mathcal{P}$ is prefix-closed and $\mathcal{S}$ is suffix-closed.*

Note that, given a string $x \in \Sigma^*$, there are exactly $|x| + 1$ decompositions $x = uv$ with $u, v \in \Sigma^*$. A direct consequence of this fact is that if $\mathcal{W} \subseteq \Sigma^*$ is prefix-closed and has $|\mathcal{W}| = n$, then $|w| < n$ for every $w \in \mathcal{W}$. The same holds for suffix-closed sets. When combined with the previous proposition, this observation yields a bound on *how far* in Σ^* one needs to look in order to find a Hankel basis for a Hankel matrix $\mathbf{H}$ with rank n.

Corollary 1. *Let $\mathbf{H}$ be Hankel with $\mathrm{rank}(\mathbf{H}) = n$. Then $\mathcal{B} = (\Sigma^{<n}, \Sigma^{<n})$ is a Hankel basis for $\mathbf{H}$.*

3.5   WFA Reconstruction from Complete Minimal Masks

In this section, we describe the class of complete minimal Hankel masks, which can be used to specify the information needed to solve a WFA reconstruction problem via the Gaussian elimination algorithm in an arbitrary field. We describe the reconstruction algorithm and show that if the given mask is a Hankel basis for some Hankel matrix $\mathbf{H}_f$, then the algorithm will reconstruct a minimal WFA computing f.

Definition 7 (Hankel sub-blocks $\mathbf{H}_a$ and $\mathbf{H}_\Sigma$). *Let $\mathcal{B} = (\mathcal{P}, \mathcal{S})$ be a Hankel mask in Σ^*. For every symbol $a \in \Sigma$, we define the Hankel mask $\mathcal{B}_a = (\mathcal{P}a, \mathcal{S})$, where $\mathcal{P}a = \{ua : u \in \mathcal{P}\}$. Given a Hankel matrix $\mathbf{H}$, we will use the shorthand $\mathbf{H}_a = \mathbf{H}_{\mathcal{B}_a} \in \mathbb{S}^{\mathcal{P} \times \mathcal{S}}$. Note the entries of $\mathbf{H}_a$ satisfy $\mathbf{H}_a(u, v) = \mathbf{H}(ua, v)$ for every $u \in \mathcal{P}$ and $v \in \mathcal{S}$. We denote by $\mathbf{H}_\Sigma$ the block-matrix obtained by stacking together the matrices $\mathbf{H}_a$ for all $a \in \Sigma$, that is $\mathbf{H}_\Sigma^\top = [\mathbf{H}_{a_1}^\top \cdots \mathbf{H}_{a_r}^\top]$ if $\Sigma = \{a_1, \ldots, a_r\}$.*

Definition 8 (Complete and minimal Hankel masks). *A Hankel mask $\mathcal{B} = (\mathcal{P}, \mathcal{S})$ is said to be complete for a Hankel matrix $\mathbf{H}$ if $\epsilon \in \mathcal{P} \cap \mathcal{S}$ and $\mathrm{rank}([\mathbf{H}_{\mathcal{B}}^\top \mid \mathbf{H}_\Sigma^\top]) = \mathrm{rank}(\mathbf{H}_{\mathcal{B}}^\top)$. A complete Hankel mask $\mathcal{B}$ is minimal if $\mathrm{rank}(\mathbf{H}_{\mathcal{B}}) = |\mathcal{P}|$. Note this last condition implies $|\mathcal{P}| \leq |\mathcal{S}|$.*

We now proceed to describe a WFA reconstruction algorithm that takes as input a complete minimal Hankel mask $\mathcal{B} = (\mathcal{P}, \mathcal{S})$ for a Hankel matrix $\mathbf{H}$ and the corresponding Hankel sub-blocks $\mathbf{H}_\mathcal{B}$ and $\mathbf{H}_\Sigma$, and returns a WFA $\mathcal{A}$ with $k = |\mathcal{P}|$ states. Let us write $\mathcal{P} = \{u_1, \ldots, u_k\}$ and $\mathcal{S} = \{v_1, \ldots, v_{k'}\}$ with $u_1 = v_1 = \epsilon$. First, let $\boldsymbol{\alpha}_\mathcal{A}^\top = [\bar{1}, \bar{0}, \ldots, \bar{0}] \in \mathbb{S}^k$ and $\boldsymbol{\beta}_\mathcal{A}^\top = [\mathbf{H}_\mathcal{B}(u_1, \epsilon), \ldots, \mathbf{H}_\mathcal{B}(u_k, \epsilon)] = (\mathbf{H}_\mathcal{B}(\cdot, \epsilon))^\top$. Second, note that since $\mathcal{B}$ is complete and minimal we have $\mathrm{rank}([\mathbf{H}_\mathcal{B}^\top \mid \mathbf{H}_a^\top]) = \mathrm{rank}(\mathbf{H}_\mathcal{B}^\top) = k$ for every $a \in \Sigma$. Thus, by the Rouché–Capelli theorem, for every $a \in \Sigma$ there exists a unique $\mathcal{A}_a \in \mathbb{S}^{k \times k}$ such that $\mathcal{A}_a \mathbf{H}_\mathcal{B} = \mathbf{H}_a$. Using the Gaussian elimination algorithm, each of these systems of equations can be solved in $O(k^2(k' + k))$ arithmetic operations in $\mathbb{S}$. Thus, the arithmetic complexity of reconstructing a WFA $\mathcal{A}$ with $|\mathcal{P}|$ states from a complete minimal basis $\mathcal{B} = (\mathcal{P}, \mathcal{S})$ is in $O(|\Sigma||\mathcal{P}|^2|\mathcal{S}|)$ [30]. If, in addition to being complete and minimal, the mask $\mathcal{B}$ is a Hankel basis for $\mathbf{H}_f$, the above procedure recovers a minimal WFA computing f.

Theorem 2. *If $\mathcal{B}$ is a complete minimal Hankel basis for $\mathbf{H}_f$, then the reconstructed WFA $\mathcal{A}$ computes f and is minimal.*

Proof. Let $\mathcal{A}'$ be a minimal WFA computing f. Recall that $\mathcal{A}'$ induces a rank factorization $\mathbf{H}_f = \mathbf{P}_{\mathcal{A}'} \mathbf{S}_{\mathcal{A}'}^\top$, which, when restricted to the Hankel basis $\mathcal{B}$ yields a rank factorization $\mathbf{H}_\mathcal{B} = \mathbf{P}' \mathbf{S}'^\top$ and associated factorizations $\mathbf{H}_a = \mathbf{P}' \mathcal{A}'_a \mathbf{S}'^\top$ for all $a \in \Sigma$. From these, using the fact that the transition weights of $\mathcal{A}$ satisfy $\mathcal{A}_a \mathbf{H}_\mathcal{B} = \mathbf{H}_a$ we get $\mathcal{A}_a \mathbf{P}' \mathbf{S}'^\top = \mathbf{P}' \mathcal{A}'_a \mathbf{S}'^\top$. Since $\mathbf{P}'$ is invertible and $\mathbf{S}'$ has full column rank, this last equation implies $\mathcal{A}_a = \mathbf{P}' \mathcal{A}'_a \mathbf{P}'^{-1}$. A similar argument with the initial and final weights shows that $\boldsymbol{\alpha} = \mathbf{P}'^{-1} \boldsymbol{\alpha}$ and $\boldsymbol{\beta} = \mathbf{P}' \boldsymbol{\beta}$. Therefore, we see that $\mathcal{A}$ and $\mathcal{A}'$ compute the same function, and in particular $\mathcal{A}$ computes f. Minimality is immediate by observing that $\mathcal{A}$ has $|Q_\mathcal{A}| = \mathrm{rank}(\mathbf{H}_\mathcal{B}) = \mathrm{rank}(\mathbf{H}_f)$ states. $\qquad\square$

If the Hankel mask $\mathcal{B} = (\mathcal{P}, \mathcal{S})$ is complete and minimal but not necessarily a Hankel basis, then the function computed by $\mathcal{A}$ will not agree with f everywhere. However, the next result shows that if $\mathcal{P}$ is prefix-free and $\mathcal{S}$ is suffix-free, then $\mathcal{A}$ will agree with f on all strings in $\mathcal{P}(\{\epsilon\} \cup \Sigma)\mathcal{S}$.

Theorem 3. *Let $\mathcal{B} = (\mathcal{P}, \mathcal{S})$ be a complete minimal Hankel mask for $\mathbf{H}_f$. Suppose that $\mathcal{P}$ is prefix-closed and $\mathcal{S}$ is suffix-closed. Then, the WFA $\mathcal{A}$ reconstructed from $\mathbf{H}_\mathcal{B}$ and $\mathbf{H}_\Sigma$ satisfies $f(uv) = \mathcal{A}(uv)$ and $f(uav) = \mathcal{A}(uav)$ for every $u \in \mathcal{P}$, $v \in \mathcal{S}$, and $a \in \Sigma$.*

Proof. Let $k = \mathrm{rank}(\mathbf{H}_\mathcal{B}) = |\mathcal{P}|$ and $\mathcal{P} = \{u_1, \ldots, u_k\}$ with $u_1 = \epsilon$ and $|u_i| \leq |u_{i+1}|$ for all i. Let $\mathbf{H}_\mathcal{A} = \mathbf{P}_\mathcal{A} \mathbf{S}_\mathcal{A}^\top$ be the factorization induced by $\mathcal{A}$. Let us write $\mathbf{P}_\mathcal{P} \in \mathbb{S}^{\mathcal{P} \times k}$ for the sub-block of $\mathbf{P}_\mathcal{A}$ containing the rows indexed by prefixes in $\mathcal{P}$. We claim that $\mathbf{P}_\mathcal{P} = \mathbf{I}$ is the identity matrix. To see this, we will show that for $1 \leq i \leq k$ we have $\mathbf{P}_\mathcal{P}(u_i, \cdot) = \mathbf{e}_i^\top$, where $\mathbf{e}_i$ is the ith indicator vector.

By construction of $\mathcal{A}$, the case $i = 1$ holds since $\mathbf{P}_\mathcal{P}(u_i, \cdot) = \mathbf{P}_\mathcal{A}(\epsilon, \cdot) = \boldsymbol{\alpha}^\top = \mathbf{e}_1^\top$. Now, suppose the claim is true for all $1 \leq j \leq i$. Since $|u_{i+1}| \geq |u_j|$ for all $1 \leq j \leq i$ and $\mathcal{P}$ is prefix-closed, we must have $u_{i+1} = u_j a$ for some $a \in \Sigma$

and $1 \leq j \leq i$. Therefore, we have $\mathbf{P}_{\mathcal{P}}(u_{i+1}, \cdot) = \mathbf{P}_{\mathcal{A}}(u_j a, \cdot) = \mathbf{P}_{\mathcal{A}}(u_j, \cdot)\boldsymbol{\mathcal{A}}_a = \mathbf{e}_j^\top \boldsymbol{\mathcal{A}}_a = \boldsymbol{\mathcal{A}}_a(j, \cdot)$. Finally, we observe that because $\boldsymbol{\mathcal{A}}_a \mathbf{H}_{\mathcal{B}} = \mathbf{H}_a$ and $\mathbf{H}_a(u_j, \cdot) = \mathbf{H}_{\mathcal{B}}(u_j a, \cdot) = \mathbf{H}_{\mathcal{B}}(u_{i+1}, \cdot)$, when solving the system of equations for $\boldsymbol{\mathcal{A}}_a$ we will obtain $\boldsymbol{\mathcal{A}}_a(j, \cdot) = \mathbf{e}_{i+1}^\top$.

Now, let $\mathbf{S}_{\mathcal{S}} \in \mathbb{S}^{\mathcal{S} \times k}$ denote the sub-block of $\mathbf{S}_{\mathcal{A}}$ corresponding to the suffixes in $\mathcal{S}$. By the previous claim, to show that $\mathcal{A}(uv) = f(uv)$ for all $u \in \mathcal{P}$ and all $v \in \mathcal{S}$ it suffices to show that $\mathbf{S}_{\mathcal{S}}^\top = \mathbf{H}_{\mathcal{B}}$. Let $k' = |\mathcal{S}|$ and assume without loss of generality that $\mathcal{S} = \{v_1, \ldots, v_{k'}\}$ with $v_1 = \epsilon$ and $|v_{i+1}|? \geq |v_i|$ for all i. Then for $i = 1$ we immediately have $\mathbf{S}_{\mathcal{S}}(v_1, \cdot) = \mathbf{S}_{\mathcal{A}}(\epsilon, \cdot) = \boldsymbol{\beta}^\top = \mathbf{H}_{\mathcal{B}}(\cdot, \epsilon)^\top = \mathbf{H}_{\mathcal{B}}(\cdot, v_1)^\top$ by the way $\boldsymbol{\beta}$ is constructed. Now, suppose we have $\mathbf{S}_{\mathcal{S}}(v_j, \cdot) = \mathbf{H}_{\mathcal{B}}(\cdot, v_j)^\top$ for all $1 \leq j \leq i$. Note we must have $v_{i+1} = a v_j$ for some $a \in \Sigma$ and some $1 \leq j \leq i$. Thus, we see that $\mathbf{S}_{\mathcal{S}}(v_{i+1}, \cdot) = \mathbf{S}_{\mathcal{A}}(a v_j, \cdot) = \mathbf{S}_{\mathcal{A}}(v_j, \cdot)\boldsymbol{\mathcal{A}}_a^\top = \mathbf{H}_a(\cdot, v_j)^\top = \mathbf{H}_{\mathcal{B}}(\cdot, a v_j)^\top = \mathbf{H}_{\mathcal{B}}(\cdot, v_{i+1})^\top$.

To complete the proof it just remains to show that $\mathcal{A}(uav) = f(uav) = \mathbf{H}_a(u, v)$ for all $u \in \mathcal{P}$, $v \in \mathcal{S}$, and $a \in \Sigma$. This follows from the previous claims by noting that $\mathcal{A}(u_i a v) = \mathbf{P}_{\mathcal{P}}(u_i, \cdot)\boldsymbol{\mathcal{A}}_a \mathbf{S}_{\mathcal{S}}(v, \cdot)^\top = \mathbf{e}_i^\top \boldsymbol{\mathcal{A}}_a \mathbf{H}_{\mathcal{B}}(\cdot, v) = \mathbf{e}_i^\top \mathbf{H}_a(\cdot, v) = \mathbf{H}_a(u_i, v)$. $\qquad\square$

3.6   WFA Reconstruction via Rank Factorizations

In this section, we show how a rank factorization of $\mathbf{H}_{\mathcal{B}}$ for a non-minimal complete Hankel mask $\mathcal{B}$ can be used to reconstruct a WFA. The main difference with the procedure presented in the previous sections is that here the number of states of the resulting WFA is not tied to the number of prefixes $|\mathcal{P}|$ in the mask, but to the rank of $\mathbf{H}_{\mathcal{B}}$, which can be small, even if $|\mathcal{P}|$ is large.

Let $\mathcal{B} = (\mathcal{P}, \mathcal{S})$ be a Hankel mask in Σ^* with $\epsilon \in \mathcal{P} \cap \mathcal{S}$. Given a Hankel matrix $\mathbf{H}$, in addition to the matrices $\mathbf{H}_a = \mathbf{H}_{\mathcal{B}_a} \in \mathbb{S}^{\mathcal{P}a \times \mathcal{S}}$ for $a \in \Sigma$ introduced in the previous section, we define vectors $\mathbf{h}_{\mathcal{P}} \in \mathbb{S}^{\mathcal{P}}$ and $\mathbf{h}_{\mathcal{S}} \in \mathbb{S}^{\mathcal{S}}$ with entries given by $\mathbf{h}_{\mathcal{P}}(u) = \mathbf{H}(u, \epsilon)$ and $\mathbf{h}_{\mathcal{S}}(v) = \mathbf{H}(\epsilon, v)$. Note that the condition $\epsilon \in \mathcal{P} \cap \mathcal{S}$ implies that $\mathbf{h}_{\mathcal{P}}$ (resp. $\mathbf{h}_{\mathcal{S}}^\top$) can be found as a column (resp. a row) in $\mathbf{H}_{\mathcal{B}}$.

Suppose $\mathcal{B}$ is a complete Hankel mask and let k be the rank of $\mathbf{H}_{\mathcal{B}}$, $\mathrm{rank}(\mathbf{H}_{\mathcal{B}}) = k$. Then, $\mathbf{H}_{\mathcal{B}}$ admits a rank factorization $\mathbf{H}_{\mathcal{B}} = \mathbf{P}_{\mathcal{B}}\mathbf{S}_{\mathcal{B}}^\top$ with $\mathbf{P}_{\mathcal{B}} \in \mathbb{S}^{\mathcal{P} \times k}$ and $\mathbf{S}_{\mathcal{B}} \in \mathbb{S}^{\mathcal{S} \times k}$. Such a rank factorization can be obtained using a Gaussian elimination algorithm [30]. Next, we show how to use this rank factorization in order to reconstruct a WFA $\mathcal{A}$ with $k = |Q_{\mathcal{A}}|$ states.

The algorithm proceeds by solving a series of systems of linear equations. For the initial and final weights we find the unique solutions to $\mathbf{S}_{\mathcal{B}}\boldsymbol{\alpha}_{\mathcal{A}} = \mathbf{h}_{\mathcal{P}}$ and $\mathbf{P}_{\mathcal{B}}\boldsymbol{\beta}_{\mathcal{A}} = \mathbf{h}_{\mathcal{S}}$. Note that $\boldsymbol{\alpha}_{\mathcal{A}}$ exists and is unique since $\mathbf{S}_{\mathcal{B}}$ contains a basis of linearly independent vectors for the column-span of $\mathbf{H}_{\mathcal{B}}$ and $\mathbf{h}_{\mathcal{P}}$ is a column of $\mathbf{H}_{\mathcal{B}}$. A similar argument holds for $\boldsymbol{\beta}_{\mathcal{A}}$. For the transition weights associated with a symbol $a \in \Sigma$, we use the unique solution to the system of linear equations $\mathbf{H}_a = \mathbf{P}_{\mathcal{B}}\boldsymbol{\mathcal{A}}_a \mathbf{S}_{\mathcal{B}}^\top$.

One way to solve this last system of equations — and to see that indeed it admits a unique solution — is to recall that the equation for $\boldsymbol{\mathcal{A}}_a$ is equivalent to $\mathrm{vec}(\mathbf{H}_a) = (\mathbf{S}_{\mathcal{B}} \otimes_{\mathrm{K}} \mathbf{P}_{\mathcal{B}}) \mathrm{vec}(\boldsymbol{\mathcal{A}}_a)$, where $\otimes_{\mathrm{K}}$ denotes the Kronecker product between matrices, and vec($\mathbf{M}$) the result of stacking the columns of $\mathbf{M}$ into a single vector. Observe that the new system of equations admits k^2 unknowns. Its coefficients satisfy rank($\mathbf{S}_{\mathcal{B}} \otimes_K \mathbf{P}_{\mathcal{B}}$) = rank($\mathbf{S}_{\mathcal{B}}$) rank($\mathbf{P}_{\mathcal{B}}$) = k^2 by a basic property of Kronecker products, and rank($[\mathbf{S}_{\mathcal{B}} \otimes_K \mathbf{P}_{\mathcal{B}} | \text{vec}(\mathbf{H}_a)]$) = rank($\mathbf{S}_{\mathcal{B}} \otimes_K \mathbf{P}_{\mathcal{B}}$) since the columns of $\mathbf{H}_a$ are linear combinations of the columns of $\mathbf{P}_{\mathcal{B}}$ because the mask $\mathcal{B}$ is complete. Thus, by the Rouché–Capelli theorem, there exists a unique solution for $\mathcal{A}_a$. Furthermore, the solution can be found using Gaussian elimination in $O(|\mathcal{P}||\mathcal{S}|k^2)$ arithmetic operations.

Overall, the cost of reconstructing the WFA $\mathcal{A}$ starting from a complete Hankel mask takes $O(|\Sigma||\mathcal{P}||\mathcal{S}|k^2)$ arithmetic operations. As in the previous section, if, in addition to being complete, the mask $\mathcal{B}$ is a Hankel basis for some Hankel matrix $\mathbf{H}_f$, then the WFA recovered is a minimal automaton for f. The proof of this result is almost identical to that of Theorem 2 and is omitted.

Theorem 4. *If $\mathcal{B}$ is a complete Hankel basis for $\mathbf{H}_f$, then the reconstructed WFA $\mathcal{A}$ computes f and is minimal.*

3.7   WFA Reconstruction from Noisy Hankel Matrices

In the two WFA reconstruction algorithms described in the previous sections, we assumed that the Hankel sub-blocks used in the reconstruction procedure are known exactly. However, that assumption is not realistic in practice, especially when we are concerned with learning problems. We now describe a variation of the WFA reconstruction algorithm from rank factorizations that works in situations where the only available information are approximations to the Hankel sub-blocks specified by a Hankel mask.

This procedure relies in a crucial manner on the computation of a singular value decomposition (SVD), which is only possible for the real case $\mathbb{S} = \mathbb{R}$, the complex case $\mathbb{S} = \mathbb{C}$, and, in general, in the case where $\mathbb{S}$ is a field obtained as the intersection of real closed fields [48]. Since $\mathbb{S} = \mathbb{R}$ is the case which occurs more frequently in applications, and is also a case for which efficient SVD algorithms are widely available, we will present the algorithm in this section only for this case. The ideas can be straightforwardly generalized to other fields admitting an SVD.

As before, we will assume that the algorithm is given as input an arbitrary Hankel mask $\mathcal{B} = (\mathcal{P}, \mathcal{S})$. The difference is that here, instead of the exact versions of the matrices and vectors $\mathbf{H}_{\mathcal{B}}$, $\mathbf{H}_a$, $\mathbf{h}_{\mathcal{P}}$, and $\mathbf{h}_{\mathcal{S}}$ that represent sub-blocks of some Hankel matrix $\mathbf{H}$, the algorithm will only have access to approximate versions of these objects. For example, we are given a matrix $\widehat{\mathbf{H}}_{\mathcal{B}} \in \mathbb{R}^{\mathcal{P} \times \mathcal{S}}$ such that $\widehat{\mathbf{H}}_{\mathcal{B}} = \mathbf{H}_{\mathcal{B}} + \mathbf{E}_{\mathcal{B}}$, where $\mathbf{E}_{\mathcal{B}} \in \mathbb{R}^{\mathcal{P} \times \mathcal{S}}$ is a *noise* matrix. Likewise, where are given $\widehat{\mathbf{H}}_a = \mathbf{H}_a + \mathbf{E}_a$ for every $a \in \Sigma$, $\widehat{\mathbf{h}}_{\mathcal{P}} = \mathbf{h}_{\mathcal{P}} + \mathbf{e}_{\mathcal{P}}$, and $\widehat{\mathbf{h}}_{\mathcal{S}} = \mathbf{h}_{\mathcal{S}} + \mathbf{e}_{\mathcal{S}}$.

The important point to note here is that even if $\mathbf{H}_{\mathcal{B}}$ has small rank, say rank($\mathbf{H}_{\mathcal{B}}$) = $k \leq$ rank($\mathbf{H}$) = n, its approximation $\widehat{\mathbf{H}}_{\mathcal{B}}$ may have a much larger rank, and thus, in this case, the straightforward rank factorization approach will yield a large WFA which does not necessarily resemble the one we would recover had we had access to the exact versions of $\mathbf{H}_{\mathcal{B}}$ and the other matrices.

For example, if the error matrix $\mathbf{E}_{\mathcal{B}}$ is in generic position, or random, then $\mathbf{H}_{\mathcal{B}}$ will have full rank.

Thus, the question is now how to use these matrices to reconstruct a WFA with less states than $\mathrm{rank}(\widehat{\mathbf{H}}_{\mathcal{B}})$, and that ideally resembles the one we would obtain in the exact case if the amount of noise is small. The key to the solution consists of using an SVD and replace the rank factorization in the previous WFA reconstruction algorithm by a low rank approximation of $\widehat{\mathbf{H}}_{\mathcal{B}}$.

Now we proceed to describe the first steps of the algorithm. As input it receives the Hankel mask $\mathcal{B}$, the number of states k' that the output WFA must have, and the approximated Hankel sub-blocks described above. We start by computing the SVD of $\widehat{\mathbf{H}}_{\mathcal{B}}$ and using it to obtain the best rank k' approximation $\widehat{\mathbf{H}}_{\mathcal{B}} \approx \widehat{\mathbf{U}}\widehat{\mathbf{D}}\widehat{\mathbf{V}}^{\top}$, where $\widehat{\mathbf{D}} = \mathrm{diag}(\widehat{s}_1, \ldots, \widehat{s}_{k'})$ is a diagonal matrix containing the top k' singular values of $\widehat{\mathbf{H}}_{\mathcal{B}}$, and $\widehat{\mathbf{U}} \in \mathbb{R}^{\mathcal{P} \times k'}$ and $\widehat{\mathbf{V}} \in \mathbb{R}^{\mathcal{S} \times k'}$ contain the associated left and right singular vectors respectively. With this notation, one can see that now $\widehat{\mathbf{P}}_{\mathcal{B}} = \widehat{\mathbf{U}}\widehat{\mathbf{D}}$ and $\widehat{\mathbf{S}}_{\mathcal{B}} = \widehat{\mathbf{V}}$ provide a rank factorization $\widehat{\mathbf{P}}_{\mathcal{B}}\widehat{\mathbf{S}}_{\mathcal{B}}$ of the best rank k' approximation to $\widehat{\mathbf{H}}_{\mathcal{B}}$.

The next natural step in the algorithms would be to solve the following systems of linear equations in order to reconstruct a WFA: $\widehat{\mathbf{S}}_{\mathcal{B}}\widehat{\alpha} = \widehat{\mathbf{h}}_{\mathcal{P}}$, $\widehat{\mathbf{P}}_{\mathcal{B}}\widehat{\beta} = \widehat{\mathbf{h}}_{\mathcal{S}}$, and $(\widehat{\mathbf{S}}_{\mathcal{B}} \otimes_{\mathrm{K}} \widehat{\mathbf{P}}_{\mathcal{B}})\,\mathrm{vec}(\widehat{\mathbf{A}}_a) = \mathrm{vec}(\widehat{\mathbf{H}}_a)$ for every $a \in \Sigma$. There is, however, an obstruction to the direct application of this strategy in this case: these equations are no longer guaranteed to admit a unique solution. Due to the errors in the Hankel sub-blocks introduced by the approximation, these equations might now be unsatisfiable or not admit a unique solution. Thus, we will follow a least-squares approach and look for a solution to these equations that minimizes the norm of the residual. A way to express these solutions in closed-form is via the Moore–Penrose pseudo-inverse $\mathbf{M}^{+} \in \mathbb{R}^{d_2 \times d_1}$ of a matrix $\mathbf{M} \in \mathbb{R}^{d_1 \times d_2}$. In particular, given a linear system of equations $\mathbf{M}\mathbf{x} = \mathbf{b}$ the pseudo-inverse yields a solution $\mathbf{x} = \mathbf{M}^{+}\mathbf{b}$ that satisfies the equation if it is satisfiable, and that minimizes the error $\|\mathbf{M}\mathbf{x} - \mathbf{b}\|$ otherwise.

Now we proceed to describe the rest of the algorithm, which essentially applies this strategy to solve the linear systems given above. For the initial and final weights this yields $\widehat{\alpha} = \widehat{\mathbf{S}}_{\mathcal{B}}^{+}\widehat{\mathbf{h}}_{\mathcal{P}}$ and $\widehat{\beta} = \widehat{\mathbf{P}}_{\mathcal{B}}^{+}\widehat{\mathbf{h}}_{\mathcal{S}}$. In our case, these are easy to compute because by properties of the pseudo-inverse it can be shown that $\widehat{\mathbf{P}}_{\mathcal{B}}^{+} = (\widehat{\mathbf{U}}\widehat{\mathbf{D}})^{+} = \widehat{\mathbf{D}}^{-1}\widehat{\mathbf{U}}^{\top}$ and $\widehat{\mathbf{S}}_{\mathcal{B}}^{+} = \widehat{\mathbf{V}}^{+} = \widehat{\mathbf{V}}^{\top}$. For the transition weights, a short algebraic calculation shows that $(\widehat{\mathbf{S}}_{\mathcal{B}} \otimes_{\mathrm{K}} \widehat{\mathbf{P}}_{\mathcal{B}})^{+} = (\widehat{\mathbf{S}}_{\mathcal{B}}^{+} \otimes_{\mathrm{K}} \widehat{\mathbf{P}}_{\mathcal{B}}^{+}) = (\widehat{\mathbf{V}}^{\top} \otimes_{\mathrm{K}} \widehat{\mathbf{D}}^{-1}\widehat{\mathbf{U}}^{\top})$. Substituting into $\mathrm{vec}(\widehat{\mathbf{A}}_a) = (\widehat{\mathbf{S}}_{\mathcal{B}} \otimes_{\mathrm{K}} \widehat{\mathbf{P}}_{\mathcal{B}})^{+}\,\mathrm{vec}(\widehat{\mathbf{H}}_a)$ and applying the equivalence between vectorized and unvectorized systems of linear equations, we obtain the expression $\widehat{\mathbf{A}}_a = \widehat{\mathbf{D}}^{-1}\widehat{\mathbf{U}}^{\top}\widehat{\mathbf{H}}_a\widehat{\mathbf{V}}$.

Overall, the complexity of this process is dominated by the low-rank SVD computation, which takes $O(|\mathcal{P}||\mathcal{S}|k')$ arithmetic operations. Hence, the arithmetic complexity of computing the WFA $\widehat{A}$ with k' states given by $\widehat{\alpha}$, $\widehat{\beta}$, and $\widehat{\mathbf{A}}_a$ for $a \in \Sigma$, is in $O(|\Sigma||\mathcal{P}||\mathcal{S}|k')$.

The main result of this section is a bound on the sensitivity of this algorithm to the magnitude of the noise. To make this more precise, we need two ingredients. The first is a precise way to quantify the error in the approximations.

Different choices lead to slightly different results, but in order to illustrate the point we will simply use the Euclidean norm for vectors and the Frobenius norm for matrices. Thus, we will define $\varepsilon_{\mathcal{B}} = \|\mathbf{E}_{\mathcal{B}}\|_F$, $\varepsilon_a = \|\mathbf{E}_a\|_F$ for every $a \in \Sigma$, $\varepsilon_{\mathcal{P}} = \|\mathbf{e}_{\mathcal{P}}\|_2$, and $\varepsilon_{\mathcal{S}} = \|\mathbf{e}_{\mathcal{S}}\|_2$. For convenience we will also write $\varepsilon = \max\{\varepsilon_{\mathcal{B}}, \varepsilon_{a_1}, \ldots, \varepsilon_{a_r}, \varepsilon_{\mathcal{P}}, \varepsilon_{\mathcal{S}}\}$. The second ingredient is to determine what would be the output of the algorithm be if the input had no noise. For that purpose, let us assume that $k' = \text{rank}(\mathbf{H}_{\mathcal{B}}) = k$ and $\varepsilon = 0$. In that case, we have $\widehat{\mathbf{H}}_{\mathcal{B}} = \mathbf{H}_{\mathcal{B}}$ and the SVD of rank k yields an exact rank factorization $\mathbf{H}_{\mathcal{B}} = \mathbf{PS}^\top = (\mathbf{UD})(\mathbf{V})^\top$. Thus, the algorithm returns a WFA with k states given by $\boldsymbol{\alpha} = \mathbf{V}^\top \mathbf{h}_{\mathcal{P}}$, $\boldsymbol{\beta} = \mathbf{D}^{-1}\mathbf{U}^\top \mathbf{h}_{\mathcal{S}}$, and $\mathcal{A}_a = \mathbf{D}^{-1}\mathbf{U}^\top \mathbf{H}_a \mathbf{V}$, where we dropped the hat notation to indicate that we are in the case $\varepsilon = 0$. For this automaton, a direct application of Theorem 4 yields the following result, which shows that this is essentially a generalization of the WFA reconstruction algorithm based on rank factorizations.

Corollary 2. *Suppose $k' = \text{rank}(\mathbf{H}_{\mathcal{B}})$ and $\varepsilon = 0$. If $\mathcal{B}$ is a complete basis for $\mathbf{H}_f$, then the reconstructed WFA $\mathcal{A}$ computes f and is minimal.*

The most important result about the WFA reconstructions algorithm based on SVD is the following, which bounds the error between the noisy and the noiseless cases.

Theorem 5. *Suppose $k' = \text{rank}(\mathbf{H}_{\mathcal{B}})$. Let $\mathcal{A}$ denote the WFA obtained in the case $\varepsilon = 0$ and $\widehat{\mathcal{A}}$ the WFA obtained in the noisy case. Then, the following approximation guarantee holds as $\varepsilon \to 0$:*

$$\Delta = \max\{\|\boldsymbol{\alpha} - \widehat{\boldsymbol{\alpha}}\|_2, \|\boldsymbol{\beta} - \widehat{\boldsymbol{\beta}}\|_2, \|\mathcal{A}_{a_1} - \widehat{\mathcal{A}}_{a_1}\|_F, \ldots, \|\mathcal{A}_{a_r} - \widehat{\mathcal{A}}_{a_r}\|_F\} = O(\varepsilon).$$

The proof of this results is technical and goes beyond the scope of the present survey. Essentially, it involves a detailed analysis using perturbation theory for singular values and vectors (see [9, Chapter 5] for details).

4 Algorithms for Learning WFAs

In this section, we show how the reconstruction techniques described in the previous section can be used in the design of algorithms for learning WFAs. We describe three WFA learning algorithms, each designed for a different learning scenario. The scenarios mainly differ by the way the data about the target function $f\colon \Sigma^* \to \mathbb{S}$ is gathered: exact learning from membership and equivalence queries (Section 4.1), PAC learning (Probably approximately correct learning) of a probability distribution represented by a WFA from i.i.d. samples (Section 4.2), and statistical learning of WFA from general string–label pairs (Section 4.3).

We also present learning guarantees in each case, thereby showcasing an important trade-off between degree of fidelity of the information collected versus quality of the learned WFA with respect to a target automaton or distribution. Of the three scenarios, only the first one can learn WFA over an arbitrary field $\mathbb{S}$; in the other two scenarios we restrict ourselves only to the case $\mathbb{S} = \mathbb{R}$.

4.1   Learning WFAs From Queries

In this section, we describe an algorithm for learning WFAs defined over an arbitrary field $\mathbb{S}$. The algorithm was first presented in [14] for the special case $\mathbb{S} = \mathbb{Q}$ and later generalized to arbitrary fields in [15]. It can be interpreted as a direct generalization of Angluin's classical algorithm for learning DFAs from membership and equivalence queries [5] and can be further applied to other learning problems (see [12,13]).

The learning scenario for this algorithm coincides with the active learning scenario defined and adopted by Angluin [5] for learning (unweighted) automata. In this scenario, given a target rational function $f \colon \Sigma^* \to \mathbb{S}$ the learner can make the following two types of queries to which an oracle responds:

- *membership queries* MQ_f: the learner requests the target value $f(w)$ of a string $w \in \Sigma^*$ and receives that value;
- *equivalence queries* EQ_f: the learner conjectures a WFA $\mathcal{A}$; he receives the response yes if f can be computed by $\mathcal{A}$, a counter-example $w \in \Sigma^*$ with $f(w) \neq \mathcal{A}(w)$ otherwise.

The objective of the learner is to determine exactly a WFA $\mathcal{A}$ representing f. We will denote by n the unknown rank of the Hankel matrix of f, $n = \mathrm{rank}(\mathbf{H}_f)$.

The main idea behind the algorithm is to build a complete minimal Hankel basis $\mathcal{B}$ for $\mathbf{H}_f$, fill the associated Hankel sub-blocks $\mathbf{H}_\mathcal{B}$ and $\mathbf{H}_\Sigma$ by making a series of calls to MQ_f, and then reconstruct the corresponding WFA using the Gaussian elimination algorithm described in Section 3.5. In order to find such a basis $\mathcal{B}$ several intermediate complete minimal Hankel masks are considered. For each, the corresponding WFA is reconstructed using information collected from membership queries, and the counter-examples supplied by the equivalence queries used to extend the current Hankel mask.

Given two bases $\mathcal{B} = (\mathcal{P}, \mathcal{S})$ and $\mathcal{B}' = (\mathcal{P}', \mathcal{S}')$, we will write in short $\mathcal{B} \subseteq \mathcal{B}'$ for $\mathcal{P} \subseteq \mathcal{P}'$ and $\mathcal{S} \subseteq \mathcal{S}'$. The algorithm constructs a sequence of complete minimal Hankel masks $\mathcal{B}_0 \subseteq \mathcal{B}_1 \subseteq \cdots \subseteq \mathcal{B}_d$, where the last mask $\mathcal{B}_d$ is a Hankel basis for $\mathbf{H}_f$. At each step, the inequality $\mathrm{rank}(\mathbf{H}_{\mathcal{B}_{i+1}}) > \mathrm{rank}(\mathbf{H}_{\mathcal{B}_i})$ holds, which guarantees that the total number of iterations is at most $d \leq n$. The starting mask is $\mathcal{B}_0 = (\{\epsilon\}, \{\epsilon\})$, which is clearly a complete and minimal mask.

The main inductive step is given by the following procedure. First, given $\mathcal{B}_i = (\mathcal{P}_i, \mathcal{S}_i)$ with $i \geq 0$, the algorithm reconstructs a WFA $\mathcal{A}_i$ by filling the corresponding Hankel sub-blocks using calls to MQ_f and then applying the reconstruction algorithm of Section 3.5. Second, it makes an equivalence query $\mathsf{EQ}_f(\mathcal{A}_i)$. If the answer is yes, the algorithm terminates. Otherwise, it receives a counter-example $w \in \Sigma^*$ such that $\mathcal{A}_i(w) \neq f(w)$. This is used to build the new Hankel mask $\mathcal{B}_{i+1} = (\mathcal{P}_{i+1}, \mathcal{S}_{i+1})$ as follows:

1. find a decomposition $w = uav$ where u is the longest prefix of w in $\mathcal{P}_i$;
2. let $\mathcal{S}_{i+1} = \mathcal{S}_i \cup \mathrm{suffs}(v)$, where $\mathrm{suffs}(v)$ is the set of all suffixes of v;
3. starting from $\mathcal{P}_{i+1} = \mathcal{P}_i$, and while $\mathrm{rank}([\mathbf{H}_{\mathcal{B}_{i+1}}^\top \mid \mathbf{H}_\Sigma^\top]) > s\,\mathrm{rank}(\mathbf{H}_{\mathcal{B}_{i+1}})$, keep adding to $\mathcal{P}_{i+1}$ prefixes $ua \in \mathcal{P}_{i+1}\Sigma$ such that $\mathrm{rank}([\mathbf{H}_{\mathcal{B}_{i+1}}^\top \mid \mathbf{H}_a(u,:)^\top]) = \mathrm{rank}(\mathbf{H}_{\mathcal{B}_{i+1}}) + 1$.

Note the resulting mask $\mathcal{B}_{i+1}$ is complete by construction and minimal because only prefixes that increase the rank of $\mathbf{H}_{\mathcal{B}_{i+1}}$ are added to $\mathcal{P}_{i+1}$. Also note that the algorithm maintains the property that $\mathcal{P}_i$ is prefix-closed and $\mathcal{S}_i$ suffix-closed. It is clear that if the algorithm terminates, it returns the correct answer. To prove that the algorithm terminates it suffices to show that at each iteration the inequality $|\mathcal{P}_{i+1}| > |\mathcal{P}_i|$ holds since this will guarantee that at each iteration the rank of $\mathbf{H}_{\mathcal{B}_i}$ increases. Since this rank can be at most $n = \text{rank}(\mathbf{H}_f)$, and since whenever $\text{rank}(\mathbf{H}_{\mathcal{B}_i}) = \text{rank}(\mathbf{H}_f)$ $\mathcal{B}_i$ is a complete minimal Hankel basis, Theorem 2 then shows that the WFA $\mathcal{A}_i$ computes f. The termination of the algorithm is guaranteed by the following result.

Lemma 1. *Let $\mathcal{B}'_i = (\mathcal{P}_i, \mathcal{S}_{i+1})$, where $\mathcal{S}_{i+1}$ is the set of suffixes obtained after processing the counter-example w received from the $(i+1)$th call to EQ_f. Then, the following inequality holds:* $\text{rank}([\mathbf{H}_{\mathcal{B}'_i}^{\top} \mid \mathbf{H}_{\Sigma}^{\top}]) > \text{rank}(\mathbf{H}_{\mathcal{B}'_i}^{\top})$.

Proof. Suppose that $\text{rank}([\mathbf{H}_{\mathcal{B}'_i}^{\top} \mid \mathbf{H}_{\Sigma}^{\top}]) = \text{rank}(\mathbf{H}_{\mathcal{B}'_i}^{\top})$ and let $\mathcal{A}'_i$ be the WFA reconstructed from $\mathcal{B}'_i$ by the algorithm in Section 3.5. Since $\mathcal{B}_i$ and $\mathcal{B}'_i$ share the same prefixes, both are minimal and complete, and $\mathcal{S}_i \subseteq \mathcal{S}_{i+1}$, then $\mathcal{A}_i$ and $\mathcal{A}'_i$ must compute the same function. Thus, we have $f(w) \neq \mathcal{A}_i(w) = \mathcal{A}'_i(w)$. On the other hand, $w = uav$ with $u \in \mathcal{P}_i$ and $v \in \mathcal{S}_{i+1}$. Thus, in the matrix $\mathbf{H}_a$ used to reconstruct $\mathcal{A}'_i$ we have $\mathbf{H}_a(u, v) = f(w)$, and by Theorem 3 it holds that $\mathcal{A}'_i(w) = f(w)$. We conclude by contradiction that $\text{rank}([\mathbf{H}_{\mathcal{B}'_i}^{\top}|\mathbf{H}_{\Sigma}^{\top}]) > \text{rank}(\mathbf{H}_{\mathcal{B}'_i}^{\top})$. □

We can now bound the number of queries made by the algorithm. First observe that the number of calls to EQ_f is $O(n)$ since one such call is made for each of the $d+1$ Hankel masks. To bound the number of calls to MQ_f, note that since we have $\mathcal{B}_i \subseteq \mathcal{B}_{i+1}$ for each i, at each stage most of the queries needed to fill $\mathbf{H}_{\mathcal{B}_{i+1}}$ have already been asked in previous iterations. Thus, it suffices to count the number of MQ_f queries needed to fill the matrices corresponding to the last Hankel mask $\mathcal{B}_d = (\mathcal{P}_d, \mathcal{S}_d)$. This number is clearly $(|\Sigma| + 1)|\mathcal{P}_d||\mathcal{S}_d|$.

Let L denote the length of the longest counter-example returned by the successive calls to EQ_f, we have $|\mathcal{S}_d| \leq 1 + dL$. This, combined with $|\mathcal{P}_d| = n$, shows that the total number of calls to MQ_f is in $O(|\Sigma|n^2 L)$. Note that this complexity is not optimal: [17] give an improved technique for processing counter-examples that yields an algorithm making only $O(|\Sigma|n^2 \log(L))$ calls to MQ_f.

4.2 Learning Stochastic WFAs from I.I.D. Samples

A *stochastic WFA* is a WFA computing a probability distribution. In this section, we consider the problem of learning a stochastic WFA and therefore assume that $\mathbb{S} = \mathbb{R}$. The learning scenario commonly adopted for stochastic WFAs is one where the learner receives a finite set of strings sampled i.i.d. from the target stochastic WFA. The objective of the learner is to use this training sample to learn a WFA computing a function close the target distribution with respect to some measure of accuracy.

In this section, we present an algorithm for this problem which consists of first using the training sample to estimate a sub-block of the Hankel matrix of the target WFA, and next of using the algorithm described in Section 3.7 to reconstruct a WFA based on those estimates. Several variants of this algorithm can be found in the literature, including [33,7] for the first such algorithms based on SVD, [10] for variants using prefix and substring statistics, and [6,9] for detailed analyses and further references.

A stochastic WFA over Σ is one that computes a probability distribution over Σ^*, that is, a WFA $\mathcal{A}$ with $\mathcal{A}(w) \geq 0$ for all $w \in \Sigma^*$ and $\sum_{w \in \Sigma^*} \mathcal{A}(w) = 1$. Probabilistic automata with stopping probabilities or absorbing states are typical examples of stochastic WFAs in this class (see [57,24] for a discussion of the relations between different finite-state machines computing probability distributions).

Let $\mathcal{A}$ be a fixed unknown target stochastic WFA $\mathcal{A}$. We assume that the learning algorithm receives a sample $S = (w_1, \ldots, w_m) \in (\Sigma^*)^m$ of m strings sampled i.i.d. from the distribution computed by $\mathcal{A}$. In addition to S, the algorithm receives the alphabet Σ, a number of states n that the output automaton should have, and a finite Hankel mask $\mathcal{B} = (\mathcal{P}, \mathcal{S})$ with $n \leq \min\{|\mathcal{P}|, |\mathcal{S}|\}$.

The first step of the algorithm is to compute empirical estimates of the matrices and vectors required by the SVD-based WFA reconstruction algorithm of Section 3.7: $\widehat{\mathbf{H}}_\mathcal{B}$, $\widehat{\mathbf{H}}_a$ for $a \in \Sigma$, $\widehat{\mathbf{h}}_\mathcal{P}$, and $\widehat{\mathbf{h}}_\mathcal{S}$. This is done by assigning to each entry in these matrices and vectors the relative frequency of the corresponding string in the sample $S = (w_1, \ldots, w_m)$. For example, for $u \in \mathcal{P}$ and $v \in \mathcal{S}$ the algorithm sets

$$\widehat{\mathbf{H}}_\mathcal{B}(u, v) = \frac{1}{m} \sum_{i=1}^{m} \mathbb{I}[w_i = uv] \ .$$

The same is done for $\widehat{\mathbf{H}}_a$, $\widehat{\mathbf{h}}_\mathcal{P}$, and $\widehat{\mathbf{h}}_\mathcal{S}$. These approximations are then used by the WFA reconstruction algorithm to obtain an automaton $\widehat{\mathcal{A}}$ with n states.

The empirical probabilities used in the estimations of the Hankel sub-blocks converge to the true probabilities as $m \to \infty$. One can also expect that the difference between the unknown probability distribution f and the function computed by $\widehat{\mathcal{A}}$ decreases as m increases. The next theorem gives a stronger guarantee which holds for finite samples, as opposed to a result holding in the limit. It is a *probably approximately correct* (PAC) learning guarantee: for a sample size m polynomial in the size of the $1/\epsilon$ where ϵ is the precision sought, $\log(1/\delta)$ where δ is the confidence parameter and several other parameters including $1/\mathfrak{s}_n(\mathbf{H}_\mathcal{B})$ where $\mathfrak{s}_n(\mathbf{H}_\mathcal{B})$ is the singular value of $\mathbf{H}_\mathcal{B}$ and the string length L, the WFA $\widehat{\mathcal{A}}$ returned by the algorithm is ϵ-close to f for the norm-1 over the set of strings of length at most L.

Theorem 6. *Let $\varepsilon > 0$. Then, for any $\delta \in (0,1)$, with probability at least $1 - \delta$ over the draw of a sample S of size $m \geq p(|\Sigma|, n, |\mathcal{P}|, |\mathcal{S}|, 1/\mathfrak{s}_n(\mathbf{H}_\mathcal{B}), L, 1/\varepsilon, \log(1/\delta))$ from the (target) probability distribution f, where p is a polynomial, the WFA $\widehat{\mathcal{A}}$ returned by the algorithm after* *receiving S, a complete Hankel basis $\mathcal{B} = (\mathcal{P}, \mathcal{S})$ for $\mathbf{H}_f$ and $n = \mathrm{rank}(\mathbf{H}_f)$ verifies the following inequality:*

$$\sum_{w \in \Sigma^{\leq L}} |f(w) - \widehat{A}(w)| \leq \varepsilon \ .$$

The proof of this result admits three components: Theorem 5, a concentration bound for the estimates $\widehat{\mathbf{H}}_\mathcal{B}$, and a bound relating accuracy in transition weights between A and $\widehat{A}$ to accuracy in the function they compute (see [33,6,9] for detailed proofs).

4.3   Learning WFAs from String–Value Pairs

In this section, we present an algorithm for learning WFAs in a more general scenario than the previous ones. This scenario was first introduced in [11]. The learning algorithm for WFAs described here is also due to [11].

Here, as in the standard supervised learning, the learner receives a labeled sample $S = ((w_1, y_1), \ldots, (w_m, y_m)) \in (\Sigma^* \times \mathbb{S})^m$ containing m string–value pairs $(w_i, y_i) \in \Sigma^* \times \mathbb{S}$, drawn i.i.d. according to some unknown distribution $\mathcal{D}$. The learning problem consists of finding a WFA A with small expected loss, that is with small $\mathbb{E}_{(w,y) \sim \mathcal{D}}[\ell(A(w), y)]$, where ℓ is a loss function defined over semiring pairs. We will consider again here the case $\mathbb{S} = \mathbb{R}$. The problem is then an instance of a regression learning problem. The loss function $\ell : \mathbb{R} \times \mathbb{R} \to \mathbb{R}_+$ is used to measure the closeness of the labels. Some common choices for ℓ are the quadratic loss defined for all $y, y' \in \mathbb{R}$ by $\ell_2(y, y') = (y - y')^2$ and the absolute loss defined by $\ell_1(y, y') = |y - y'|$.

Note that in this formulation we did not assume that the labels y in pairs (x, y) drawn from $\mathcal{D}$ are computed by some WFA. Thus, in learning-theoretic terms, we consider an *agnostic setting*.

Note also that one could find a WFA A consistent with the labeled sample, that is such that $A(w_i) = y_i$ for all $i \in [1, m]$. But, such a WFA could be large and might not benefit from a favorable expected loss. Furthermore, it was recently shown in [39] that the problem of finding the smallest WFA A consistent with the labeled sample is computationally hard.

A WFA minimizing the empirical loss $\frac{1}{m} \sum_{i=1}^m \ell(A(w_i), y_i)$ could *overfit* the training sample and typically would not benefit from favorable learning guarantees unless it is selected out of a less *complex* sub-family of WFAs. The algorithm we describe here avoids overfitting by constraining the choice of a WFA in two ways: by restricting the number of states, and by controlling the norm of a certain Hankel matrix.

The algorithm works in two stages. In the first stage, the sample S is used to find a sub-block of a Hankel matrix on a given mask. The second stage uses this Hankel block to reconstruct a WFA with a given number of states using the SVD-based method from Section 3.7. The algorithm receives as input the sample S, the alphabet Σ, a Hankel mask $\mathcal{B} = (\mathcal{P}, \mathcal{S})$ with $\epsilon \in \mathcal{P} \cap \mathcal{S}$, a number of states $k \leq \min\{|\mathcal{P}|, |\mathcal{S}|\}$, a convex loss function $\ell : \mathbb{R} \times \mathbb{R} \to \mathbb{R}_+$, and a regularization parameter $\lambda > 0$.

The first stage builds a basis $\mathcal{B}' = (\mathcal{P}', \mathcal{S})$ with $\mathcal{P}' = \mathcal{P} \cup \mathcal{P}\Sigma$ and a modified sample S' containing only those $(w_i, y_i) \in S$ such that $w_i \in \mathcal{P}'\mathcal{S}$. Then, the algorithm solves the convex optimization problem

$$\widehat{\mathbf{H}}_{\mathcal{B}'} \in \operatorname*{argmin}_{\mathbf{H} \in \mathbb{H}_{\mathcal{B}'}} \frac{1}{|S'|} \sum_{(w,y) \in S'} \ell(\mathbf{H}(w), y) + \lambda \|\mathbf{H}\|_* \ ,$$

where $\mathbb{H}_{\mathcal{B}'}$ denotes the set of all Hankel matrices $\mathbf{H} \in \mathbb{R}^{\mathcal{P}' \times \mathcal{S}}$, $\mathbf{H}(w)$ denotes $\mathbf{H}(u, v)$ for some arbitrary decomposition $w = uv$ with $u \in \mathcal{P}$ and $v \in \mathcal{S}$, and where $\|\mathbf{H}\|_*$ denotes the nuclear norm of $\mathbf{H}$ defined as the sum of the singular values of $\mathbf{H}$.

The second stage of the algorithm starts by extracting from $\widehat{\mathbf{H}}_{\mathcal{B}'}$ the Hankel sub-blocks associated with the Hankel mask $\mathcal{B}$: $\widehat{\mathbf{H}}_{\mathcal{B}}, \widehat{\mathbf{H}}_a$ for $a \in \Sigma$, $\mathbf{h}_{\mathcal{P}}$, and $\mathbf{h}_{\mathcal{S}}$. Then, it uses the SVD-based WFA reconstruction algorithm of Section 3.7 to obtain a WFA $\widehat{\mathcal{A}}$ with k states.

The design of the algorithm, and in particular the choice of the nuclear norm as a regularization term for finding the Hankel matrix $\widehat{\mathbf{H}}_{\mathcal{B}'}$ is supported by several properties. First, the nuclear norm is a convex surrogate for the rank function commonly used in machine learning algorithms [27]. By Theorem 1, low-rank Hankel matrices correspond to WFAs with small numbers of states, thus it favors the selection of smaller WFAs by the algorithm. A second justification is given by the following theorem, which provides a guarantee for learning with WFAs in terms of the nuclear norm of the associated Hankel matrix.

Let $M > 0$ and define $\tau_M : \mathbb{R} \to \mathbb{R}$ as the function defined by $\tau_M(y) = \operatorname{sign}(y)M$ if $|y| > M$, $\tau_M(y) = y$ otherwise. Let $S = ((w_1, y_1), \ldots, (w_m, y_m)) \in (\Sigma^* \times \mathbb{R})^m$. Given a decomposition $w_i = u^i v^i$, for any $1 \le i \le m$, we define $U_S = \max_{u \in \Sigma^*} |\{i : u^i = u\}|$ and $V_S = \max_{v \in \Sigma^*} ?|\{i : v^i = v\}|$. A measure of the complexity of S that will appear in the next theorem is $W_S = \min \max\{U_S, V_S\}$, where the minimum is taken over all possible decompositions of the strings w_i in S. For any $R > 0$, let $\mathcal{F}_R$ denote the following class of functions

$$\mathcal{F}_R = \{f(w) = \tau_M(\mathcal{A}(w)) : \mathcal{A} \text{ WFA}, \|\mathbf{H}_f\|_* \le R\} \ .$$

The following gives a learning bound for the algorithm just discussed.

Theorem 7. *Let ℓ_1 denote the absolute loss. Assume that there exists $M > 0$ such that $\mathbb{P}_{(w,y) \sim D}[|y| \le M] = 1$. Then, for any $\delta > 0$, with probability at least $1 - \delta$ over the draw of an i.i.d. sample S of size m from D, the following inequality holds simultaneously for all $f \in \mathcal{F}_R$:*

$$\mathop{\mathbb{E}}_{(w,y) \sim D}[\ell_1(f(w), y)] \le \frac{1}{m} \sum_{i=1}^{m} \ell_1(f(w_i), y_i) + 3M \sqrt{\frac{\log(\frac{2}{\delta})}{2m}}$$

$$+ O\left(\frac{R(\log(m+1) + \sqrt{W_S \log(m+1)})}{m} \right).$$

A similar result was first proven in [11] using a Frobenius norm regularizer instead of a nuclear norm. The analysis in [11] was based on a stability argument, and it is not clear how to extend it to the nuclear norm case, which is known to perform better than the Frobenius norm in some applications [51]. Theorem 7 is proven using a Rademacher complexity analysis of WFAs recently given by [8].

5   Conclusion

We presented a detailed survey of modern algorithms for learning WFAs. We highlighted the key role played by the notion of Hankel matrix and its properties in the design of these learning algorithms which are designed for different scenarios. These properties and the algorithms we described could inspire other variants of these algorithms as well as other algorithms.

Acknowledgments. This work was partly funded by the NSF award IIS-1117591 and NSERC.

References

1. Allauzen, C., Mohri, M., Riley, M.: Statistical modeling for unit selection in speech synthesis. In: Proceedings of ACL (2004)
2. Allauzen, C., Mohri, M., Talwalkar, A.: Sequence kernels for predicting protein essentiality. In: Proceedings of ICML (2008)
3. Aminof, B., Kupferman, O., Lampert, R.: Formal analysis of online algorithms. In: Bultan, T., Hsiung, P.-A. (eds.) ATVA 2011. LNCS, vol. 6996, pp. 213–227. Springer, Heidelberg (2011)
4. Angluin, D.: On the complexity of minimum inference of regular sets. Information and Control 3(39) (1978)
5. Angluin, D.: Learning regular sets from queries and counterexamples. Information and Computation 75(2) (1987)
6. Bailly, R.: Méthodes spectrales pour l'inférence grammaticale probabiliste de langages stochastiques rationnels. Ph.D. thesis, Aix-Marseille Université (2011)
7. Bailly, R., Denis, F., Ralaivola, L.: Grammatical inference as a principal component analysis problem. In: Proceedings of ICML (2009)
8. Balle, B., Mohri, M.: On the Rademacher complexity of weighted automata. In: Proceedings of ALT (2015)
9. Balle, B.: Learning Finite-State Machines: Statistical and Algorithmic Aspects. Ph.D. thesis, Universitat Politecnica de Catalunya (2013)
10. Balle, B., Carreras, X., Luque, F.M., Quattoni, A.: Spectral learning of weighted automata. Machine Learning 96(1–2) (2014)
11. Balle, B., Mohri, M.: Spectral learning of general weighted automata via constrained matrix completion. In: Proceedings of NIPS (2012)
12. Beimel, A., Bergadano, F., Bshouty, N.H., Kushilevitz, E., Varricchio, S.: On the applications of multiplicity automata in learning. In: Proceeding FOCS (1996)
13. Beimel, A., Bergadano, F., Bshouty, N.H., Kushilevitz, E., Varricchio, S.: Learning functions represented as multiplicity automata. Journal of the ACM 47(3) (2000)

14. Bergadano, F., Varricchio, S.: Learning behaviors of automata from multiplicity and equivalence queries. In: Bonuccelli, M.A., Crescenzi, P., Petreschi, R. (eds.) CIAC 1994. LNCS, vol. 778, pp. 54–62. Springer, Heidelberg (1994)
15. Bergadano, F., Varricchio, S.: Learning behaviors of automata from multiplicity and equivalence queries. SIAM Journal on Computing 25(6) (1996)
16. Berstel, J., Reutenauer, C.: Rational Series and Their Languages. Springer (1988)
17. Bisht, L., Bshouty, N.H., Mazzawi, H.: On optimal learning algorithms for multiplicity automata. In: Lugosi, G., Simon, H.U. (eds.) COLT 2006. LNCS (LNAI), vol. 4005, pp. 184–198. Springer, Heidelberg (2006)
18. Breuel, T.M.: The OCRopus open source OCR system. In: Proceedings of IS&T/SPIE (2008)
19. Cardon, A., Crochemore, M.: Détermination de la représentation standard d'une série reconnaissable. ITA 14(4), 371–379 (1980)
20. Carlyle, J.W., Paz, A.: Realizations by stochastic finite automata. J. Comput. Syst. Sci. 5(1) (1971)
21. Chalermsook, P., Laekhanukit, B., Nanongkai, D.: Pre-reduction graph products: Hardnesses of properly learning dfas and approximating edp on dags. In: Proceedings of FOCS (2014)
22. Cortes, C., Haffner, P., Mohri, M.: Rational kernels: Theory and algorithms. Journal of Machine Learning Research 5 (2004)
23. Droste, M., Kuich, W. (eds.): Handbook of weighted automata. EATCS Monographs on Theoretical Computer Science. Springer (2009)
24. Dupont, P., Denis, F., Esposito, Y.: Links between probabilistic automata and hidden markov models: probability distributions, learning models and induction algorithms. Pattern Recognition (2005)
25. Durbin, R., Eddy, S.R., Krogh, A., Mitchison, G.J.: Biological Sequence Analysis: Probabilistic Models of Proteins and Nucleic Acids. Cambridge University Press (1998)
26. Eilenberg, S.: Automata, Languages and Machines. Academic Press (1974)
27. Fazel, M.: Matrix rank minimization with applications. Ph.D. thesis, Stanford University (2002)
28. Fliess, M.: Matrices de Hankel. Journal de Mathématiques Pures et Appliquées 53 (1974)
29. Gold, E.M.: Complexity of automaton identification from given data. Information and Control 3(37) (1978)
30. Golub, G., Loan, C.V.: Matrix Computations. Johns Hopkins University Press (1983)
31. Haussler, D., Littlestone, N., Warmuth, M.K.: Predicting {0, 1}-functions on randomly drawn points. In: Proceedings of COLT (1988)
32. Hsu, D., Kakade, S.M., Zhang, T.: A spectral algorithm for learning hidden markov models. In: Proceedings of COLT (2009)
33. Hsu, D., Kakade, S.M., Zhang, T.: A spectral algorithm for learning hidden markov models. Journal of Computer and System Sciences 78(5) (2012)
34. II, K.C., Kari, J.: Image compression using weighted finite automata. Computers & Graphics 17(3) (1993)
35. Kaplan, R.M., Kay, M.: Regular models of phonological rule systems. Computational Linguistics 20(3) (1994)
36. Karttunen, L.: The replace operator. In: Proceedings of ACL (1995)
37. Kearns, M.J., Valiant, L.G.: Cryptographic limitations on learning boolean formulae and finite automata. Journal of ACM 41(1) (1994)

38. Kearns, M.J., Vazirani, U.V.: An Introduction to Computational Learning Theory. MIT Press (1994)
39. Kiefer, S., Marusic, I., Worrell, J.: Minimisation of multiplicity tree automata. In: Pitts, A. (ed.) FOSSACS 2015. LNCS, vol. 9034, pp. 297–311. Springer, Heidelberg (2015)
40. Kuich, W., Salomaa, A.: Semirings, Automata. Springer, Languages (1986)
41. Mohri, M.: Finite-state transducers in language and speech processing. Computational Linguistics 23(2) (1997)
42. Mohri, M.: Weighted automata algorithms. In: Handbook of Weighted Automata. Springer (2009)
43. Mohri, M., Pereira, F., Riley, M.: Weighted automata in text and speech processing. In: Proceedings of ECAI 1996 Workshop on Extended finite state models of language (1996)
44. Mohri, M., Pereira, F., Riley, M.: Speech recognition with weighted finite-state transducers. In: Handbook on Speech Processing and Speech Comm. Springer (2008)
45. Mohri, M., Pereira, F.C.N.: Dynamic compilation of weighted context-free grammars. In: Proceedings of COLING-ACL (1998)
46. Mohri, M., Rostamizadeh, A., Talwalkar, A.: Foundations of Machine Learning. MIT Press (2012)
47. Mohri, M., Sproat, R.: An efficient compiler for weighted rewrite rules. In: Proceedings of ACL (1996)
48. Mornhinweg, D., Shapiro, D.B., Valente, K.: The principal axis theorem over arbitrary fields. American Mathematical Monthly (1993)
49. Pereira, F., Riley, M.: Speech recognition by composition of weighted finite automata. In: Finite-State Language Processing. MIT Press (1997)
50. Pitt, L., Warmuth, M.K.: The minimum consistent DFA problem cannot be approximated within any polynomial. J. ACM 40(1) (1993)
51. Quattoni, A., Balle, B., Carreras, X., Globerson, A.: Spectral regularization for max-margin sequence tagging. In: Proceedings of ICML (2014)
52. Salomaa, A., Soittola, M.: Automata-Theoretic Aspects of Formal Power Series. Springer (1978)
53. Schützenberger, M.P.: On a special class of recurrent events. The Annals of Mathematical Statistics 32(4) (1961)
54. Schützenberger, M.P.: On the definition of a family of automata. Information and Control 4 (1961)
55. Sproat, R.: A finite-state architecture for tokenization and grapheme-to-phoneme conversion in multilingual text analysis. In: Proceedings of the ACL SIGDAT Workshop. ACL (1995)
56. Valiant, L.G.: A theory of the learnable. Commun. ACM 27(11) (1984)
57. Vidal, E., Thollard, F., de la Higuera, C., Casacuberta, F., Carrasco, R.C.: Probabilistic finite-state machines - part I. PAMI (2005)

More Than 1700 Years of Word Equations

Volker Diekert[✉]

Institut Für Formale Methoden der Informatik,
Universität Stuttgart, Stuttgart, Germany
diekert@fmi.uni-stuttgart.de

Abstract. Geometry and Diophantine equations have been ever-present in mathematics. According to the existing literature the work of Diophantus of Alexandria was mentioned before 364 AD, but a systematic mathematical study of word equations began only in the 20th century. So, the title of the present article does not seem to be justified at all. However, a Diophantine equation can be viewed as a special case of a system of word equations over a unary alphabet, and, more importantly, a word equation can be viewed as a special case of a Diophantine equation. Hence, the problem WordEquations: "Is a given word equation solvable?", is intimately related to Hilbert's 10th problem on the solvability of Diophantine equations. This became clear to the Russian school of mathematics at the latest in the mid 1960s, after which a systematic study of that relation began.

Here, we review some recent developments which led to an amazingly simple decision procedure for WordEquations, and to the description of the set of all solutions as an EDT0L language.

Word Equations

A word equation is easy to describe: it is a pair (U, V) where U and V are strings over finite sets of constants A and variables Ω. A *solution* is mapping $\sigma : \Omega \to A^*$ which is extended to homomorphism $\sigma : (A \cup \Omega)^* \to A^*$ such that $\sigma(U) = \sigma(V)$. Word equations are studied in other algebraic structures and frequently one is not interested only in satisfiability. For example, one may be interested in all solutions, or only in solutions satisfying additional criteria like *rational constraints* for free groups [6]. Here, we focus on the simplest case of word equations over free monoids; and by *WordEquations* we understand the formal language of all word equations (over a given finite alphabet A) which are satisfiable, that is, for which there exists a solution.

History

The problem WordEquations is closely related to the theory of Diophantine equations. The publication of Hilbert's 1900 address at the International Congress of Mathematicians listed 23 problems. The tenth problem (Hilbert 10) is:

© Springer International Publishing Switzerland 2015
A. Maletti (Ed.): CAI 2015, LNCS 9270, pp. 22–28, 2015.
DOI: 10.1007/978-3-319-23021-4_2

"Given a Diophantine equation with any number of unknown quantities and with rational integral numerical coefficients: To devise a process according to which it can be determined in a finite number of operations whether the equation is solvable in rational integers."

There is a natural encoding of a word equation as a Diophantine problem. It is based on the fact that two 2×2 integer matrices $\left(\begin{smallmatrix}1 & 0 \\ 1 & 1\end{smallmatrix}\right)$ and $\left(\begin{smallmatrix}1 & 1 \\ 0 & 1\end{smallmatrix}\right)$ generate a free monoid. Moreover, these matrices generate exactly those matrices on $\mathrm{SL}(2, \mathbb{Z})$ where all coefficients are natural numbers. This is actually easy to show, and also used in fast the "fingerprint" algorithm by Karp and Rabin [12]. A reduction from WordEquations to Hilbert 10 is now straightforward. For example, the equation $abX = Yba$ is solvable if and only if the following Diophantine system in unknowns $X_1, \ldots, Y_4$ is solvable over integers:

$$\left(\begin{smallmatrix}1 & 0 \\ 1 & 1\end{smallmatrix}\right) \cdot \left(\begin{smallmatrix}1 & 1 \\ 0 & 1\end{smallmatrix}\right) \cdot \left(\begin{smallmatrix}X_1 & X_2 \\ X_3 & X_4\end{smallmatrix}\right) = \left(\begin{smallmatrix}Y_1 & Y_2 \\ Y_3 & Y_4\end{smallmatrix}\right) \cdot \left(\begin{smallmatrix}1 & 1 \\ 0 & 1\end{smallmatrix}\right) \cdot \left(\begin{smallmatrix}1 & 0 \\ 1 & 1\end{smallmatrix}\right)$$

$$X_1 X_4 - X_2 X_3 = 1$$

$$Y_1 Y_4 - Y_2 Y_3 = 1$$

$$X_i \geq 0 \quad \& \quad Y_i \geq 0 \quad \text{for } 1 \leq i \leq 4$$

The reduction of a Diophantine system to a single Diophantine equation is classic. It is based on the fact that every natural number can be written as a sum of four squares. In the mid 1960s the following mathematical project was launched: show that Hilbert 10 is undecidable by showing that WordEquations is undecidable. The hope was to encode the computations of a Turing machine into a word equation. The project failed greatly, producing two great mathematical achievements. In 1970 Matiyasevich showed that Hilbert 10 is undecidable, based on number theory and previous work by Davis, Putnam, and Robinson, see the textbook [17]. A few years later, in 1977 Makanin showed that WordEquations is decidable [15].

In the 1980s, Makanin showed that the existential and positive theories of free groups are decidable [16]. In 1987 Razborov gave a description of all solutions for an equation in a free group via "Makanin-Razborov" diagrams [21, 22]. Finally, in a series of papers ending in [13] Kharlampovich and Myasnikov proved Tarski's conjectures dating back to the 1940s:

1. The elementary theory of free groups is decidable.
2. Free non-abelian groups are elementary equivalent.

The second result has also been shown independently by Sela [24].

It is not difficult to see (by encoding linear Diophantine systems over the naturals) that WordEquations is NP-hard, but the first estimations of Makanin's algorithm was something like

$$\mathrm{DTIME}\left(2^{2^{2^{2^{2^{2^{\mathrm{poly}(n)}}}}}}\right).$$

Over the years Makanin's algorithm was modified to bring the complexity down to EXPSPACE [9], see also the survey in [5]. For equations in free groups the complexity seemed to be much worse. Kościelski and Pacholski published a result that the scheme of Makanin's algorithm for free groups is not primitive recursive [14]. However, a few years later Plandowski and Rytter showed in [20] that solutions of word equations can be compressed by Lempel-Ziv encodings (actually by straight-line programs); and the conjecture was born that WordEquations is in NP; and, moreover, the same should be true for word equations over free groups. The conjecture has not yet been proved, but in 1999 Plandowski showed that WordEquations is in PSPACE [18,19]. The same is true for equations in free groups and allowing rational constraints we obtain a PSPACE-complete problem [6,10].

In 2013 Jeż applied *recompression* to WordEquations and simplified all (!) known proofs for decidability [11]. Actually, using his method he could describe all solutions of a word equation by a finite graph where the labels are of two types. Either the label is a compression $c \mapsto ab$ where a, b, c where letters or the label is a linear Diophantine system. His method copes with free groups and with rational constraints: this was done in [7].

Moreover, the method of Jeż led Ciobanu, Elder, and the present author to an even simpler description for the set of all solutions: it is an EDT0L language [3]. Such a simple structural description of solution sets was known before only for quadratic word equations by [8].

The notion of an *EDT0L system* refers to **E**xtended, **D**eterministic, **T**able, **0** *interaction, and* **L**indenmayer. There is a vast literature on Lindenmayer systems, see [23], but actually we need very little from the "Book of **L**".

Rational Sets of Endomorphisms

The starting point is a word equation (U, V) of length n over a set of constants A and set of variables $X_1, \ldots, X_k$ (without restriction, $|A| + k \leq n$). There is an nondeterministic algorithm which takes (U, V) as input and which works in space $\mathsf{NSPACE}(n \log n)$. The output is an extended alphabet $C \supseteq A$ of linear size in n and a finite trim nondeterministic automaton $\mathcal{A}$ where the arc labels are endomorphisms over C^*. The automaton $\mathcal{A}$ accepts therefore a rational set $\mathcal{R} = L(\mathcal{A}) \subseteq \mathrm{End}(C^*)$, and enjoys various properties which are explained next. The arc labels are restricted. An endomorphism used for an arc label is defined by mapping $c \mapsto u$ where $c \in C$ is a letter and u is some word of length at most 2. The monoid $\mathrm{End}(C^*)$ is neither free nor finitely generated, but $\mathcal{R}$ lives inside a finitely generated submonoid $H^* \subseteq \mathrm{End}(C^*)$ where H is finite. Thus, we can think of $\mathcal{R}$ as a rational (or regular) expression over a finite set of endomorphisms H as we are used to in standard formal language theory. For technical reasons it is convenient to assume that C contains a special symbol $\#$ whose main purpose is serve as a marker. The algorithm is designed in such a way that it yields an automaton $\mathcal{A}$ accepting a rational set $\mathcal{R}$ such that

$$\{h(\#) \mid h \in \mathcal{R}\} \subseteq \underbrace{A^* \# \cdots \# A^*}_{k-1 \text{ symbols } \#}.$$

Thus, applying the set of endomorphisms to the special symbol # we obtain a formal language in $(A^* \{\#\})^{k-1} A^*$. The set $\{h(\#) \mid h \in \mathcal{R}\}$ encodes a set of k-tuples over A^*. Due to Asfeld [1] we can take a description like $\{h(\#) \mid h \in \mathcal{R}\}$ as the very *definition* for EDT0L. Now, the result by Ciobanu et al. in [3] is the following equality:

$$\{h(\#) \mid h \in \mathcal{R}\} = \{\sigma(X_1)\# \cdots \#\sigma(X_k) \mid \sigma(U) = \sigma(V)\}.$$

Here, σ runs over all solutions of the equation (U, V). Hence, the set of all solutions for a given word equation is an EDT0L language.

The results stated in [3] are more general.[1] They cope with the existential theory of equations with rational constraints in finitely generated free products of free groups, finite groups, free monoids, and free monoids with involution. For example, they cover the existential theory of equations with rational constraints in the modular group PSL$(2, \mathbb{Z})$.

The NSPACE$(n \log n)$ algorithm produces some $\mathcal{A}$ whether or not (U, V) has a solution. (If there is no solution then the trimmed automaton $\mathcal{A}$ has no states accepting the empty set.) This shifts the viewpoint on how to solve equations. The idea is that $\mathcal{A}$ answers basic questions about the solution set of (U, V). Indeed, the construction in [3] is such that the following assertions hold.

- The equation (U, V) is solvable if and only if $L(\mathcal{A}) \neq \emptyset$.
- The equation (U, V) has infinitely many solutions if and only if $L(\mathcal{A})$ is infinite.

In particular, decision problems like "Is (U, V) satisfyable?" or "Does (U, V) have infinitely many solutions" can be answered in NSPACE$(n \log n)$ for finitely generated free products over free groups, finite groups, free monoids, and free monoids with involution. Actually, we conjecture that NSPACE$(n \log n)$ is the best complexity bound for WordEquations with respect to space. This conjecture might hold even if the problem WordEquations was in NP.

How to Solve a Linear Diophantine System

Many of the aspects of our method of solving word equations are present in the special case of solving a system of word equations over a unary alphabet. In this particular case Jeż's recompression is closely related to [2]. There are many other places where the following is explained, so in some sense we can view the rest of this section as folklore.

Assume that Alice wants to explain to somebody, say Bob, in a very short time, say 15 minutes, that the set of solvable linear Diophantine systems over integers is decidable. Assume that this fundamental insight is entirely new to Bob. Alice might start to explain something with Cramer's rule, determinants or Gaussian elimination, but Bob does not know any of these terms, so better not to start with a course on linear algebra within a time slot of 15 minutes.

---

[1] Full proofs are in [4].

What Bob knows is basic matrix operations and the notion of a linear Diophantine system:

$$AX = c, \text{ where } A \in \mathbb{Z}^{n \times n}, \ X = (X_1, \ldots, X_n)^T \text{ and } c \in \mathbb{Z}^{n \times 1}.$$

Here, the X_i are variables over natural numbers. (This is not essential, and actually makes the problem more difficult than looking for a solution over integers.)

The complexity of the problem depends on the or values n, $\|c\|_1 = \sum_i |c_i|$ and $\|A\|_1 = \sum_{i,j} |a_{ij}|$. Without restriction (by adding dummies) we have

$$\|c\|_1 \le \|A\|_1 . \tag{1}$$

Alice explains the compression algorithm with respect to a given solution $x \in \mathbb{N}^n$. Of course, the algorithm does not know the solution, so the algorithm uses nondeterministic guesses. This is allowed provided two properties are satisfied: soundness and completeness. Soundness means that a guess can never transform a unsolvable system into a solvable one. Completeness means that for every solution x, there is some choice of correct guesses such that the procedure terminates with a system which has a trivial solution.

So we begin by guessing a solution $x \in \mathbb{N}^n$. First, we can check whether $x = 0$ is a solution by looking at c. Indeed, $x = 0$ is a solution if and only if $c = 0$.

Hence, let us assume $x \ne 0$ (this might be possible even if $c = 0$.) We define a vector $b = c$. The vector b (and the solution x) will be modified during the procedure. Perform the following while-loop.

while $x \ne 0$

1. For all i define $x_i' = x_i - 1$ if x_i is odd and $x_i' = x_i$ otherwise. Thus, all x_i' are even. Rewrite the system with a new vector b' such that $Ax' = b'$. Note that

$$\|b'\|_1 \le \|b\|_1 + \|A\|_1 . \tag{2}$$

2. Now, all b_i' must be even. Otherwise we made a mistake and x was not a solution.
3. Define $b_i'' = b_i'/2$ and $x_i'' = x_i'/2$. We obtain a new system $AX = b''$ with solution $Ax'' = b''$.
4. Rename b'' and x'' as b and x.

end while.

The clue is that, since $\|b\|_1 \le \|A\|_1$ by Equation (1), we obtain by Equation (2) and the third step an invariant:

$$\|b''\|_1 = \|b'\|_1 /2 \le \|b\|_1 /2 + \|A\|_1 /2 \le \|A\|_1 .$$

The procedure is obviously sound. It is complete because in each round $\|x\|_1$ decreases and therefore termination is guaranteed for every solution as long as we make correct guesses. The final observation is that the procedure defines a finite graph. The vertices are the vectors $b \in \mathbb{Z}^n$ with $\|b\|_1 \leq \|A\|_1$. There are at most $\|A\|_1^{2n+1}$ such vectors. We are done! It is reported that the explanation of Alice took less than 15 minutes. It is not reported whether Bob understood.

Alice explanation has a bonus: there is more information. We can label the arcs according to our guesses with affine mappings of two types: either $x \mapsto x+1$ or $x \mapsto 2x$. Thus, we have a finite graph of at most exponential size where the arc labels are affine mappings of $x \mapsto \lambda x + \varepsilon$ with $\lambda \in \{1, 2\}$ and $\varepsilon \in \{0, 1\}^n$. Letting $b = 0$ be the initial state and the initial vector c the final state, we have a nondeterministic finite automaton which accepts a rational set $\mathcal{R}$ of affine mappings from $\mathbb{N}^n$ to itself. By construction, we obtain

$$\{x \in \mathbb{N}^n \mid Ax = c\} = \{h(0) \mid h \in \mathcal{R}\}.$$

References

1. Asveld, P.R.: Controlled iteration grammars and full hyper-AFL's. Information and Control **34**(3), 248–269 (1977)
2. Boudet, A., Comon, H.: Diophantine equations, Presburger arithmetic and finite automata. In: Kirchner, H. (ed.) CAAP 1996. LNCS, vol. 1059, pp. 30–43. Springer, Heidelberg (1996)
3. Ciobanu, L., Diekert, V., Elder, M.: Solution Sets for equations over free groups are EDT0L languages. In: Halldórsson, M.M., Iwama, K., Kobayashi, N., Speckmann, B. (eds.) ICALP 2015, Part II. LNCS, vol. 9135, pp. 134–145. Springer, Heidelberg (2015)
4. Ciobanu, L., Diekert, V., Elder, M.: Solution sets for equations over free groups are EDT0L languages. ArXiv e-prints, abs/1502.03426 (2015)
5. Diekert, V.: Makanin's algorithm. In: Lothaire, M., (eds.) Algebraic Combinatorics on Words. Encyclopedia of Mathematics and its Applications, vol. 90, chapter 12, pp. 387–442. Cambridge University Press (2002)
6. Diekert, V., Gutiérrez, C., Hagenah, Ch.: The existential theory of equations with rational constraints in free groups is PSPACE-complete. Information and Computation **202**, 105–40 (2005). Conference version in STACS 2001. LNCS 2010, pp. 170–182. Springer, Heidelberg (2001)
7. Diekert, V., Jeż, A., Plandowski, W.: Finding all solutions of equations in free groups and monoids with involution. In: Hirsch, E.A., Kuznetsov, S.O., Pin, J.É., Vereshchagin, N.K. (eds.) CSR 2014. LNCS, vol. 8476, pp. 1–15. Springer, Heidelberg (2014)
8. Ferté, J., Marin, N., Sénizergues, G.: Word-mappings of level 2. Theory Comput. Syst. **54**, 111–148 (2014)
9. Gutiérrez, C.: Satisfiability of word equations with constants is in exponential space. In: Proc. 39th Ann. Symp. on Foundations of Computer Science (FOCS 1998), pp. 112–119. IEEE Computer Society Press, Los Alamitos (1998)
10. Gutiérrez, C.: Satisfiability of equations in free groups is in PSPACE. In: Proceedings 32nd Annual ACM Symposium on Theory of Computing, STOC 2000, pp. 21–27. ACM Press (2000)
11. Jeż, A.: Recompression: a simple and powerful technique for word equations. In: Portier, N., Wilke, T. (eds.) STACS. LIPIcs, vol. 20, pp. 233–244. Schloss Dagstuhl-Leibniz-Zentrum für Informatik. To appear in JACM, Dagstuhl, Germany (2013)

12. Karp, R.M., Rabin, M.O.: Efficient randomized pattern-matching algorithms. IBM Journal of Research and Development **31**, 249–260 (1987)
13. Kharlampovich, O., Myasnikov, A.: Elementary theory of free non-abelian groups. J. of Algebra **302**, 451–552 (2006)
14. Kościelski, A., Pacholski, L.: Complexity of Makanin's algorithm. Journal of the Association for Computing Machinery **43**(4), 670–684 (1996)
15. Makanin, G.S.: The problem of solvability of equations in a free semigroup. Math. Sbornik **103**, 147–236 (1977). English transl. in Math. USSR Sbornik 32 (1977)
16. Makanin, G.S.: Decidability of the universal and positive theories of a free group. Izv. Akad. Nauk SSSR, Ser. Mat. **48** 735–749 (1984) (in Russian). nglish translation. In: Math. USSR Izvestija **25**(75–88) (1985)
17. Matiyasevich, Yu.: Hilbert's Tenth Problem. MIT Press, Cambridge (1993)
18. Plandowski, W.: Satisfiability of word equations with constants is in PSPACE. In: Proc. 40th Ann. Symp. on Foundations of Computer Science, FOCS 1999, pp. 495–500. IEEE Computer Society Press (1999)
19. Plandowski, W.: Satisfiability of word equations with constants is in PSPACE. Journal of the Association for Computing Machinery **51**, 483–496 (2004)
20. Plandowski, W., Rytter, W.: Application of lempel-ziv encodings to the solution of word equations. In: Larsen, K.G., Skyum, S., Winskel, G. (eds.) ICALP 1998. LNCS, vol. 1443, pp. 731–742. Springer, Heidelberg (1998)
21. Razborov, A.A.: On Systems of Equations in Free Groups. PhD thesis, Steklov Institute of Mathematics (1987) (in Russian)
22. Razborov, A.A.: On systems of equations in free groups. In: Combinatorial and Geometric Group Theory, pp. 269–283. Cambridge University Press (1994)
23. Rozenberg, G., Salomaa, A.: The Book of L. Springer (1986)
24. Sela, Z.: Diophantine geometry over groups VIII: Stability. Annals of Math. **177**, 787–868 (2013)

An Algebraic Geometric Approach
to Multidimensional Words

Jarkko Kari[(✉)] and Michal Szabados

Department of Mathematics and Statistics,
University of Turku, 20014 Turku, Finland
jkari@utu.fi

Abstract. We apply linear algebra and algebraic geometry to study infinite multidimensional words of low pattern complexity. By low complexity we mean that for some finite shape, the number of distinct sub-patterns of that shape that occur in the word is not more than the size of the shape. We are interested in discovering global regularities and structures that are enforced by such low complexity assumption. We express the word as a multivariate formal power series over integers. We first observe that the low pattern complexity assumption implies that there is a non-zero polynomial whose formal product with the power series is zero. We call such polynomials the annihilators of the word. The annihilators form an ideal, and using Hilbert's Nullstellensatz we construct annihilators of simple form. In particular, we prove a decomposition of the word as a sum of finitely many periodic power series. We consider in more details a particular interesting example of a low complexity word whose periodic decomposition contains necessarily components with infinitely many distinct coefficients. We briefly discuss applications of our technique in the Nivat's conjecture and the periodic tiling problem. The results reported here have been first discussed in a paper that we presented at ICALP 2015.

1 Introduction

A *multidimensional infinite word*, or simply a *configuration*, $c \in A^{\mathbb{Z}^d}$ is a d-dimensional infinite array filled with symbols from a (usually finite) *alphabet* A. For each *cell* $v \in \mathbb{Z}^d$, we denote by $c_v \in A$ the symbol in position v. Suppose that for some finite observation window $D \subseteq \mathbb{Z}^d$, the number of distinct patterns of shape D that exist in c is small, at most the cardinality $|D|$ of D. We investigate global regularities and structures in c that are enforced by such low local complexity assumption.

Suppose that the alphabet A is a subset of $\mathbb{Z}$. This can be established by renaming the symbols if A is finite. It is then possible to perform arithmetics on configurations; for example the sum of two configurations is defined cell wise. The main result that we report (Theorem 3) is that c can be expressed as a finite sum $c = c_1 + \cdots + c_m$ of *periodic* $c_1, \ldots, c_m \in \mathbb{Z}^{(\mathbb{Z}^d)}$. Recall that a configuration e is called *periodic* if it is invariant under some translation, so that there is

© Springer International Publishing Switzerland 2015
A. Maletti (Ed.): CAI 2015, LNCS 9270, pp. 29–42, 2015.
DOI: 10.1007/978-3-319-23021-4_3

a vector $u \in \mathbb{Z}^d \setminus \{0\}$ such that $\forall v \in \mathbb{Z}^d : e_v = e_{v+u}$. Note that the periodic components c_i in the decomposition $c = c_1 + \cdots + c_m$ are not necessarily over any finite alphabet, but they are allowed to contain infinitely many distinct integer values. After the main result we present and analyze an example of a low local complexity configuration c over two letters, whose periodic decomposition uses *necessarily* an infinite alphabet. Finally, we briefly discuss applications of our results on two open problems: Nivat's conjecture [Niv97] and the periodic tiling problem [LW96].

To prove our main Theorem 3 we proceed in two steps.

(1) We show how the low complexity assumption on c implies that there is a non-trivial *filter* that annihilates c to the zero configuration. The filtering operation is the usual convolution of c with a finite mask, which we conveniently express in terms of multiplication by a multivariate polynomial. This step is based on basic linear algebra.

(2) We analyze configurations annihilated by non-trivial filtering, that is, by multiplying them with some non-zero polynomial. The set of annihilating polynomials is an ideal of the polynomial ring. Using Hilbert's Nullstellensatz we show that the annihilator ideal contains polynomials of simple form. In particular, we show that the configuration can be annihilated by a product of difference filters $(X^v - 1)$ that subtract from a configuration its translated copy. This in turn implies a decomposition of the configuration into a sum of periodic components.

The result reported here have been presented in [KS15], except for the proofs related to the example in Section 5.

2   Preliminaries

Classically, configurations are just assignments $c : \mathbb{Z}^d \longrightarrow A$ of symbols of a (finite or infinite) alphabet A on an infinite grid. We use the subscript notation c_v for the symbol assigned in cell $v \in \mathbb{Z}^d$. In order to apply algebra it is convenient to let the symbols in A be numbers, and to represent c as a *formal power series* over d variables $x_1, \ldots, x_d$ and with coefficients in A:

$$c(x_1, \ldots, x_d) = \sum_{v_1=-\infty}^{\infty} \cdots \sum_{v_d=-\infty}^{\infty} c_{v_1, \ldots, v_d} x_1^{v_1} \ldots x_d^{v_d}.$$

As usual, we abbreviate the vector $(x_1, \ldots, x_d)$ of variables as X, and write monomial $x_1^{v_1} \ldots x_d^{v_d}$ as X^v for $v = (v_1, \ldots, v_d) \in \mathbb{Z}^d$. Configuration c can now be expressed compactly as

$$c(X) = \sum_{v \in \mathbb{Z}^d} c_v X^v. \tag{1}$$

Usually we let $A \subseteq \mathbb{Z}$ so that configurations are power series with integer coefficients, but to use Nullstellensatz we need an algebraically closed field, so that frequently we consider multivariate power series and polynomials over $\mathbb{C}$. Anyway, for $R = \mathbb{Z}$ or $R = \mathbb{C}$, we denote by $R[[X^{\pm 1}]]$ the set of formal power series as in (1) with coefficients c_v in domain R. Note that we include negative exponents in the series. We call power series (1) *integral* if all coefficients c_v are integers, and it is *finitary* if there are only finitely many distinct coefficients c_v. In our usual setup $A \subseteq \mathbb{Z}$ is finite so that the corresponding power series is finitary and integral.

A *polynomial* over R is a formal sum $a(X) = \sum a_v X^v$ where $a_v \in R$ and the sum is over a finite set of d-tuples $v = (v_1, \ldots, v_d)$ with non-negative coordinates $v_i \geq 0$. If the coordinates are also allowed be negative we get a *Laurent polynomial* over R. We denote by $R[X]$ and $R[X^{\pm 1}]$ the sets of polynomials and Laurent polynomials over R. We sometimes use the term *proper* polynomial when we want to emphasize that $a(X)$ is a polynomial and not only a Laurent polynomial.

Here are some notational remarks: We use both notations $a(X)$ and a to denote (Laurent) polynomials and power series, that is, we may or may not explicitly write the formal variable in the notation. For any formal polynomial, Laurent polynomial or power series a we denote by a_v the value in cell v, that is, the coefficient of monomial X^v. Sometimes we may wish to write the coefficients explicitly differently, e.g., we may write $f(X) = \sum a_v X^v$.

The *support* of a polynomial or a Laurent polynomial $a(X)$ is the set

$$\mathrm{supp}(a) = \{v \in \mathbb{Z}^d \mid a_v \neq 0\} \tag{2}$$

of cells with non-zero value.

The formal product between a power series and a (Laurent) polynomial is defined the usual way, as a convolution. This is a filtering operation, and the result is again a power series. Note that multiplying a power series with monomial X^v is equivalent to translating it by the vector v. It follows that power series $c(X)$ is *periodic* with period v if and only if $(X^v - 1)c = 0$. We say that $(X^v - 1)$ annihilates $c(X)$.

3   Step 1: From Low Local Complexity to an Annihilating Filter

We are studying configurations in which the number of distinct patterns of some finite shape D is at most the size $|D|$ of the shape. More precisely, for any finite $D \subseteq \mathbb{Z}^d$ we denote by π_D the *projection* operator on $R[[X^{\pm 1}]]$ defined by

$$\pi_D(c) = \sum_{v \in D} c_v X^v,$$

and define the D-patterns of c to be the elements of

$$\mathrm{Patt}_D(c) = \{\pi_D(X^u c) \mid u \in \mathbb{Z}^d \}.$$

Configuration c has *low complexity* with respect to a finite $D \subseteq \mathbb{Z}^d$ if

$$| \operatorname{Patt}_D(c)| \leq |D|, \tag{3}$$

and we say that c has low complexity if (3) is satisfied for some finite D.

We say that a Laurent polynomial $f(X)$ *annihilates* configuration $c(X)$ if $f(X)c(X) = 0$. The following lemma guarantees that each low complexity configuration is annihilated by some non-zero Laurent polynomial, and hence also by a non-zero proper polynomial.

Lemma 1. *Let R be a field or $R = \mathbb{Z}$. Let $c(X) \in R[[X^{\pm 1}]]$ be a configuration and $D \subset \mathbb{Z}^d$ a finite set such that $| \operatorname{Patt}_D(c)| \leq |D|$. Then there exists a non-zero polynomial $f(X) \in R[X]$ such that $f(X)c(X) = 0$.*

Proof. Let R be a field. We use elementary linear algebra. Let $D = \{u_1, \ldots, u_n\}$. By the low complexity assumption, the set

$$\{(1, c_{u_1 + v}, \ldots, c_{u_n + v}) \mid v \in \mathbb{Z}^d\}$$

of vectors in R^{n+1} contains at most $n = |D|$ elements. There exists hence a non-zero vector $(a_0, a_1, \ldots, a_n)$ orthogonal to the set. Consider the product of $c(X)$ and the Laurent polynomial $g(X) = a_1 X^{-u_1} + \cdots + a_n X^{-u_n}$. In any position v, the coefficient in the product $g(X)c(X)$ is

$$a_1 c_{u_1 + v} + \cdots + a_n c_{u_n + v} = -a_0.$$

Hence the product is a constant configuration, so that $(X^v - 1)g(X)c(X) = 0$ for any v. We conclude that $c(X)$ is annihilated by all non-zero Laurent polynomials $h(X) = (X^v - 1)g(X)$.

To obtain a non-zero proper polynomial that annihilates c, notice that if $h(X)$ is an annihilator of $c(X)$, so is $a(X)h(X)$ for any Laurent polynomial $a(X)$. In particular, by choosing $a(X) = X^u$ for $u \in \mathbb{Z}^d$ with sufficiently large coordinates, we have that $f(X) = X^u h(X) \in R[X]$ is a polynomial.

Consider then the case $R = \mathbb{Z}$. By the proof above (for $R = \mathbb{Q}$) we see that there exists a non-zero polynomial $f(X) \in \mathbb{Q}[X]$ such that $f(X)c(X) = 0$. There is a positive integer m such that $m \cdot f(X) \in \mathbb{Z}[X]$, so that $m \cdot f(X)$ satisfies the claim. $\qquad \square$

As a first application of this simple observation we infer the classical Morse-Hedlund theorem [MH38]. Consider the case $d = 1$, and hence a one-variable configuration $c(x) \in \mathbb{C}[[x^{\pm 1}]]$ that satisfies the low complexity assumption. By Lemma 1, there is a (one variable) polynomial $f(x)$ that annihilates $c(x)$. Multiplying by a suitable monomial, we can take an annihilating $f(x)$ with the constant term one:

$$f(x) = 1 + a_1 x + a_2 x^2 + \ldots a_n x^n.$$

Now $f(x)c(x) = 0$ means that, for all $i \in \mathbb{Z}$,

$$c_i = a_1 c_{i-1} + a_2 c_{i-2} + \cdots + a_n c_{i-n},$$

so that the symbol in position i is determined by the n symbols on its left. A deterministic process on a finite set is necessarily periodic, so clearly c has to be a periodic configuration. We have established

Theorem 1 (Morse, Hedlund 1938). *If a one-dimensional bi-infinite word contains at most n distinct subwords of length n then the word is periodic.*

4   Step 2: From an Annihilating Filter to a Periodic Decomposition

Let c be a configuration. We define

$$\mathrm{Ann}(c) = \big\{\, f \in \mathbb{C}[X] \mid fc = 0 \,\big\}$$

to be the set of polynomials that annihilate it. Note that $\mathrm{Ann}(c)$ contains proper polynomials only. Note also that we take complex polynomials so that we can apply Hilbert's Nullstellensatz that requires an algebraically closed field.

It is easy to see that $\mathrm{Ann}(c)$ is an ideal of the polynomial ring $\mathbb{C}[X]$, the *annihilator ideal* of configuration c. We always have $0 \in \mathrm{Ann}(c)$ where 0 is the zero polynomial with zero coefficients. If $\mathrm{Ann}(c) = \{0\}$ then the annihilator ideal is *trivial*; if $\mathrm{Ann}(c)$ contains also some non-zero polynomial then it is *non-trivial*. By Lemma 1, the annihilator ideal of a low complexity configuration is always non-trivial. It is also easy to see that if c is integral and $\mathrm{Ann}(c)$ is non-trivial then $\mathrm{Ann}(c)$ contains a non-zero polynomial from $\mathbb{Z}[X]$, that is, a polynomial with integer coefficients.

More generally, if $\mathcal{C}$ is a set of configurations (e.g., a subshift), we let

$$\mathrm{Ann}(\mathcal{C}) = \big\{\, f \in \mathbb{C}[X] \mid fc = 0 \text{ for all } c \in \mathcal{C} \,\big\}$$

be the set of common annihilators. Again, $\mathrm{Ann}(\mathcal{C})$ is an ideal of the polynomial ring.

If $Z = (z_1, \ldots, z_d) \in \mathbb{C}^d$ is a complex vector then it can be plugged into a polynomial, producing a complex value. In particular, plugging into a monomial X^v results in $Z^v = z_1^{v_1} \cdots z_d^{v_d}$.

In this section we use Hilbert's Nullstellensatz as a tool to infer other elements of the ideal $\mathrm{Ann}(c)$. Recall the statement of the Nullstellensatz: Suppose $g(X)$ is a polynomial such that $g(Z) = 0$ for all common roots Z of $\mathrm{Ann}(c)$, that is, for all $Z \in \mathbb{C}^d$ such that $f(Z) = 0$ for all $f \in \mathrm{Ann}(c)$. Then $g^n(X) \in \mathrm{Ann}(c)$ for some n.

First we show that annihilating integral polynomials can be spatially "blown-up":

Lemma 2. *Let $c(X)$ be a finitary integral configuration and $f(X) \in \mathrm{Ann}(c)$ a non-zero integral polynomial, that is, $f(X) \in \mathrm{Ann}(c) \cap \mathbb{Z}[X]$. Then there exists an integer r such that for every positive integer n relatively prime to r we have $f(X^n) \in \mathrm{Ann}(c)$.*

Proof. Denote $f(X) = \sum a_v X^v$. First we prove the claim for the case when n is a large enough prime.

Let p be a prime, then we have $f^p(X) \equiv f(X^p) \pmod{p}$. Because f annihilates c, multiplying both sides by $c(X)$ results in

$$0 \equiv f(X^p)c(X) \pmod{p}.$$

The coefficients in $f(X^p)c(X)$ are bounded in absolute value by

$$s = c_{max} \sum |a_v|,$$

where c_{max} is the maximum absolute value of coefficients in c. Therefore if $p > s$ we have $f(X^p)c(X) = 0$.

For the general case, set $r = s!$. Now every n relatively prime to r is of the form $p_1 \cdots p_k$ where each p_i is a prime greater than s. Note that we can repeat the argument with the same bound s also for polynomials $f(X^m)$ for arbitrary m – the bound s depends only on c and the (multi)set of coefficients a_v, which is the same for all $f(X^m)$. Thus we have $f(X^{p_1 \cdots p_k}) \in \mathrm{Ann}(c)$.   □

The next lemma establishes a polynomial $g(X)$ of simple form that becomes zero at all common roots of $\mathrm{Ann}(c)$:

Lemma 3. *Let c be a finitary integral configuration and $f(X) = \sum a_v X^v$ a non-trivial integral polynomial annihilator. Let $S = \mathrm{supp}(f)$ be the support of $f(X)$. Define*

$$g(X) = x_1 \cdots x_d \prod_{\substack{v \in S \\ v \neq v_0}} (X^{rv} - X^{rv_0})$$

where r is the integer from Lemma 2 and $v_0 \in S$ arbitrary. Then $g(Z) = 0$ for any common root $Z \in \mathbb{C}^d$ of $\mathrm{Ann}(c)$.

Proof. Fix Z such that $h(Z) = 0$ for all $h \in \mathrm{Ann}(c)$. If any of its complex coordinates is zero then clearly $g(Z) = 0$. For this reason we included $x_1 \cdots x_d$ as a factor of $g(X)$.

Assume then that all coordinates of Z are non-zero. Let us define for $\alpha \in \mathbb{C}$

$$S_\alpha = \{ v \in S \mid Z^{rv} = \alpha \},$$
$$f_\alpha(X) = \sum_{v \in S_\alpha} a_v X^v.$$

Because S is finite, there are only finitely many non-empty sets $S_{\alpha_1}, \ldots, S_{\alpha_m}$ and they form a partitioning of S. In particular we have $f = f_{\alpha_1} + \cdots + f_{\alpha_m}$.

Numbers of the form $1 + ir$ are relatively prime to r for all non-negative integers i, therefore by Lemma 2, $f(X^{1+ir}) \in \mathrm{Ann}(c)$. Plugging in Z we obtain $f(Z^{1+ir}) = 0$. Now compute:

$$f_\alpha(Z^{1+ir}) = \sum_{v \in S_\alpha} a_v Z^{(1+ir)v} = \sum_{v \in S_\alpha} a_v Z^v \alpha^i = f_\alpha(Z)\alpha^i$$

Summing over $\alpha = \alpha_1, \ldots, \alpha_m$ gives

$$0 = f(Z^{1+ir}) = f_{\alpha_1}(Z)\alpha_1^i + \cdots + f_{\alpha_m}(Z)\alpha_m^i.$$

Let us rewrite the last equation as a statement about orthogonality of two vectors in $\mathbb{C}^m$:

$$(f_{\alpha_1}(Z), \ldots, f_{\alpha_m}(Z)) \perp (\alpha_1^i, \ldots, \alpha_m^i)$$

By Vandermode determinant, for $i \in \{0, \ldots, m-1\}$ the vectors on the right side span the whole $\mathbb{C}^m$. Therefore the left side must be the zero vector, and especially for α such that $v_0 \in S_\alpha$ we have

$$0 = f_\alpha(Z) = \sum_{v \in S_\alpha} a_v Z^v.$$

Because Z does not have zero coordinates, each term on the right hand side is non-zero. But the sum is zero, therefore there are at least two vectors $v_0, v \in S_\alpha$. From the definition of S_α we have $Z^{rv} = Z^{rv_0} = \alpha$, so Z is a root of $X^{rv} - X^{rv_0}$. $\square$

Now we are ready to apply the Nullstellensatz to obtain a simple annihilator:

Theorem 2. *Let c be a finitary integral configuration with a non-trivial annihilator. Then there are non-zero $v_1, \ldots, v_m \in \mathbb{Z}^d$ such that the Laurent polynomial*

$$(X^{v_1} - 1) \cdots (X^{v_m} - 1)$$

annihilates c.

Proof. This is an easy corollary of Lemma 3. First notice that the non-trivial annihilator can be taken so that it has integer coefficients. The polynomial $g(X)$ provided by Lemma 3 vanishes on all common roots of $\mathrm{Ann}(c)$, therefore by Hilbert's Nullstellensatz there is n such that $g^n(X) \in \mathrm{Ann}(c)$. Note that any monomial multiple of an annihilator is again an annihilator. Therefore also

$$\frac{g^n(X)}{x_1^n \cdots x_d^n X^{nrv_0(|S|-1)}}$$

is, and it is a Laurent polynomial of the desired form. $\square$

Multiplying a configuration by $(X^v - 1)$ is a "difference operator" on the configuration. Theorem 2 then says that there is a sequence of difference operators which annihilates the configuration. We can reverse the process: let us start by the zero configuration and step by step "integrate" until we obtain the original configuration. This idea gives the Decomposition theorem:

Theorem 3 (Decomposition theorem [KS15]). *Let c be a finitary integral configuration with a non-trivial annihilator. Then there exist periodic integral configurations $c_1, \ldots, c_m$ such that $c = c_1 + \cdots + c_m$.*

5   An Example

In this section we illustrate how the theory applies to a concrete example. Its properties were briefly mentioned in [KS15], without proofs. Recall that configurations are not assumed to be finitary or integral unless explicitly stated so.

Fix $\alpha \in \mathbb{R}$ irrational and define two-dimensional configurations $c^{(1)}, c^{(2)}, c^{(3)}$ and s by

$$c^{(1)}_{ij} = -\lfloor i\alpha \rfloor, \qquad c^{(2)}_{ij} = -\lfloor j\alpha \rfloor, \qquad c^{(3)}_{ij} = \lfloor (i+j)\alpha \rfloor,$$

$$s = c^{(1)} + c^{(2)} + c^{(3)}.$$

Then s is a finitary integral configuration over the alphabet $\{0, 1\}$. Obviously, $c^{(1)}, c^{(2)}, c^{(3)}$ are periodic in directions $(0, 1), (1, 0), (-1, 1)$ respectively, but they are not finitary. In the following we prove that s cannot be expressed as a finite sum of finitary periodic configurations.

There is a certain symmetry in s which becomes apparent when the configuration is affinely transformed such that these three directions become symmetric. In that case, it is natural to show the coefficients in a hexagonal grid, see Figure 1.

Fig. 1. The configuration s from Section 5 when α is the golden ratio is shown on the left. On the right the configuration is skewed such that the three directions $(0, 1), (1, 0)$ and $(1, -1)$ became symmetrical, the bottom left corner is preserved.

For any Laurent polynomials $f_1, \ldots, f_n \in \mathbb{C}[X^{\pm 1}]$, we let

$$\langle f_1, \ldots, f_n \rangle = \{g_1 f_1 + \cdots + g_n f_n \mid g_1, \ldots, g_n \in \mathbb{C}[X^{\pm 1}]\}$$

be the Laurent polynomial ideal they generate. Note that in this notation we let all involved polynomials be Laurent so that this is not a polynomial ideal. For Laurent polynomials $f(X)$ and $g(X)$, we denote $f \equiv g \bmod \langle f_1, \ldots, f_n \rangle$ if and only if $f(X) - g(X) \in \langle f_1, \ldots, f_n \rangle$.

A (Laurent) polynomial $a(X)$ is called a *line (Laurent) polynomial* if the support supp(a) defined by (2) contains at least two points and all the points of the support lie on a single line. If $u, v \in \mathbb{Z}^d$ are such that $\{u + tv \mid t \in \mathbb{R}\}$ contains the support of a line (Laurent) polynomial $a(X)$ then we say that v is a *direction* of $a(X)$. By *rational directions* we mean elements of $\mathbb{Z}^d \setminus \{0\}$. We say that two line (Laurent) polynomials are *parallel* if they have the same directions.

Let configuration $c \in \mathbb{C}[[X^{\pm 1}]]$ be such that Ann(c) contains a line polynomial. We call such c *directed*. This terminology applies to both finitary and non-finitary configurations. Notice that for any line Laurent polynomial that annihilates c there is a parallel line polynomial in Ann(c), obtained by multiplying it with a monomial. If all line polynomials in Ann(c) are parallel to each other, we say that c is *one-directed*, and if c has non-parallel annihilating line polynomials we say that c is *multi-directed*. Non-finitary configurations can be directed without being periodic, but if c is finitary then it is one-directed if and only if it is periodic in one direction only, and it is multi-directed if and only if it has several directions of periodicity. In the two-dimensional setting $d = 2$, such configurations are sometimes called singly periodic and doubly periodic, respectively.

It is well known that a doubly periodic configuration is periodic in all rational directions. An analogous statement holds more generally for two-dimensional directed configurations:

Lemma 4. *If $a(X)$ and $b(X)$ are non-parallel two-dimensional line Laurent polynomials then $\langle a, b \rangle$ contains line Laurent polynomials in all rational directions. In particular, in two dimensions, if Ann(c) contains two non-parallel line polynomials then it contains a line polynomial in every rational direction.*

Proof. The proof is easy using simple algebraic geometry and zero dimensionality of $\langle a, b \rangle$. Here we give it as an elementary linear algebraic reasoning. It is easy to see that there is a finite domain $D \subseteq \mathbb{Z}^2$ (a parallelogram determined by the supports of a and b) such that for any Laurent polynomial f there is a Laurent polynomial $f' \equiv f \bmod \langle a, b \rangle$ with support supp(f') $\subseteq D$.

Let $u \in \mathbb{Z}^2 \setminus \{(0,0)\}$ be any rational direction. Consider the monomials $X^0, X^u, X^{2u}, \ldots$ and, for each $k = 0, 1, 2 \ldots$, let $f_k(X) \equiv X^{ku} \bmod \langle a, b \rangle$ be the representative with supp(f_k) $\subseteq D$. It follows from the finiteness of the support that $f_1, f_2, \ldots$ are linearly dependant, and hence there is a non-zero vector $(a_0, a_1, \ldots, a_n)$ of coefficients such that $a_0 f_0(X) + \cdots + a_n f_n(X) = 0$. But then $f(X) = a_0 X^0 + a_1 X^u + \cdots + a_n X^{nu}$ is in $\langle a, b \rangle$. If $f(X)$ is a monomial then $1 \in \langle a, b \rangle$ and hence $\langle a, b \rangle$ contains all Laurent polynomials. Otherwise $f(X)$ has at least two non-zero coefficients and it is then a line Laurent polynomial in direction u. □

The next lemma states that one-directed configurations in different directions are linearly independent.

Lemma 5. *Let $c_1(X), \dots, c_n(X)$ be two-dimensional configurations that are one-directed and pairwise non-parallel. Then $a_1, \dots, a_n \in \mathbb{C}$ satisfy $a_1 c_1(X) + \cdots + a_n c_n(X) = 0$ if and only if $a_1 = \cdots = a_n = 0$.*

Proof. We prove the claim by induction on n. Case $n = 1$: since $c_1(X)$ is one-directed, it is not the zero power series. Hence $a_1 c_1(X) = 0$ if and only if $a_1 = 0$.

Suppose then the claim has been proved for $n - 1$, and let $a_1, \dots, a_n$ be such that

$$a_1 c_1(X) + \cdots + a_n c_n(X) = 0. \tag{4}$$

Because $c_n(X)$ is one-directed it is annihilated by some line Laurent polynomial $a(X)$. We multiply (4) by $a(X)$.

Let $1 \le i \le n - 1$ and consider $a(X) c_i(X)$. It is annihilated by the same line Laurent polynomial that annihilates $c_i(X)$ so it is directed. If it were multi-directed then, by Lemma 4, it would be annihilated by some line Laurent polynomial $b(X)$ that is parallel to $a(X)$. Then $c_i(X)$ would be annihilated by the line Laurent polynomial $a(X) b(X)$ that is parallel to $a(X)$, a contradiction with the fact that $c_i(X)$ and $c_n(X)$ are one-directed in different directions. We conclude that $a(X) c_i(X)$ is one-directed in the same direction as $c_i(X)$.

Multiplying (4) by $a(X)$ implies that

$$a_1 a(X) c_1(X) + \cdots + a_{n-1} a(X) c_{n-1}(X) = 0.$$

By the inductive hypothesis, $a_1 = \cdots = a_{n-1} = 0$. Case $n = 1$ applied to $a_n c_n(X) = 0$ shows that also $a_n = 0$. $\qquad\square$

Now we are ready to analyze the configuration $s = c^{(1)} + c^{(2)} + c^{(3)}$ defined at the beginning of this section. We want to show that it is not a sum of finitely many periodic finitary configurations. Suppose the contrary: $c^{(1)} + c^{(2)} + c^{(3)} = f_1 + \cdots + f_n$ for some periodic finitary $f_i(X)$. By moving the terms on the same side, and combining terms that are directed in the same direction, we obtain that

$$(c^{(1)} + p_1) + (c^{(2)} + p_2) + (c^{(3)} + p_3) + p_4 + \cdots + p_m = 0, \tag{5}$$

for some directed finitary $p_i(X)$ with the following properties:

– Configurations $p_1(X), p_2(X)$ and $p_3(X)$ have line Laurent polynomial annihilators in the same directions $(0, 1)$, $(1, 0)$ and $(-1, 1)$ as $c^{(1)}(X), c^{(2)}(X)$ and $c^{(3)}(X)$, respectively. They may have line annihilators also in other directions so that any doubly periodic $f_i(X)$ in the original bounded periodic decomposition may be added in them.
– Configurations $p_4(X), \dots, p_m(X)$ are one-directed in pairwise non-parallel directions. These directions are also not parallel to the directions $(0, 1)$, $(1, 0)$ and $(-1, 1)$ of the line annihilators of $c^{(1)}(X), c^{(2)}(X)$ and $c^{(3)}(X)$.

Lemma 6. *In* (5), *configurations* $c^{(k)} + p_k$ *are one-directed, for* $k = 1, 2, 3$.

Proof. It is clear that $c^{(k)} + p_k$ is directed in the same direction as $c^{(k)}$, so it is enough to show that it is not multi-directed. For $k = 1$ or $k = 3$ let us read the coefficients of $c^{(k)} + p_k$ horizontally along cells $\ldots, (-1, 0), (0, 0), (1, 0), \ldots$, and in the case $k = 2$ along the vertical line $\ldots, (0, -1), (0, 0), (0, 1), \ldots$. In each case we obtain a one-dimensional configuration $d(x) = c(x) + p(x)$ with $c_i = \lfloor i\alpha \rfloor$ for all $i \in \mathbb{Z}$, and with $p(x)$ finitary. (Note that in the cases $k = 1$ and $k = 2$ we negate the coefficients to get from $-\lfloor i\alpha \rfloor$ to $\lfloor i\alpha \rfloor$.)

If $c^{(k)} + p_k$ is multi-directed then by Lemma 4 it has an annihilating line Laurent polynomial in every direction and then, in particular, in the horizontal and vertical directions. This means that the one-dimensional configuration $d(x)$ has a non-trivial annihilator $b(x)$. Then also

$$(1 - x)d(x) = c'(x) + p'(x)$$

is annihilated by $b(x)$, where $c'(x) = (1 - x)c(x)$ has coefficient $c'_i = \lfloor i\alpha \rfloor - \lfloor (i - 1)\alpha \rfloor \in \{0, 1\}$ in cell i, and also $p'(x) = (1 - x)p(x)$ is finitary. A one-dimensional finitary configuration with a non-trivial annihilator is periodic by the determinism argument we used in the proof of Theorem 1, so that $d'(x) = (1 - x)d(x)$ is n-periodic for some $n > 0$. Let $h = d'_1 + \cdots + d'_n$ be the sum over one period. Notice that $d_i - d_0 = d'_1 + \cdots + d'_i$ for all $i > 0$, so that $d_{jn} = d_0 + jh$ for all $j > 0$. As $d(x) = c(x) + p(x)$ we have

$$p_{jn} = d_{jn} - c_{jn} = d_0 + jh - \lfloor jn\alpha \rfloor. \tag{6}$$

Because $p(x)$ is finitary, there are $j_1 < j_2$ such that $p_{j_1n} = p_{j_2n}$. By (6) this means $(j_2 - j_1)h = \lfloor j_2n\alpha \rfloor - \lfloor j_1n\alpha \rfloor$, so that h is a rational number and cannot hence be equal to irrational $n\alpha$. But then, using (6) again, $\lim_{j\to\infty} p_{jn} = \pm\infty$ so that $p(x)$ cannot be finitary, a contradiction. $\square$

Now it is clear that (5) is a non-trivial linear dependency among one-directed configurations in pairwise non-parallel directions. This is impossible by Lemma 5 so (5) cannot hold. We have proved the following result:

Theorem 4. *Let* $\alpha > 0$ *be irrational. The two-dimensional configuration* s *over the binary alphabet* $\{0, 1\}$ *defined by*

$$s_{ij} = \lfloor (i + j)\alpha \rfloor - \lfloor i\alpha \rfloor - \lfloor j\alpha \rfloor$$

is a sum of three periodic integral configurations but not a sum of finitely many finitary periodic configurations.

6 Conclusions and Applications

We have proved that multidimensional configurations of low local complexity can be expressed as a sum of periodic configurations. We have also demonstrated that sometimes the periodic components are necessarily non-finitary. We believe that the periodic decomposition will be useful in tackling a number of questions in multidimensional symbolic dynamics and combinatorics of words. Here we present two open problems whose setup is amenable to our approach.

Nivat's Conjecture

Nivat's conjecture (proposed by M. Nivat in his keynote address in ICALP 1997 [Niv97]) claims that in the two-dimensional case $d = 2$, the low complexity assumption (3) for a *rectangle* D implies that c is periodic. The conjecture is a natural generalization of the one-dimensional Morse-Hedlund theorem that we presented as Theorem 1. In the two-dimensional setting, for $m, n \in \mathbb{N}$, let us denote by $\text{Patt}_{m \times n}(c)$ the set of $m \times n$ rectangles in configuration c.

Conjecture 1 (Nivat's conjecture). If for some m, n we have $|\text{Patt}_{m \times n}(c)| \le mn$ then c is periodic.

The conjecture has recently raised wide interest, but it remains unsolved. In [EKM03] it was shown $P_c(m, n) \le mn/144$ is enough to guarantee the periodicity of c. This bound was improved to $P_c(m, n) \le mn/16$ in [QZ04], and recently to $P_c(m, n) \le mn/2$ in [CK13b]. Also the cases of narrow rectangles have been investigated: it was shown in [ST02] and recently in [CK13a] that $P_c(2, n) \le 2n$ and $P_c(3, n) \le 3n$, respectively, imply that c is periodic.

The analogous conjecture in the higher dimensional setups is false [ST00]. The following example recalls a simple counter example for $d = 3$.

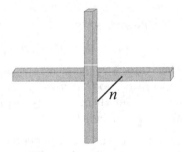

Fig. 2. A non-periodic three-dimensional configuration where two infinite stripes in orthogonal orientations are at distance n of each other. The number of distinct $n \times n \times n$ patterns in the configuration is $2n^2 + 1$.

Example 1. Fix $n \ge 3$, and consider the following $c \in \{0, 1\}^{\mathbb{Z}^3}$ consisting of two perpendicular lines of 1's on a 0-background, at distance n from each other: $c(i, 0, 0) = c(0, i, n) = 1$ for all $i \in \mathbb{Z}$, and $c(i, j, k) = 0$ otherwise. See Figure 2 for a picture of the configuration. For D equal to the $n \times n \times n$ cube we have $|\text{Patt}_D(c)| = 2n^2 + 1$ since the D-patterns in c have at most a single 1-line piercing a face of the cube. Clearly c is not periodic although $2n^2 + 1 < n^3 = |D|$. Notice that c is the sum of two periodic configurations. Our results imply that any counter example must decompose into a sum of periodic components.   □

In [KS15] we reported the following asymptotic result, using the approach discussed in the present paper. The detailed proof of the result will be published elsewhere.

Theorem 5 ([KS15]). *Let c be a two-dimensional non-periodic configuration. Then* $|\mathrm{Patt}_{m \times n}(c)| > mn$ *for all but finitely many pairs* m, n.

Periodic tiling problem

Another related open problem is the *periodic (cluster) tiling problem* by Lagarias and Wang [LW96]. A (cluster) tile is a finite $D \subset \mathbb{Z}^d$. Its co-tiler is any subset $C \subseteq \mathbb{Z}^d$ such that

$$D \oplus C = \mathbb{Z}^d. \tag{7}$$

The co-tiler can be interpreted as the set of positions where copies of D are placed so that they together cover the entire $\mathbb{Z}^d$ without overlaps. Note that the tile D does not need to be connected — hence the term "cluster tile" is sometimes used. The tiling is by translations of D only: the tiles may not be rotated.

It is natural to interpret any $C \subseteq \mathbb{Z}^d$ as the binary configuration $c \in \{0,1\}^{\mathbb{Z}^d}$ with $c_v = 1$ if and only if $v \in C$. Then the tiling condition (7) states that C is a co-tiler for D if and only if the $(-D)$-patterns in the corresponding configuration c contain exactly a single 1 in the background of 0's. In fact, as co-tilers of D and $-D$ coincide [Sze98], this is equivalent to all D-patterns having a single 1.

We see that the set $\mathcal{C}$ of all co-tiler configurations for D is a *subshift of finite type* [LM95]. We also see that the low local complexity assumption (3) is satisfied, even for the entire subshift of valid tilings so that $|\mathrm{Patt}_D(\mathcal{C})| \leq |D|$.

Conjecture 2 (Periodic Tiling Problem). If tile D has a co-tiler then it has a periodic co-tiler.

This conjecture was first formulated in [LW96]. In the one-dimensional case it is easily seen true, but already for $d = 2$ it is open. Interestingly, it is known that if $|D|$ is a prime number then *every* co-tiler of D is periodic [Sze98]. (See also [KS15] for an alternative proof that uses power series and polynomials.). The same is true if D is connected, that is, a polyomino [BN91].

References

[BN91] Beauquier, D., Nivat, M.: On Translating One Polyomino to Tile the Plane. Discrete & Computational Geometry **6** (1991)

[CK13a] Cyr, V., Kra, B.: Complexity of short rectangles and periodicity (2013) (submitted). arXiv: 1307.0098 [math.DS]

[CK13b] Cyr, V., Kra, B.: Nonexpansive $\mathbb{Z}^2$-subdynamics and Nivat's conjecture. Trans. Amer. Math. Soc. (2013) (to appear)

[EKM03] Epifanio, C., Koskas, M., Mignosi, F.: On a conjecture on bidimensional words. Theor. Comput. Sci. (1–3), 299 (2003)

[KS15] Kari, Jarkko, Szabados, Michal: An algebraic geometric approach to nivat's conjecture. In: Halldórsson, Magnús M., Iwama, Kazuo, Kobayashi, Naoki, Speckmann, Bettina (eds.) ICALP 2015. LNCS, vol. 9135, pp. 273–285. Springer, Heidelberg (2015)

[LW96] Lagarias, J.C., Wang, Y.: Tiling the Line with Translates of One Tile. Inventiones Mathematicae **124**, 341–365 (1996)

[LM95] Lind, D., Marcus, B.: An Introduction to Symbolic Dynamics and Coding. Cambridge University Press (1995)

[MH38] Morse, M., Hedlund, G.A.: Symbolic Dynamics. American Journal of Mathematics **60**(4), 815–866 (1938)

[Niv97] Nivat, M.: Invited talk at ICALP, Bologna (1997)

[QZ04] Quas, A., Zamboni, L.Q.: Periodicity and local complexity. Theor. Comput. Sci. **319**(1–3), 229–240 (2004)

[ST00] Sander, J.W., Tijdeman, R.: The complexity of functions on lattices. Theor. Comput. Sci. **246**(1–2), 195–225 (2000)

[ST02] Sander, J.W., Tijdeman, R.: The rectangle complexity of functions on two-dimensional lattices. Theor. Comput. Sci. **270**(1–2), 857–863 (2002)

[Sze98] Szegedy, M.: Algorithms to tile the infinite grid with finite clusters. In: FOCS, pp. 137–147. IEEE Computer Society (1998)

Why We Need Semirings in Automata Theory
(Extended Abstract)

Werner Kuich[(✉)]

Technische Universität Wien, Vienna, Austria
kuich@tuwien.ac.at

In this lecture we will report on generalizations of some classical results on formal languages. These generalizations are achieved by an algebraic treatment using semirings, formal power series, fixed point theory and matrices. By the use of these mathematical constructs, definitions, constructions, and proofs are obtained that are very satisfactory from a mathematical point of view. The use of these mathematical constructs yields the following advantages:

(i) The constructions needed in the proofs are mainly the usual ones.
(ii) The descriptions of the constructions by formal series and matrices do not need as much indexing as the usual descriptions.
(iii) The proofs are separated from the constructions and do not need the intuitive contents of the constructions. Often they are shorter than the usual proofs.
(iv) The results are more general than the usual ones. Depending on the semiring used, the results are valid for classical grammars and automata, classical grammars and automata with ambiguity considerations, probabilistic grammars or automata, etc.
(v) The use of semirings, formal power series and matrices in formal language and automata theory gives insight into the mathematical structure of problems and yields new results and solutions to unsolved problems that are difficult, if not impossible to obtain by other means.

In our lecture we will concentrate on item (v) and show the validity of this statement by presenting several of these problems.

We first discuss the formal power series $S\langle\langle M \rangle\rangle$ over a graded monoid M according to Sakarovitch [9] and show some decidability questions for these power series. One of them is a weak version of the famous decidability result of Harju and Karhumäki [5] on the equivalence of deterministic finite multitape automata. (See Esik, Kuich [3].)

Then we discuss the following problem: It is well known that, given a context-free grammar G and a regular language R, the problem $L(G) = R$ is undecidable. But if G is unambiguous this problem becomes decidable. The known proofs of this result all use formal power series. The given outline of the proof follows Kuich, Salomaa [8].

© Springer International Publishing Switzerland 2015
A. Maletti (Ed.): CAI 2015, LNCS 9270, pp. 43–44, 2015.
DOI: 10.1007/978-3-319-23021-4_4

Next we show how formal power series can be used to show that certain context-free languages are inherently ambiguous. (See Baron, Kuich [1] and Flajolet [4].)

Then we give a condition on a context-free language L which implies that L can not be generated by an unambiguous nonexpansive context-free grammar. (See Kuich [6].)

The last part of the lecture gives an answer to a problem raised by Jean Berstel [2]: We give the characterization of Kuich [7] of the families of context-free grammars that generate the languages of a cone.

References

1. Baron, G., Kuich, W.: The characterization of nonexpansive grammars by rational power series. Inf. Control **48**, 109–118 (1981)
2. Berstel, J.: Transductions and Context-Free Languages. Teubner (1979)
3. Ésik, Z., Kuich, W.: On power series over a graded monoid. In: Calude, C.S., Freivalds, R., Kazuo, I. (eds.) Gruska Festschrift. LNCS, vol. 8808, pp. 49–55. Springer, Heidelberg (2014)
4. Flajolet, P.: Ambiguity and transcendence. In: Brauer, W. (ed.) Automata, Languages and Programming. LNCS, pp. 179–188. Springer, Heidelberg (1985)
5. Harju, T., Karhumäki, J.: The equivalence problem of multitape finite automata. Theoretical Computer Science **78**, 347–355 (1991)
6. Kuich, W.: On the entropy of context-free languages. Inf. Control **16**, 173–200 (1970)
7. Kuich, W.: Forty years of formal power series in automata theory. In: Salomaa, A., Wood, D., Yu, S. (eds.) Half Century of Automata Theory, pp. 49–71. World Scientific (2001)
8. Kuich, W., Salomaa, A.: Semirings, Automata, Languages. Springer (1986)
9. Sakarovitch, J.: Rational and recognisable power series. In: Droste, M., Kuich, W., Vogler, H. (eds.) Handbook of Weighted Automata, Chapter 4, pp. 105–174. Springer (2009)

Unbordered Pictures: Properties and Construction

Marcella Anselmo[1], Dora Giammarresi[2], and Maria Madonia[3]([✉])

[1] Dipartimento di Informatica, Università di Salerno,
Via Giovanni Paolo II, 132-84084 Fisciano, SA, Italy
anselmo@dia.unisa.it
[2] Dipartimento di Matematica, Università Roma "Tor Vergata",
via Della Ricerca Scientifica, 00133 Roma, Italy
giammarr@mat.uniroma2.it
[3] Dipartimento di Matematica e Informatica,
Università di Catania, Viale Andrea Doria 6/a, 95125 Catania, Italy
madonia@dmi.unict.it

Abstract. The notion of unbordered picture generalizes to two dimensions the notion of unbordered (or bifix-free) string. We extend to two dimensions Nielsen's construction of unbordered strings ([23]) and describe an algorithm to construct the set $U(m,n)$ of unbordered pictures of fixed size (m,n). The algorithm recursively computes the set of quasi-unbordered pictures $Q(m,n)$, i.e. pictures that can possibly have some "large" borders.

Keywords: Bifix-free strings · Unbordered pictures

1 Introduction

The study of the structure and special patterns of the strings plays an important role in combinatorics of strings, both from theoretical and applicative side. Given a string s, a *bifix* or a *border* of s is a substring x that is both prefix and suffix of s. A string s is *bifix-free* or *unbordered* if it has no other bifixes besides the empty string and s itself.

Bifix-free strings are connected with the theory of codes [9] and are involved in the data structures for pattern matching algorithms [15,19]. From a more applicative point of view, bifix-free strings are suitable as synchronization patterns in digital communications and similar communications protocols [23]. The combinatorial structure of bifix-free strings over a given alphabet was studied by P.T. Nielsen in [23]: he provided an algorithm to enumerate recursively all bifix-free strings of the same length n over a given alphabet. A set of strings X in which no prefix of any string is the suffix of any other string in X is

Partially supported by MIUR Projects *"Formal Languages and Automata: Mathematical Structures and Applicative Directions"* and *"PRISMA PON04a2 A/F"*, and by FARB Projects of University of Catania, Roma "Tor Vergata", Salerno.

A. Maletti (Ed.): CAI 2015, LNCS 9270, pp. 45–57, 2015.
DOI: 10.1007/978-3-319-23021-4_5

called a *cross-bifix-free code*. Constructive methods for cross-bifix-free codes are investigated in [7,10,13].

The increasing interest for pattern recognition and image processing has motivated the research on two-dimensional languages of pictures. A two dimensional string is called *picture* and it is given by a rectangular array of symbols taken from a finite alphabet Σ. The set of all pictures over Σ is usually denoted by Σ^{**}. Extending results from the formal (string) languages theory to two dimensions is a very challenging task. The two-dimensional structure in fact imposes some intrinsic difficulties even in the basic concepts. For example, between two pictures we can define two concatenation operations (horizontal and vertical concatenations) but they are only partial operations and do not induce a monoid structure to the set Σ^{**}. The definition of "prefix" can be extended to a picture by considering its rectangular portion in the top-left corner: nevertheless, if one deletes a prefix from a picture, the remaining part is not a picture anymore.

Several results from string language theory can be worthy extended to pictures. Many researchers have investigated how the notion of recognizability by finite state automata can be transferred to two dimensions to accept picture languages ([2,4–6,11,17,18,20,24,25]). Two dimensional codes were studied in different contexts ([1,8,12,21] and recently two-dimensional prefix codes were introduced as the two-dimensional counterpart of prefix string codes ([3,6]). Matrix periodicity plays a fundamental role in two-dimensional pattern matching (see e.g. [15,22]), while two-dimensional quasi-periodicity was very recently studied in [16].

In this paper we investigate the notion of *unbordered* picture that is somehow connected both to picture codes and to two-dimensional pattern matching. Observe that the notion of border extends very naturally from strings to pictures since it is not related to any scanning direction. Informally we can say that a picture p is *bordered* if a copy p' of p can be overlapped on p by putting a corner of p' somewhere on some position in p. The border of p will be the subpicture corresponding to the portion where p and p' match. The two dimensions of the structure allow several possibilities to specialize this notion. The simplest one is when the matching is checked only by sliding the two picture copies with a horizontal or a vertical move: in this case we allow only borders with the same number of columns or rows of the picture p itself. Notice that this case is not really interesting, since pictures can be handled as they were thick strings on the alphabet either of the columns or of the rows: then the string algorithm by Nielsen can be directly applied to calculate all unbordered pictures. A more intriguing case is taking square pictures and allow only overlaps that put a corner of p' on positions of the diagonals of p. This corresponds to consider only square borders as defined in ([14]). Also in this special case some properties of string borders still hold for pictures.

We consider the more general situation when the overlaps can be made on any position in p and therefore the borders can be of any size. This leads to a different scenario with respect to the string case. It can be proved that if a string

s of lenght n has a border, then it can be written in the form $s = xvx$, i.e. s admits also a "small" border of length less than or equals to $\frac{n}{2}$. Unfortunately this property does not hold in general in two dimensions. Borders of a picture p of size (m, n), can be of three types: borders with dimensions both greater or both smaller than the half of the corresponding dimensions of p (say "large" or "small" borders) and borders with only one dimension greater than the half of the corresponding dimension of p (say a "medium" border). We can only prove that the presence of a "large" border implies also a medium or a small border. For this reason it is not possible to directly generalize Nielsen's construction for unbordered strings to pictures. In this paper we use *quasi-unbordered* pictures as intermediate concept: they can have only certain types of borders that become unlikely when the size of the pictures grows. We describe a recursive procedure to calculate all quasi-unbordered pictures of a given size: the (pure) unbordered pictures can then be easily extracted from this set.

The paper is organized as follows: Section 2 reports the recursive construction of bifix-free strings given by Nielsen in [23], together with all the needed notations and definitions on pictures. In Section 3 the notion of unbordered picture is introduced as two-dimensional extension from the string case. Some related properties are stated together with some examples. Section 4 contains the recursive construction for the set of all unbordered pictures of a given size. Some conclusions together with a table of experimental results are given in Section 5.

2 Preliminaries

In this section we first report the formal definition of unbordered strings together with their recursive construction given by Nielsen. Then, we recall all definitions on pictures needed for the main results of the paper.

2.1 Unbordered Strings and Nielsen's Construction

A string is a sequence of zero or more symbols from an alphabet Σ. A string w of length h is a substring of s if $s = uwv$ for $u, v \in \Sigma^*$. Moreover we say that a string w occurs at position j of s if and only if $w = s_j \ldots s_{j+h}$. A string x of length $m < n$ is a *prefix* of s if x is a substring that occurs in s at position 1; a string y is a *suffix* of s if it is a substrings that occurs in s at position $n - m + 1$. A string x that is both prefix and suffix of s is called a *border* or a *bifix* of s. The empty string and s itself are *trivial* borders of s. A string s is *unbordered* or *bifix-free* if it has no borders unless the trivial ones.

Unbordered strings have received very much attention since they occur in many applications as message synchronization or string matching. In [23] P. T. Nielsen proposed a recursive procedure to generate all bifix-free strings of a given length that is based on a property of string borders. We report briefly the main steps that will be used as base for the results of this paper.

The *bifix indicator* h_i of a string s of length n, $1 \le i < n$, is equal to 1 if s has a border of size i, and $h_i = 0$ otherwise. Then the following results hold.

Lemma 1. *A string $s \in \Sigma^*$ is unbordered if and only if $h_i = 0$ for $1 \leq i \leq \lfloor n/2 \rfloor$.*

Saying differently, the previous lemma states that if a string is *not* unbordered, then it must have a "short" border, i.e. of length less than the half of the length of the string. Let $s = s_1 s_2 \ldots s_n \in \Sigma^*$ be a unbordered string of even length n, $s_L = s_1 s_2 \ldots s_{n/2}$ and $s_R = s_{n/2+1} \ldots s_n$. Consider now the strings $s' = s_L a s_R$ and $s'' = s_L a b s_R$, with $a, b \in \Sigma$. Then Lemma 1 is used to prove the following one.

Lemma 2. *The string s is unbordered if and only if s' is unbordered. If s'' is unbordered then s is unbordered. If s is unbordered, then s'' has a border if and only if the following conditions are satisfied: $a = s_n$, $b = s_1$ and $s_2 \ldots s_{n/2} = s_{n/2+1} \ldots s_{n-1}$ for $n \geq 4$.*

Lemma 2 is then exploited to construct all bifix-free strings of length n from bifix-free strings of shorter length, by inserting extra symbols in the central positions. The starting set of bifix-free strings of length 2 is simply the set of all strings ab with $a, b \in \Sigma$ and $a \neq b$. Remark that Lemmas 1 and 2 and the deriving construction hold for alphabets of any cardinality.

2.2 Basic Notations on Pictures

We recall some definitions about pictures (see [18]). A *picture* over a finite alphabet Σ is a two-dimensional rectangular array of elements of Σ. Given a picture p, $|p|_{row}$ and $|p|_{col}$ denote the number of rows and columns, respectively while $size(p) = (|p|_{row}, |p|_{col})$ denotes the picture *size*. The pictures of size $(m, 0)$ or $(0, n)$ for all $m, n \geq 0$, called *empty* pictures, will be never considered in this paper. The set of all pictures over Σ of fixed size (m, n) is denoted by $\Sigma^{m,n}$, while the set of all pictures over Σ is denoted by Σ^{**}.

Let p be a picture of size (m, n). The set of coordinates $dom(p) = \{1, 2, \ldots, m\} \times \{1, 2, \ldots, n\}$ is referred to as the *domain* of a picture p. We let $p(i, j)$ denote the symbol in p at coordinates (i, j). We assume the top-left corner of the picture to be at position $(1, 1)$. Moreover, to easily detect border positions of pictures, we use initials of words "top", "bottom", "left" and "right": then, for example, the *tl-corner* of p refers to position $(1, 1)$ while the *br-corner* refers to position (m, n).

A *subdomain* of $dom(p)$ is a set d of the form $\{i, i+1, \ldots, i'\} \times \{j, j+1, \ldots, j'\}$, where $1 \leq i \leq i' \leq m$, $1 \leq j \leq j' \leq n$, also specified by the pair $[(i, j), (i', j')]$. The portion of p corresponding to positions in subdomain $[(i, j), (i', j')]$ is denoted by $p[(i, j), (i', j')]$. Then a non-empty picture x is *subpicture* of p if $x = p[(i, j), (i', j')]$, for some $1 \leq i \leq i' \leq m$, $1 \leq j \leq j' \leq n$; we say that x *occurs* at position (i, j) (its tl-corner).

Several operations can be defined on pictures (cf. [18]). Let $p, q \in \Sigma^{**}$ be pictures of size (m, n) and (m', n'), respectively, the *column concatenation* of p and q ($p \oplus q$) and the *row concatenation* of p and q ($p \ominus q$) are partial operations, defined only if $m = m'$ and if $n = n'$, respectively, as:

$$p \oplus q = \boxed{p \mid q} \qquad\qquad p \ominus q = \dfrac{\boxed{p}}{\boxed{q}}$$

The reverse operation on strings can be generalized to pictures and give rise to two different mirror operations (called *row*- and *col-mirror*) obtained by reflecting with respect to a vertical and a horizontal axis, respectively. Another operation that has no counterpart in one dimension is the *rotation*. The rotation of a picture p of size (m,n), is the clockwise rotation of p by $90°$, denoted by $p^{90°}$. Note that $p^{90°}$ has size (n,m). All the operations defined on pictures can be extended in the usual way to sets of pictures.

We conclude by remarking that any string $s = y_1 y_2 \cdots y_n$ can be identified either with a single-row or with a single-column picture, i.e. a picture of size $(1,n)$ or $(n,1)$. In the sequel it will be used the notation $[y_1 y_2 \cdots y_n]$ to indicate a single-row picture, while a single-column picture will be denoted by $[y_1 y_2 \cdots y_n]^{90°}$.

3   Bordered and Unbordered Pictures

We first generalize the notion of border from strings to pictures. Note that the notions of prefix and suffix of a string implicitly assume the left-to-right reading direction. On the other hand the notion of border is completely independent from any preferred direction. A string has a border when we can find the same substring at the two ends of the string. We extend these concepts to two dimensions.

Informally we say that a picture p is bordered when we can find the same rectangular portion at two opposite corners. Remark that there are two different kinds of borders depending on the pair of opposite corners that hold the border.

More formally we state the following definition.

Definition 3. *Given pictures $p \in \Sigma^{m,n}$ and $x \in \Sigma^{m',n'}$, with $1 \le m' \le m$ and $1 \le n' \le n$, the picture x is a tl-border of p, if x is a subpicture of p occurring at position $(1,1)$ and at position $(m-m'+1, n-n'+1)$; picture x is a bl-border of p, if x is a subpicture of p occurring at position $(m-m'+1, 1)$ and at position $(1, n-n'+1)$ Moreover x is a border of p if it is either a tl- or a bl-border.*

As special cases, p is a *trivial border* of itself, and x is a *proper border* of p if it is not trivial. A tl-border is called a diagonal border in [14]. Notice that a tl-border x of a picture p of size (m,n) can be univocally detected either by giving the position where it occurs in p (besides position $(1,1)$) or by giving its size. The same holds for bl-borders. Examples of pictures together with their borders are given below.

$$
p = \begin{array}{ccc|ccc}
0 & 1 & 0 & 0 & 0 & 0 \\
1 & 1 & 0 & 1 & 1 & 1 \\ \hline
0 & 0 & 1 & 1 & 1 & 0 \\
0 & 1 & 1 & 0 & 1 & 0 \\
1 & 1 & 1 & 1 & 1 & 0
\end{array}
\qquad
q = \begin{array}{ccc|cc}
1 & 0 & 0 & 1 & 0 \\
1 & 1 & 0 & 1 & 1 \\
1 & 1 & 1 & 0 & 0 \\ \hline
1 & 0 & 1 & 1 & 0 \\
1 & 1 & 1 & 1 & 0 \\
0 & 0 & 0 & 1 & 0
\end{array}
\qquad
r = \begin{array}{c|cc}
0 & 0 & 1 \\
1 & 1 & 1 \\
0 & 1 & 1
\end{array}
\qquad
s = \begin{array}{cccc}
0 & 1 & 0 & 0 \\
1 & 1 & 1 & 1 \\
0 & 0 & 1 & 1 \\
0 & 1 & 0 & 0 \\
1 & 1 & 1 & 1
\end{array}
$$

Note that if a picture p has a tl-border x, then the rotation $p^{90°}$ has a bl-border (that coincides with $x^{90°}$). In the figure above $q = p^{90°}$.

Definition 4. *A picture $p \in \Sigma^{m,n}$ is bordered if there exists a picture x that is a proper border of p. Picture p is unbordered (or border-free) if it is not bordered.*

The set of all unbordered pictures of size (m,n) over an alphabet Σ is denoted by $U_\Sigma(m,n)$, or simply $U(m,n)$, when the alphabet can be omitted.

Few simple results can be immediately listed.

Proposition 5. *Let Σ be an alphabet. For any $m, n \geq 1$, the set $U_\Sigma(m,n)$ is closed with respect to the rotation, col- and row-mirror operations, and with respect to permutation or renaming of symbols in Σ. Moreover, $U_\Sigma(m,n)^{90°} = U_\Sigma(n,m)$.*

Remark 6. The opposite corners of an unbordered picture p of size (m,n) must contain different symbols otherwise p would have a border of size $(1,1)$. Moreover the first row (column, resp.) must be different from the last one: otherwise p would have a border of size $(1,n)$ $((m,1)$, respectively).

The aim of the rest of the paper will be to construct all the unbordered pictures of a fixed size (m,n). The unbordered pictures of size $(1,n)$ or $(m,1)$ coincide with the unbordered strings and therefore can be calculate using techniques described in Section 2.1. Before studying the general case let us consider the case of the binary alphabet $\Sigma = \{0,1\}$ and of pictures of "small" size. It is immediate to see that there are no unbordered pictures of size $(2,2)$: there is no way to have different opposite corners and different first and last row (see Remark 6). For similar reasons there are no unbordered pictures of sizes $(2,3), (3,2)$ and $(3,3)$. The "smallest" unbordered pictures are of size $(4,2)$ and are all listed below:

$$
\begin{array}{|cc|} \hline 0 & 0 \\ 1 & 0 \\ 0 & 1 \\ 1 & 1 \\ \hline \end{array},\quad
\begin{array}{|cc|} \hline 0 & 0 \\ 0 & 1 \\ 1 & 0 \\ 1 & 1 \\ \hline \end{array},\quad
\begin{array}{|cc|} \hline 1 & 1 \\ 0 & 1 \\ 1 & 0 \\ 0 & 0 \\ \hline \end{array},\quad
\begin{array}{|cc|} \hline 1 & 1 \\ 1 & 0 \\ 0 & 1 \\ 0 & 0 \\ \hline \end{array}.
$$

Notice that they can be obtained from the first one by applying mirror operations.

Then the 40 unbordered pictures of size $(4,3)$ can be obtained by somehow generalizing Nielsen's construction of unbordered strings: it is possible to construct them by inserting a suitable middle column in the unbordered pictures of size $(4,2)$ listed above. Unfortunately this procedure does not work anymore when the size of pictures grows, as shown by the following example.

Example 7. The picture of size $(5,4)$ below is unbordered. Nevertheless all the pictures obtained by deleting some columns in the "middle" of the picture (the second column or the third one or both) are all bordered ones. Note that also by deleting the middle (the third) row, one obtains a bordered picture.

0	1	0	1
0	1	0	0
0	1	1	1
0	0	0	0
0	0	1	1

The main reason why Nielsen's construction of unbordered strings can not be directly generalized to pictures (as in Example 7), is that it is based on Lemma 1, that does not hold in two dimensions. For pictures we have the following weaker result.

Lemma 8. *Let $p \in \Sigma^{m,n}$. If p has a border of size (i,j) with $i \geq \lfloor m/2 \rfloor + 1$ and $j \geq \lfloor n/2 \rfloor + 1$ then p has a border of size (h,k) with $h \leq \lfloor m/2 \rfloor$ or $k \leq \lfloor n/2 \rfloor$.*

Proof. Let b be a border of size (i,j) with $i \geq \lfloor m/2 \rfloor + 1$ and $j \geq \lfloor n/2 \rfloor + 1$. Then p has a border x of size $(h,k) = (2i - m, 2j - n)$. The border x is given by the "intersection" of the two occurrences of the border b in p. More formally, if b is a tl-border then $x = p[(r,s),(r',s')]$ where $[(r,s),(r',s')] = [(1,1),(i,j)] \cap [(m - i + 1, n - j + 1),(m,n)]$. The case of bl-border is analogous.

Note that $h < i$ and $k < j$. Now, if x is still "large" (i.e. $h \geq \lfloor m/2 \rfloor + 1$ and $k \geq \lfloor n/2 \rfloor + 1$) one can iterate the reasoning until a border, with at least one of the dimension that satisfies the desired inequality, is obtained. □

Informally Lemma 8 claims that if a picture has a "large" border then it necessarily has a "small" or a "middle" border. Indeed, according to its size, a border of a picture p can be of three types: a border with both dimensions greater (smaller, resp.) than the half of the corresponding dimensions of p, say a "large" ("small", resp.) border; or a border with only one dimension greater than the half of the corresponding dimension of p, say a "medium" border. It is the presence of these medium borders that does not allow a simple generalization.

4   Construction of Unbordered Pictures

In this section we present a construction of the class $U(m,n)$ of all unbordered pictures of given size (m,n), that takes inspiration from Nielsen's construction of unbordered strings given in [23] (see Section 2.1). With this aim we introduce the class of *quasi-unbordered* pictures and present its recursive construction. The set $U(m,n)$ will be extracted from the set of quasi-unbordered pictures.

Informally a picture is quasi-unbordered if it has no border occurring in its right side.

Definition 9. *A picture $p \in \Sigma^{m,n}$ is quasi-unbordered if p has no border at position (i,j) with $1 \leq i \leq m$ and $\lceil n/2 \rceil + 1 \leq j \leq n$.*

The set of all quasi-unbordered pictures of size (m, n) over an alphabet Σ is denoted by $Q_\Sigma(m, n)$, or simply $Q(m, n)$, when the alphabet can be omitted. Examples of quasi-unbordered pictures can be found in Example 13. Observe that $U(m, n) \subseteq Q(m, n)$.

In the following the set $Q(m, n)$ is constructed in a recursive way by the insertion of one column in the middle of pictures in $Q(m, n - 1)$. We introduce first some formal notations. For any picture $p \in \Sigma^{m,n}$, the *left side* of p is the subpicture $p_L = p[(1, 1), (m, \lceil n/2 \rceil)]$, containing the first $\lceil n/2 \rceil$ columns of p, and the *right side* of p is the subpicture $p_R = p[(1, \lceil n/2 \rceil + 1), (m, n)]$ containing the remaining columns. Hence $p = p_L \oplus p_R$.

The picture obtained by inserting in the "middle" of p a column $c \in \Sigma^{m,1}$ is denoted $p^{\parallel c} = p_L \oplus c \oplus p_R$. We also define the inverse operation of removing the central column in a picture. More exactly, if n is odd, then $p^{\#}$ denotes the picture obtained by removing the $\lceil n/2 \rceil$-th column; if n is even, then $p^{\#}$ denotes the picture obtained by removing the $(n/2 + 1)$-th column.

Let us now focus on quasi-unbordered pictures and show the properties used for their recursive construction.

Proposition 10. *Let* $p \in \Sigma^{m,n}$. *If* p *is quasi-unbordered then* $p^{\#}$ *is quasi-unbordered.*

Proof. Suppose by contradiction that $p^{\#}$ has a tl-border x that occurs at position (i, j) in its right side. It is easy to see that the same tl-border x occurs at position $(i, j+1)$ of p, contradicting the hypothesis that p is quasi-unbordered (note that $\lceil n/2 \rceil + 1 \leq j + 1 \leq n$). The case of bl-borders is analogous.   □

Proposition 11. *Let* p *be a quasi-unbordered picture,* $p \in Q(m, n)$, *and* c *be a column,* $c \in \Sigma^{m,1}$.
1. *If* n *is even then* $p^{\parallel c} \in Q(m, n + 1)$
2. *If* n *is odd then* $p^{\parallel c}$ *has a border in its right side if and only if the border occurs at a position in* c.

Proof. 1. Arguing by contradiction, suppose that there exist i and j, with $1 \leq i \leq m$ and $\lceil (n+1)/2 \rceil + 1 \leq j \leq n+1$, such that $p^{\parallel c}$ has a tl-border x that occurs at the position (i, j). It is easy to see that the same tl-border border x occurs at the position $(i, j-1)$ of p contradicting the hypothesis that p is quasi-unbordered (note that $\lceil n/2 \rceil + 1 \leq j - 1 \leq n$). The case of bl-borders is analogous.

2. Suppose first that $p^{\parallel c}$ has a border x in its right side, and suppose w.l.o.g. that x is a tl-border. If x occurs at a position (i, j) not in c, then we can find the same tl-border x at position $(i, j - 1)$ of p, that is a position in the right side of p, and this contradicts the assumption p quasi-unbordered. Suppose now that $p^{\parallel c}$ has a border that occurs at a position in c. Since n is odd, then all the positions of c belong to the right side of $p^{\parallel c}$ and this concludes the proof.   □

Consider now the basis case of the recursion, that is quasi-unbordered pictures with one or two columns. Quasi-unbordered pictures with one column are indeed unbordered strings. Quasi-unbordered pictures with two columns can be characterized in terms of special unbordered strings, that we call *heart-free*. An unbordered string of even length $w \in \Sigma^{2m}$ is *heart-free* if $w = w_1 w_2$, with $|w_1| = |w_2| = m$ and there exists no $x \in \Sigma^*$ that is a suffix of w_1 and a prefix of w_2. In other words both $w_1 w_2$ and $w_2 w_1$ are unbordered.

Proposition 12. *Let $p \in \Sigma^{m,2}$ for some $m \geq 2$, $p = c_1 \oplus c_2$ with $c_1, c_2 \in \Sigma^{m,1}$. Then p is quasi-unbordered if and only if $(c_1 \ominus c_2)^{90°}$ is a heart-free unbordered string.*

Proof. By definition p is quasi-unbordered if p has no border in its right side, i.e. c_2. Then p has a tl-border of size $(i, 1)$ iff $(c_1 \ominus c_2)^{90°}$ is a string with a border of length i; and p has a bl-border of size $(i, 1)$ iff $(c_2 \ominus c_1)^{90°}$ has a border of length i. $\qquad\square$

Thanks to Proposition 12, all quasi-unbordered pictures in $\Sigma^{m,2}$, for any $m \geq 2$, can be constructed as follows. Use Nielsen's construction to obtain all unbordered strings over Σ of length $2m-2$. For any unbordered string $w = w_1 w_2$, with $w_1, w_2 \in \Sigma^{m-1}$, insert in the middle only pairs of symbols (a, b) with $a \neq b$, that satisfy the heart-free and unbordered requirements. Then $w_1 a$ and $b w_2$ are the columns of the pictures.

We are now ready to sketch the algorithm that provides the set $Q(m, n)$ of quasi-unbordered pictures of a given size (m, n). It consists in the following two steps.

1. Construct $Q(m, 2)$ (following Proposition 12).
2. Recursively construct $Q(m, n)$ from $Q(m, n-1)$ as follows.

If n is odd then define $Q(m, n)$ as the set of all pictures $p^{\|c}$ for all $p \in Q(m, n-1)$, $c \in \Sigma^{m,1}$.

If n is even then define $Q(m, n)$ as the set of all pictures $p^{\|c}$ for all $p \in Q(m, n-1)$, $c \in \Sigma^{m,1}$, such that $p^{\|c}$ has no border occurring at a position in c.

Let us roughly estimate the complexity of the algorithm. Observe that Step 2 when n is odd requires no comparisons. On the other hand, for $k = 2, \cdots \lfloor m/2 \rfloor$, the pictures in $Q(m, 2k)$ are obtained by inserting in any $p \in Q(m, 2k-1)$ a column $c = [c_m c_{m-1} \ldots c_1]^{90°}$; symbols in c must be taken so that no border occurs at c. First consider tl-borders. To avoid a tl-border of size (i, k), for $i = 1, \cdots, m$, the algorithm does ik comparisons at most. The same number of comparisons is then necessary to avoid also bl- borders at positions in c. Hence the algorithm does $2 \sum_{i=1,\cdots,m} ik \leq 2km^2$ comparisons for any picture in $Q(m, 2k-2)$. The construction of $Q(m, n)$ from $Q(m, 2)$, needs in total a number of comparisons $C(m, n) \leq \sum_{k=1,\cdots,n/2} |Q(m, 2k-2)| 2km^2$.

A simple bound on $|Q(m, n)|$ is $|Q(m, n)| \leq 1/4 |\Sigma^{m,n}|$, for any $m, n \geq 2$, since opposite corners in quasi-unbordered pictures must be different (in an analogous way as for unbordered ones, Remark 6). Applying this bound and some mathematical formulas on summations, one can obtain $C(m, n) \leq \frac{1}{2|\Sigma^{2m}|} m^2 \sum_{k=1,\cdots,n/2} |\Sigma^{2m}|^k k$ and finally $C(m, n) = O(m^2 n |\Sigma|^{mn})$.

Example 13. As an example of the algorithm sketched above, let us show how to obtain some pictures in $Q(3,4)$ for $\Sigma = \{0,1\}$. Note that $|Q(3,4)| = 196$ (see Section 5).

The basis case is the construction of $Q(3,2)$. Applying Proposition 12, take the 20 unbordered binary strings, extract the 6 heart-free unbordered strings and obtain:

$$Q(3,2) = \left\{ \begin{array}{cc} 0 & 1 \\ 0 & 0 \\ 0 & 1 \end{array} , \begin{array}{cc} 0 & 1 \\ 0 & 1 \\ 0 & 1 \end{array} , \begin{array}{cc} 0 & 1 \\ 1 & 1 \\ 0 & 1 \end{array} , \begin{array}{cc} 1 & 0 \\ 1 & 1 \\ 1 & 0 \end{array} , \begin{array}{cc} 1 & 0 \\ 1 & 1 \\ 1 & 0 \end{array} , \begin{array}{cc} 1 & 0 \\ 0 & 0 \\ 1 & 0 \end{array} \right\}.$$

Then, using Proposition 11 (case 1) we have:

$$Q(3,3) = \left\{ \boxed{p_L \ c \ p_R} \text{ with } p = p_L \oplus p_R \in Q(3,2) \text{ and } c \in \Sigma^{3,1} \right\}.$$

Let us give now the construction of some pictures in $Q(3,4)$ from pictures in $Q(3,3)$. Consider for example pictures $p = \begin{array}{ccc} 0 & 0 & 1 \\ 0 & 0 & 0 \\ 0 & 0 & 1 \end{array}$ and $q = \begin{array}{ccc} 0 & 0 & 1 \\ 0 & 0 & 0 \\ 0 & 1 & 1 \end{array}$ in $Q(3,3)$ that show a different behavior. From Proposition 11 (case 2), we know that, for any $c \in \Sigma^{3,1}$, $p^{\|c}$ has a border in its right side if and only if the border occurs at a position in c. Observing the picture $p^{\|c}$, one notes that no border can occur at a position in c. Hence, $p^{\|c} \in Q(3,4)$, for any $c \in \Sigma^{3,1}$, i.e. $\begin{array}{cccc} 0 & 0 & x & 1 \\ 0 & 0 & y & 0 \\ 0 & 0 & z & 1 \end{array} \in Q(3,4)$, for any $x, y, z \in \Sigma$.

Consider now q and $q^{\|c} = \begin{array}{cccc} 0 & 0 & x & 1 \\ 0 & 0 & y & 0 \\ 0 & 1 & z & 1 \end{array}$ with $c = \begin{array}{c} x \\ y \\ z \end{array} \in \Sigma^{3,1}$. No tl-border of size $(1,2)$, $(2,2)$, and $(3,2)$ can occur in $q^{\|c}$, for any choice of z, y, x. On the other hand, in order to have no bl-border of size $(1,2)$, necessarily $x = 1$, while y, z can be chosen arbitrarily.

Let us now come back to unbordered pictures. The unbordered pictures of a given size can be obtained from the quasi-unbordered ones of the same size. All pictures in $Q(m,n)$ have no border in their right side. Then the bordered pictures in $Q(m,n)$ to be removed are the ones with a border in their left side. From Lemma 8, it can be argued that it is sufficient to remove pictures with borders of size (i,j), with $i \leq \lceil m/2 \rceil$ and $j > \lfloor n/2 \rfloor$. So only a limited number of comparisons are needed on pictures in the set $Q(m,n)$.

Example 14. (continued) Unbordered pictures in $U(3,4)$ are obtained from pictures in $Q(3,4)$. Consider again pictures p and q in Example 13. We noted that

$p^{\|c} \in Q(3,4)$, for any $c \in \Sigma^{3,1}$, i.e. $\begin{array}{|cccc|} 0 & 0 & x & 1 \\ 0 & 0 & y & 0 \\ 0 & 0 & z & 1 \end{array} \in Q(3,4)$, for any $x, y, z \in \Sigma$.

For $x \neq z$, these pictures in $Q(3,4)$ are unbordered pictures. Moreover, the pictures $\begin{array}{|cccc|} 0 & 0 & 1 & 1 \\ 0 & 0 & y & 0 \\ 0 & 1 & z & 1 \end{array}$ obtained from q belong to $U(3,4)$ if and only if $z = 0$.

We conclude the section with a simple result that in fact can be obtained as corollary of Proposition 10, but it sheds light on the original motivation to introduce the class of quasi-unbordered pictures, when interested in the construction of unbordered pictures.

Proposition 15. *Let $p \in \Sigma^{m,n}$. If p is unbordered then $p^{\|t}$ is quasi-unbordered.*

Moreover it is worthwhile to remark that in the special case of one-row pictures, that are strings, $Q(1,n) = U(1,n)$ for any $n \geq 1$, thanks to Lemma 1. Hence the construction presented here coincides with Nielsen's construction when applied to strings.

5   Final Remarks

We presented general definitions for unbordered pictures by imposing that all possible overlaps between two copies of such pictures are forbidden. This exploits the "bi-dimensionality" of the structures. As a result, the definition imposes many constrains to the pictures. Below we present a table reporting the cardinality of the sets of unbordered and quasi-unbordered pictures over a 2-letters alphabet. The rate with respect to the whole set of pictures of corresponding size is also shown.

Few considerations can be done. First of all, notice that unbordered pictures of size (m, n) are very few with respect to the whole set $\Sigma^{m,n}$ (remember that we had only a rough estimation of $1/4$ given by Remark 6). Regarding the basic step of our recursive construction, Proposition 12 states an interesting bijection between quasi-unbordered pictures with two columns and heart-free unbordered strings. In particular it allows to estimate $|Q(m,2)|$ as the cardinality of heart-free unbordered strings of length $2m$. By some clever considerations on Nielsen's construction it can be observed that the heart-free unbordered strings of given length are at most $1/2$ than the unbordered strings of same size. Moreover denote $v_n = \frac{|U(1,n)|}{|\Sigma^{1,n}|}$ and recall that v_n is a not increasing sequence with $v_4 = \frac{3}{8}$ ([23]). Hence $\frac{|Q(m,2)|}{|\Sigma^{m,2}|} \leq \frac{1}{2} \cdot v_{2m} \leq \frac{1}{2} \cdot \frac{3}{8} = \frac{3}{16}$. This bound is completely reflected in the table.

Finally, observe that the table reports also the rate $\frac{|U(m,n)|}{|Q(m,n)|}$: this is important to estimate the overhead complexity of calculating set $Q(m,n)$ as intermediate step for $U(m,n)$. Notice that, already for those small values of n, the two sets are not so different in size. This can be easily understood if we think that the probability that a picture has a border with more than $n/2$ columns sensibly decreases when n grows.

m	n	$\|U(m,n)\|$	$\|Q(m,n)\|$	$\|\Sigma^{m,n}\|$	$\frac{\|U(m,n)\|}{\|\Sigma^{m,n}\|}$	$\frac{\|Q(m,n)\|}{\|\Sigma^{m,n}\|}$	$\frac{\|U(m,n)\|}{\|Q(m,n)\|}$
2	2	0	2	16	0,00%	12,50%	0,00%
2	3	0	8	64	0,00%	12,50%	0,00%
2	4	4	18	256	1,56%	7,03%	22,2%
2	5	24	72	1024	2,34%	7,03%	33,3%
2	6	120	200	4096	2,93%	4,88%	60,0%
2	7	528	800	16384	3,22%	4,88%	66,0%
2	8	2220	2734	65536	3,39%	4,17%	81,2%
⋮	⋮	⋮	⋮	⋮	⋮	⋮	⋮
3	2	0	6	64	0,00%	9,38%	0,00%
3	3	0	48	512	0,00%	9,38%	0,00%
3	4	40	196	4096	0,98%	4,79%	20,4%
3	5	512	1568	32768	1,56%	4,79%	32,7%
3	6	5048	8542	262144	1,93%	3,26%	59,1%
3	7	44880	68336	2097152	2,14%	3,26%	65,7%
3	8	376768	465266	16777216	2,25%	2,77%	81,0%
⋮	⋮	⋮	⋮	⋮	⋮	⋮	⋮
4	2	4	22	256	1,56%	8,59%	18,2%
4	3	40	352	4096	0,98%	8,59%	11,4%
4	4	864	2720	65536	1,32%	4,15%	31,8%
4	5	16712	42920	1048576	1,59%	4,09%	38,9%
4	6	303976	472990	16777216	1,81%	2,82%	64,3%
4	7	5164176	7567840	268435456	1,92%	2,82%	68,2%
4	8	85346944	103001874	4294967296	1,99%	2,40%	82,9%
⋮	⋮	⋮	⋮	⋮	⋮	⋮	⋮
5	2	24	80	1024	2,34%	7,81%	30,0%
5	3	512	2560	32768	1,56%	7,81%	20,0%
5	4	16712	39646	1048576	1,59%	3,78%	42,2%
5	5	563584	1268672	33554432	1,68%	3,78%	44,4%
5	6	19057664	27609768	1073741824	1,77%	2,57%	69,0%

References

1. Aigrain, P., Beauquier, D.: Polyomino tilings, cellular automata and codicity. Theoretical Computer Science **147**, 165–180 (1995)
2. Anselmo, M., Giammarresi, D., Madonia, M.: Deterministic and unambiguous families within recognizable two-dimensional languages. Fund. Inform. **98**(2–3), 143–166 (2010)
3. Anselmo, M., Giammarresi, D., Madonia, M.: Strong prefix codes of pictures. In: Muntean, T., Poulakis, D., Rolland, R. (eds.) CAI 2013. LNCS, vol. 8080, pp. 47–59. Springer, Heidelberg (2013)

4. Anselmo, M., Giammarresi, D., Madonia, M., Restivo, A.: Unambiguous recognizable two-dimensional languages. RAIRO -ITA **40**(2), 227–294 (2006)
5. Anselmo, M., Giammarresi, D., Madonia, M.: A computational model for tiling recognizable two-dimensional languages. Theor. Comput. Sci. **410**(37), 3520–3529 (2009)
6. Anselmo, M., Giammarresi, D., Madonia, M.: Prefix picture codes: a decidable class of two-dimensional codes. Int. J. Found. Comput. Sci. **25**(8), 1017–1032 (2014)
7. Bajic, D., Loncar-Turukalo, T.: A simple suboptimal construction of cross-bifix-free codes. Cryptography and Communications **6**(1), 27–37 (2014)
8. Beauquier, D., Nivat, M.: A codicity undecidable problem in the plane. Theoret. Comp. Sci **303**, 417–430 (2003)
9. Berstel, J., Perrin, D., Reutenauer, C.: Codes and Automata. Cambridge University Press (2009.)
10. Bilotta, S., Pergola, E., Pinzani, R.: A new approach to cross-bifix-free sets. IEEE Transactions on Information Theory **58**(6), 4058–4063 (2012)
11. Blum, M., Hewitt, C.: Automata on a 2-dimensional tape. In: SWAT (FOCS), pp. 155–160 (1967)
12. Bozapalidis, S., Grammatikopoulou, A.: Picture codes. RAIRO - ITA **40**(4), 537–550 (2006)
13. Chee, Y.M., Kiah, H.M., Purkayastha, P., Wang, C.: Cross-bifix-free codes within a constant factor of optimality. IEEE Transactions on Information Theory **59**(7), 4668–4674 (2013)
14. Crochemore, M., Iliopoulos, C.S., Korda, M.: Two-dimensional prefix string matching and covering on square matrices. Algorithmica **20**(4), 353–373 (1998)
15. Crochemore, M., Rytter, W.: Jewels of stringology. World Scientific (2002). http://www-igm.univ-mlv.fr/mac/JOS/JOS.html
16. Gamard, G., Richomme, G.: Coverability in two dimensions. In: Dediu, A.-H., Formenti, E., Martín-Vide, C., Truthe, B. (eds.) LATA 2015. LNCS, vol. 8977, pp. 402–413. Springer, Heidelberg (2015)
17. Giammarresi, D., Restivo, A.: Recognizable picture languages. Int. Journal. Pattern Recognition and Artificial Intelligence **6**(2–3), 241–256 (1992)
18. Giammarresi, D., Restivo, A.: Two-dimensional languages. In: Rozenberg, G.(ed.) Handbook of Formal Languages, vol. III, pp. 215–268. Springer Verlag (1997)
19. Gusfield, D.: Algorithms on Strings, Trees, and Sequences - Computer Science and Computational Biology. Cambridge University Press (1997)
20. Kari, J., Salo, V.: A survey on picture-walking automata. In: Kuich, W., Rahonis, G. (eds.) Algebraic Foundations in Computer Science. LNCS, vol. 7020, pp. 183–213. Springer, Heidelberg (2011)
21. Kolarz, M., Moczurad, W.: Multiset, set and numerically decipherable codes over directed figures. In: Smyth, B. (ed.) IWOCA 2012. LNCS, vol. 7643, pp. 224–235. Springer, Heidelberg (2012)
22. Na, J.C., Ferragina, P., Giancarlo, R., Park, K.: Indexed two-dimensional string matching. In: Encyclopedia of Algorithms (2015)
23. Nielsen, P.T.: A note on bifix-free sequences (corresp.). IEEE Transactions on Information Theory **19**(5), 704–706 (1973)
24. Otto, F., Mráz, F.: Extended two-way ordered restarting automata for picture languages. In: Dediu, A.-H., Martín-Vide, C., Sierra-Rodríguez, J.-L., Truthe, B. (eds.) LATA 2014. LNCS, vol. 8370, pp. 541–552. Springer, Heidelberg (2014)
25. Pradella, M., Cherubini, A., Crespi-Reghizzi, S.: A unifying approach to picture grammars. Inf. Comput. **209**(9), 1246–1267 (2011)

Effective Invariant Theory of Permutation Groups Using Representation Theory

Nicolas Borie[(⊠)]

Laboratoire d'Informatique Gaspard Monge,
Université Paris Est á Marne-La-Vallée, Champs-sur-marne, France
nicolas.borie@u-pem.fr

Abstract. Using the representation theory of the symmetric group, we propose an algorithm to compute the invariant ring of a permutation group in the non modular case. Our approach has the advantage of reducing the amount of linear algebra computations and exploits a finer combinatorial description of the invariant ring. We build explicit generators for invariant rings by means of the higher Specht polynomials of the symmetric group.

Keywords: Computational invariant theory · Representation theory · Permutation group · Specht polynomials

1 Introduction

The purpose of invariant theory is exploiting symmetries. Problems admitting a large number of symmetries can be reduced to a problem dealing with a smaller domain. When a real function of a real variable is even, we only study its graph on the positive side as we can deduce the look of its graph on the negative side by symmetry. The goal of algebraic invariant theory is to establish general results when we consider abstract groups of symmetries acting on some formal variables. From this general algebraic approach, abstract variables can be specialized to solve practical applications [7, § 5] as the resolution of polynomial systems with symmetries [4], [11], [19, § 2.6] and [8]), in effective Galois theory [1], [5], [12], or in discrete mathematics [16].

The literature contains deep and explicit results for special classes of groups, like complex reflection groups or classical reductive groups, as well as general results applicable to any group. Given the level of generality, one cannot hope for such results to be simultaneously explicit and tight in general. Thus the subject was effective early on: given a group, one wants to *calculate* the properties of its invariant ring. Under the impulsion of modern computer algebra, computational methods, and their implementations, have largely expanded in the last twenty years [7,13,14,19,21]. However, much progress is still needed to go beyond toy examples and enlarge the spectrum of applications.

In this study, we will focus on groups of symmetries permuting a finite number of formal variables, thus subgroups of a symmetric group, and in the non modular

© Springer International Publishing Switzerland 2015
A. Maletti (Ed.): CAI 2015, LNCS 9270, pp. 58–69, 2015.
DOI: 10.1007/978-3-319-23021-4_6

case (our invariants will be invariant polynomials whose coefficients lie in a field $\mathbb{K}$ of characteristic zero.)

Given a finite permutation group G, subgroup of the symmetric group $\mathfrak{S}_n$, acting by permutation on a finite number of variables $\mathbf{x} = x_1, x_2, \ldots, x_n$, an *invariant polynomial* P under the action of G is a multivariate polynomial in $\mathbb{K}[\mathbf{x}]$ such that $P = \sigma \cdot P$ for all $\sigma \in G$. As the sum and the product of invariant polynomials are still stabilized under the action of G, the set of all invariant polynomials form a ring called the ring of invariants under the action of G and denoted $\mathbb{K}[\mathbf{x}]^G$. It is known since Hilbert and Noether that the ring of invariants $\mathbb{K}[\mathbf{x}]^G$ is finitely generated for any finite group G of matrices with an explicit bound for the degree of the generators ($|G|$, the cardinality of the group G). However the proofs of these results didn't tell how to build explicitly a set of generators.

For any permutation group G, subgroup of the symmetric group $\mathfrak{S}_n$, it is know that the ring $\mathbb{K}[\mathbf{x}]^G$ has for Krull dimension n. A set of n homogeneous polynomials $\theta_1, \ldots, \theta_n$ invariant under the action of G, algebraically independent is called a *homogeneous set of parameters*. Such a family forms the *primary invariants* which constitutes a first key element for building generators of the invariant ring. In the non modular case, the ring $\mathbb{K}[\mathbf{x}]^G$ has the Cohen-Macaulay property, this means that for any family of primary invariants, one can build a second family of invariant polynomials S called *secondary invariants* such that:

$$\mathbb{K}[\mathbf{x}]^G = \bigoplus_{\eta \in S} \eta \mathbb{K}[\theta_1, \ldots, \theta_n]. \tag{1}$$

A couple of families, primary invariants and secondary invariants give a thin description of the structure of the invariant ring, in particular, their union generates the ring of invariants.

Classical approaches computing generators of invariant rings use elimination techniques (like (SAGBI-)Gröbner basis) in vector spaces of very high dimensions. Current approaches work degree by degree and end up with a Gauss reduction over some polynomials of degree d over n variables. As the cost linear reduction is conditioned by the cube the dimension and that there are $\binom{n+d-1}{d}$ monomials of degree d over n variables, all current approaches and their practical implementation are limited to groups acting on 8, 9 or 10 variables with modern computers. The evaluation approach proposed by the author in [3] localizes the Gauss reduction in an abstract vector space whose basis is indexed by the right cosets of the symmetric group $\mathfrak{S}_n$ of order n by a permutation group G, thus with the cube of the ambient dimension, this approach is still limited to permutation groups whose index $|\mathfrak{S}_n|/|G|$ in the symmetric group is relatively controlled (around 1000 for modern computers).

We propose in this article an approach following the idea that adding more combinatorics in invariant theory would help to produce more efficient algorithms. Moreover, such algorithms outputs could also reveal some combinatorics; the long time goal being having a combinatorial description of invariant rings (generators or pairs of primary-secondary invariant families). Since Hilbert, this problem has been solved only in very restrictive and special cases (for example, in [10], the authors give secondary invariants for Young subgroups of symmetric groups).

We focus on the problem of computing secondary invariants of finite permutation groups in the non modular case. In this context, we will show how to localize computations inside selected irreducible representations of the symmetric group. These spaces are smaller than the ones used in classical approaches and we can largely take advantage of combinatorial results coming from the theory of representations of the symmetric group. Once these computations are done, we will use the higher Specht polynomials [20] as basis of the coinvariants of the symmetric group to rebuild explicit generators of invariant rings.

2   Invariant Ring and Representations of the Symmetric Group

2.1   Invariant Ring of Permutation Group and Application to Combinatorics

Our approach starts from a result in a key article of invariant theory [17, Proposition 4.9] mixing invariants of finite group and combinatorics. We recall this general result:

Proposition 1. *Let $\theta_1, \ldots, \theta_n$ be an homogeneous set of parameters for $\mathbb{K}[\mathbf{x}]^G$, where G is any finite subgroup of $GL(\mathbb{K}^n)$ of order $|G|$. Set $d_i = \deg(\theta_i)$ and $t = d_1 \cdots d_n / |G|$. Then the action of G on the quotient ring $\mathbb{K}[\mathbf{x}]/(\theta_1, \ldots, \theta_n)$ is isomorphic to t times the regular representation of G.*

Applying this result to $\mathfrak{S}_n$, the symmetric group of degree n, with $\theta_i = e_i$ (elementary symmetric polynomial) we recover the well-known result that the ring $\mathbb{K}[\mathbf{x}]/(e_1, \ldots, e_n)$ is isomorphic to the *regular representation* $RR(\mathfrak{S}_n)$ of the symmetric group (here $t = \frac{n!}{n!} = 1$). This well-known quotient $\mathbb{K}[\mathbf{x}]/(e_1, \ldots, e_n)$ is called the coinvariant ring of the symmetric group in the algebraic combinatorics world and several bases of this ring have been explicitly built (Harmonic polynomials [2], Schubert Polynomials [15], Descent monomials [10] and more).

Let G be a group of permutations, subgroup of $\mathfrak{S}_n$. We now reapply the result of Stanley to G with the same homogeneous set of parameters formed with the elementary symmetric polynomials. Then, the ring of coinvariants of the symmetric group is now also isomorphic to $t = n!/|G|$ times the regular representation of the group G:

$$\mathbb{K}[\mathbf{x}]/(e_1, \ldots, e_n) \sim_G \bigoplus_{i=1}^{n!/|G|} RR(G). \tag{2}$$

We know that for any permutation group G, in the non modular case, the ring of invariants under the action of G is a Cohen-Macaulay algebra. This implies that there exists a family of generators η_i making the ring of invariants under the action of G a free module of rank $n!/|G|$ over the ring of symmetric polynomials:

$$\mathbb{K}[\mathbf{x}]^G = \bigoplus_{i=1}^{n!/|G|} \eta_i \mathbb{K}[e_1, \ldots, e_n]. \qquad (3)$$

Taking the quotient on both sides by the ideal $(e_1, \ldots, e_n)$ and keeping η_i as representative of its equivalent class inside the quotient, we have

$$\mathbb{K}[\mathbf{x}]^G/(e_1, \ldots, e_n) = \bigoplus_{i=1}^{n!/|G|} \mathbb{K} \cdot \eta_i. \qquad (4)$$

As $\mathbb{K}[\mathbf{x}]^G$ is, by definition, the subspace of $\mathbb{K}[\mathbf{x}]$ on which the action of G is trivial, the result of Stanley implies in particular that the polynomials η_i span the subspace of the coinvariants of the symmetric group on which the action of G is trivial. A way to construct the η_i is thus to search them as G-fixed point inside the ring of coinvariants of the symmetric group and that could be done irreducible representation by irreducible representation.

Representation theory of the symmetric group has been largely studied, and allows us to formulate the following problem:

Problem 1. Let n be a positive integer and G be a permutation group, subgroup of $\mathfrak{S}_n$. Construct an explicit basis of the trivial representations of G appearing in each irreducible representation of $\mathfrak{S}_n$ inside the quotient $\mathbb{K}[\mathbf{x}]/(e_1, \ldots, e_n)$.

A first step to solve this problem consists in having a basis of the coinvariants for the symmetric group respecting the action of $\mathfrak{S}_n$ such that this basis can be partitioned by irreducible representations. We present such a basis in Section 3.

2.2 Representations of the Symmetric Group

We recall in this section some results describing the irreducible representations of the symmetric group.

For a positive integer n, we will call λ a *partition* of n (denoted by $\lambda \vdash n$) if λ is a non increasing sequence of positive integers $\lambda = (\lambda_1, \ldots, \lambda_r)$ whose entries sum up to n.

For a positive integer n, irreducible representations of the symmetric group $\mathfrak{S}_n$ are indexed by the partitions of n. Since we have a finite group, the multiplicity of an irreducible representation inside the regular representation is equal to its dimension. This information can be collected counting *standard tableaux* of a given shape.

Let n be a positive integer and $\lambda = (\lambda_1, \ldots, \lambda_r)$ a partition of n. A *tableau* of shape λ is a diagram of square boxes disposed in rows such that the first row contains λ_1 boxes, on top of it, a second row contains λ_2 boxes and so on. A *standard tableau* of shape λ is a filling of a *tableau* of shape λ with integers from

1 to n such that the integers are increasing in row and column. We will denote by $STab(\lambda)$ the set of standard tableaux of shape λ. For example,

$$STab([2,2,1]) = \{\; \begin{array}{|c|c|}\hline 3 \\\hline 2 & 5 \\\hline 1 & 4 \\\hline\end{array}\;,\; \begin{array}{|c|c|}\hline 4 \\\hline 2 & 5 \\\hline 1 & 3 \\\hline\end{array}\;,\; \begin{array}{|c|c|}\hline 4 \\\hline 3 & 5 \\\hline 1 & 2 \\\hline\end{array}\;,\; \begin{array}{|c|c|}\hline 5 \\\hline 2 & 4 \\\hline 1 & 3 \\\hline\end{array}\;,\; \begin{array}{|c|c|}\hline 5 \\\hline 3 & 4 \\\hline 1 & 2 \\\hline\end{array}\;\}. \tag{5}$$

The number f^λ of standard tableaux of a given shape λ can be easily computed using the hook-length formula. As standard tableaux of shape λ can index the basis of the vector space associated with the irreducible representation of the symmetric group indexed by λ, and since this same representation must have for multiplicity its dimension inside the regular representation of $\mathfrak{S}_n$, we have:

$$\sum_{\lambda \vdash n} (f^\lambda)^2 = |\mathfrak{S}_n| = n!. \tag{6}$$

3   Higher Specht Polynomials for the Symmetric Group

Algorithms in invariant theory must, at some point, construct invariant polynomials. Most of actual approaches use the Reynolds operator or an orbit sum of some special monomials. When the group becomes large, such invariants become very large, and even when they are stored in a sparse manner inside a computer, the number of terms can easily fit with $n!$ when G is a permutation group with a small index in $\mathfrak{S}_n$.

Our approach focuses on the combinatorics of the quotient $\mathbb{K}[\mathbf{x}]/(e_1, \ldots, e_n)$, and the *higher Specht polynomials* [20] constitute the perfect family to get an explicit answer to Problem 1.

The quotient $\mathbb{K}[\mathbf{x}]/(e_1, \ldots, e_n)$ is isomorphic to the regular representation of $\mathfrak{S}_n$ in which we have several copies of irreducible representations following their dimension. The *Specht polynomials*, which are associated with standard tableaux, allow to construct an explicit subspace of $\mathbb{K}[\mathbf{x}]$ isomorphic to an irreducible representation of the symmetric group.

Now, we will see that the higher Specht polynomials take care of the multiplicities of an irreducible representation inside the coinvariant. They are indexed by pair of standard tableaux of the same shape and they constitute a basis of the $\mathfrak{S}_n$-module $\mathbb{K}[\mathbf{x}]/(e_1, \ldots, e_n)$.

Let λ be a partition of n and S, T be two standard tableaux of shape λ. We define the word $w(S)$ by reading the tableau S from top to bottom in consecutive columns, starting from the left. The number 1 in the word $w(S)$ has for index 0. Now, recursively, if the number k in the word has index p, then $k+1$ has index $p+1$ if it lies to the left of k in the word, and it has index p otherwise. For example, with the two tableaux

$$S = \begin{array}{|c|c|}\hline 3 & 5 \\\hline 1 & 2 & 4 \\\hline\end{array} \qquad T = \begin{array}{|c|c|}\hline 2 & 4 \\\hline 1 & 3 & 5 \\\hline\end{array} .$$

The reading of the Tableau S gives 31524, now placing step by step the indices, we get

$$\begin{array}{ll} 3\ 1_0 5\ 2\ 4 & \textit{initialization} \\ 3\ 1_0 5\ 2_0 4 & \textit{right} : 0 \to 0 \\ 3_1 1_0 5\ 2_0 4 & \textit{left} : 0 \to 1 \\ 3_1 1_0 5\ 2_0 4_1 & \textit{right} : 1 \to 1 \\ w(S) = 3_1 1_0 5 2 2_0 4_1 & \textit{left} : 1 \to 2 \end{array}$$

Filling the indices in corresponding cells of the tableau S, we obtain $i(S)$, the index tableau of S.

$$i(S) = \begin{array}{|c|c|}\hline 1 & 2 \\\hline 0 & 0 & 1 \\\hline\end{array}$$

Now, using the tableaux T and $i(S)$, where cells of T are giving variable indices and the corresponding cells of $i(S)$ are giving exponents, we build a monomial $\mathbf{x}_T^{i(S)}$ as follows

$$T = \begin{array}{|c|c|}\hline 2 & 4 \\\hline 1 & 3 & 5 \\\hline\end{array} \qquad\qquad\qquad i(S) = \begin{array}{|c|c|}\hline 1 & 2 \\\hline 0 & 0 & 1 \\\hline\end{array}$$

$$\mathbf{x}_T^{i(S)} = x_1^0 x_2^1 x_3^0 x_4^2 x_5^1.$$

For T a standard tableau of shape λ, let $R(T)$ and $C(T)$ denote respectively the row stabilizer and the column stabilizer of T and consider the Young symmetrizer

$$\epsilon_T := \frac{f^\lambda}{n!} \sum_{\sigma \in R(T)} \sum_{\tau \in C(T)} sign(\tau) \tau \sigma \qquad (7)$$

which is an idempotent of the group algebra $\mathbb{Q}[\mathfrak{S}_n]$. We now define the polynomial F_T^S by

$$F_T^S(x_1, \ldots, x_n) := \epsilon_T(\mathbf{x}_T^{i(S)}). \qquad (8)$$

Theorem 1. *[20] Let n be a positive integer, the family of $n!$ polynomials F_T^S, for S, T running over standard tableaux of the same shape, forms a basis of the $Sym(\mathbf{x})$-module $\mathbb{K}[\mathbf{x}]$.*

Terasoma and Yamada proved it using the usual bilinear form in this context: the divided difference associated to the longest element of the symmetric group.

In three variables, here is the basis of $\mathbb{K}[x_1, x_2, x_3]$ as a $Sym(x_1, x_2, x_3)$-module,

S	T	$\mathbf{x}_T^{i(S)}$	F_T^S
$\boxed{1}\boxed{2}\boxed{3}$	$\boxed{1}\boxed{2}\boxed{3}$	1	1
$\begin{smallmatrix}3\\1\,2\end{smallmatrix}$	$\begin{smallmatrix}3\\1\,2\end{smallmatrix}$	x_3	$x_3 - x_1$
$\begin{smallmatrix}3\\1\,2\end{smallmatrix}$	$\begin{smallmatrix}2\\1\,3\end{smallmatrix}$	x_2	$x_2 - x_1$
$\begin{smallmatrix}2\\1\,3\end{smallmatrix}$	$\begin{smallmatrix}3\\1\,2\end{smallmatrix}$	$x_2 x_3$	$x_2(x_3 - x_1)$
$\begin{smallmatrix}2\\1\,3\end{smallmatrix}$	$\begin{smallmatrix}2\\1\,3\end{smallmatrix}$	$x_2 x_3$	$x_3(x_2 - x_1)$
$\begin{smallmatrix}3\\2\\1\end{smallmatrix}$	$\begin{smallmatrix}3\\2\\1\end{smallmatrix}$	$x_2 x_3^2$	$(x_3 - x_1)(x_3 - x_2)(x_2 - x_1)$

We will now try to solve Problem 1 by searching linear combinations of higher Specht polynomials stabilized by the action of a permutation group.

4   Combinatorial Description of the Invariant Ring

In this section, we show how to slice invariant rings finer than degree by degree. As irreducible representations of the symmetric group are homogeneous, we will build series mixing degree statistic and partitions.

4.1   A Refinement of the Moliens Series

Let $G \subset \mathfrak{S}_n$ be a permutation group. Any $\mathfrak{S}_n$-stable module is also G-stable, thus any representation of $\mathfrak{S}_n$ is also a representation of G. We will denote by $\mathcal{C}(G)$ the set of conjugacy classes of G.

Proposition 2. *Let $\lambda \vdash n$ be a partition of the positive integer n. Let $G \subset \mathfrak{S}_n$ be a permutation group. The multiplicity of the trivial representation of G inside the irreducible representation of $\mathfrak{S}_n$ indexed by λ is given by $m_\lambda(G, \mathfrak{S}_n)$ with*

$$m_\lambda(G, \mathfrak{S}_n) := \frac{1}{|G|} \sum_{C \in \mathcal{C}(G)} |C| \chi_\lambda(\sigma), \qquad (\sigma \text{ chosen arbitrary in } C) \qquad (9)$$

where $\chi_\lambda(\sigma)$ is the evaluation at σ of the character of $\mathfrak{S}_n$ associated with the irreducible representation indexed by the partition λ.

Definition 1. *Let $G \subset \mathfrak{S}_n$ be a permutation group. Using a formal set of variables $\mathbf{s} = (s_\lambda)_{\lambda \vdash n}$ indexed by partitions of n, we define the* trivial multiplicities enumerator $S(G, \mathbf{s})$ *as follow*

$$S(G, \mathbf{s}) := \sum_{\lambda \vdash n} m_\lambda(G, \mathfrak{S}_n) s_\lambda. \qquad (10)$$

For T a standard tableau, we denote by *cocharge(T)* the sum of the entries of $i(T)$ the index tableau of T.

Definition 2. *Let $\lambda \vdash n$ be a partition of a positive integer n and z a formal variable. We will denote by $\phi(\lambda, z)$ the* representation appearance polynomial *defined as follow*

$$\phi(\lambda, z) := \sum_{T \in STab(\lambda)} z^{cocharge(T)} \tag{11}$$

where the sum run over all standard tableaux T of shape λ.

$\phi(\lambda, z)$ makes the link between the degree and the irreducible representations of $\mathfrak{S}_n$ isomorphic to the abstract one indexed by λ appearing inside the quotient $\mathbb{K}[\mathbf{x}]/Sym^+(\mathbf{x})$.

If $\phi(\lambda, z)$ has for coefficient the integer k for a term in z^d, this means that the isotypical component, associated with the irreducible representation of $\mathfrak{S}_n$ indexed by λ, inside the graded quotient $\mathbb{K}[\mathbf{x}]/Sym^+(\mathbf{x})$ at degree d has multiplicity k at this degree. The higher Specht Polynomials realize explicitly these representations because the *cocharge* is exactly the sum of the entries of tableau $i(S)$ (or the degree of the corresponding polynomial).

Proposition 3. *Let $G \subset \mathfrak{S}_n$ be a permutation group. The trivial multiplicities enumerator $S(G,t)$ and the Hilbert series $H(G,z)$ are related by the following alphabet specialization:*

$$H(G, z) = \frac{S(G, s_\lambda \rightarrow \phi(\lambda, z))}{(1-z)(1-z^2)\cdots(1-z^n)}. \tag{12}$$

4.2 Secondary Invariants Built from Higher Specht Polynomials

Let $G \subset \mathfrak{S}_n$ be a permutation group and $\lambda \vdash n$ be a partition of n. Let us suppose that we have calculated $m_\lambda(G, \mathfrak{S}_n)$. We are at a stage in which we have an homogeneous G-stable space inside which we want to construct a finite and known number of independent invariant polynomials under the action of G.

The usual way to deal with this problem is to build an explicit family spanning the ambient space, to apply the Reynolds operator and to use some linear algebra to get a free family of the wanted dimension. Knowing this dimension gives a stopping criteria often very important since computations are extremely heavy even for a small number of variables.

In our context, even this usual approach would work, as permutation groups are often given by a list of generators, we can even forget the Reynolds operator.

Proposition 4. *Let $G \in \mathfrak{S}_n$ be a permutation group given by some generators: $G = \langle \sigma_1, \ldots, \sigma_r \rangle$. Let $\lambda \vdash n$ be a partition. The G-trivial abstract space inside the abstract representation of $\mathfrak{S}_n$ indexed by λ is given by the intersection of the eigenspaces of the representation matrices of $\sigma_1, \ldots, \sigma_r$ associated to the eigenvalue 1 (i.e. the common point-wise stabilized space of these matrices).*

5 Algorithm Building Secondary Invariants

We now present an effective algorithm exploiting the approach using the representation of the symmetric group.

Algorithm 1. Compute secondary invariants using representations of the symetric group

---

Input : $\sigma_1, \sigma_2, \ldots \sigma_r$ a set of permutations of size n generating a group G.

```
1     def SecondaryInvariants(σ₁, σ₂, ... σᵣ) :
2         G ← PermutationGroup(σ₁, σ₂, ... σᵣ)
4         Sec_inv ← {}
5         for λ ∈ Partitions(size = n) :
6             if mλ(G, 𝔖ₙ) ≠ 0 :
7                 V ← VectorSpace(field = ℚ, dimension = fλ)
8                 for i ∈ {1, 2, ..., r} :
9                     V ← V ∩ kernel(Mλ(σᵢ) − Id)
10                    for S ∈ StandardTableaux(shape = λ) :
11                        for P ∈ basis(V) :
12                            new_sec = 0
13                            for (coef, T) ∈ P :
14                                new_sec = new_sec + coef ∗ Fₜˢ
15                    Sec_inv ← Sec_inv ∪ {new_sec}
16        return Sec_inv
```

The returned set is composed by linear combinations of higher Specht polynomials. These polynomials can be easily evaluated but, as they contains a lot of Vandermonde type factors, there expansion on a set of n formal variables as a huge sum of monomials is an heavy computation.

The loop line(5) iterates over all irreducible representations of $\mathfrak{S}_n$, it is followed by a test selecting only these which contains G-trivial elements. line(7) initializes a full vector space whose basis vectors should be seen as standard tableaux of shape λ. $M_\lambda(\sigma_i)$ is the matrix of the permutation σ_i inside the irreducible representation of $\mathfrak{S}_n$ indexed by λ. After the lines(8 and 9), V contains abstract combinations of standard tableaux of shape λ which correspond to G-trivial elements inside the irreducible representation indexed by λ, a basis of V is an explicit solution of problem 1. The rest of the algorithm rebuilds secondary invariants from abstract G-trivial elements using the higher Specht polynomials F_T^S.

Let G be the group generated by permutations $(1,2)(3,4)$ and $(1,4)(2,3)$ (notation as product of disjoints cycles). G is of cardinality 4, its contains 3 elements of cycle type $(2,2)$ and the identity. Using Formula (9), we get $m_{(4)}(G, \mathfrak{S}_4) = 1$, $m_{(2,2)}(G, \mathfrak{S}_4) = 2$ and $m_{(1,1,1,1)}(G, \mathfrak{S}_4) = 1$. There is only a single standard tableau of shape (4) which is associated with the trivial representation of $\mathfrak{S}_4$ and the higher Specht polynomial 1 (which is a secondary invariant for any permutation group). The dimension associated to $(1, 1, 1, 1)$ is also 1. The only standard tableau of the associated shape corresponds to the Vandermonde in four variables $\Delta(x_1, x_2, x_3, x_4)$.

Finally, we have $m_{(2,2)}(G, \mathfrak{S}_4) = 2$. As there are two standard tableaux of this shape, the whole irreducible representation indexed by $(2, 2)$ and its other realization in the coinvariants of $\mathfrak{S}_4$ are invariant under the action of G. We thus obtain the following higher Specht polynomials for $T = \begin{array}{|c|c|}\hline 3 & 4 \\\hline 1 & 2 \\\hline\end{array}$ and $S = \begin{array}{|c|c|}\hline 2 & 4 \\\hline 1 & 3 \\\hline\end{array}$:

$$\begin{aligned}
F_T^T &= (x_3 - x_1)(x_4 - x_2) \\
F_S^T &= (x_2 - x_1)(x_4 - x_3) \\
F_T^S &= (x_3 - x_1)(x_4 - x_2)(x_1 x_3 + x_2 x_4) \\
F_S^S &= (x_2 - x_1)(x_4 - x_3)(x_1 x_2 + x_3 x_4)
\end{aligned} \tag{13}$$

Hence, the family $1, F_T^T, F_S^T, F_T^S, F_S^S, \Delta(x_1, x_2, x_3, x_4)$ forms a set of secondary invariants associated with the symmetric polynomials in 4 variables.

5.1 A Large Trace of the Algorithm

A never done before computational challenge consists in computing a generating family of the ring of invariants of the group acting on the edges of graphs over 5 nodes. This group is a subgroup of the symmetric group of degree $10 = \binom{5}{2}$ and has for cardinality $5! = 120$. We tried our approach on this group and we got the following verbose.

The trace have should be read has the following pattern:

```
[partition]  ambient dimension -->  (number of standard tableaux)
rank in S_n repr : (dimension of the G-trivial space)

sage: load("invariants.py")
sage: G = TransitiveGroup(10,12)
sage: Specht_basis_of_trivial_representations(G, verbose=True)
[3, 2, 2, 1, 1, 1]  ambient dimension -->  315
rank in S_n repr :  2
[6, 1, 1, 1, 1]  ambient dimension -->  126
rank in S_n repr :  3
[6, 4]  ambient dimension -->  90
rank in S_n repr :  3
...
[4, 3, 2, 1]  ambient dimension -->  768
rank in S_n repr :  6
...
total :  30240
n! / |G| :  30240
TOTAL CPU TIME :  414.837207
```

Our algorithm took 415 seconds to generate the 30240 secondary invariants as linear combinations of higher Specht polynomials. We still believe that the computation of a pair of primary and secondary invariants for this group is unreachable for Magma, Singular and the evaluation approach of Sage in less than 24 hours.

5.2   Complexity

The rich literature about effective invariant theory does not provide a lot a sharp complexity bounds for algorithms. Gröbner bases admit very general complexity bounds (in worst case $2^{2^{O(n)}}$ for n variables) which appears to be overestimated compared to their practical behavior. The thesis [3] of the author present an evaluation approach to compute the invariants inside a quotient of a reduced dimension. The algorithm computing secondary with this technique has a complexity in $O((n!)^2 + \frac{(n!)^3}{|G|^2})$. Using the representation of the symmetric group, it is still very hard to establish a sharp bound. However, we can produce better bounds.

Theorem 2. *Let G be a permutation group, subgroup of $\mathfrak{S}_n$, given by r generators. The complexity of the linear algebra reduction in algorithm 1 computing the secondary invariants of G is bounded by*

$$r \cdot \left(\sum_{\lambda \vdash n} m_\lambda(|G|, \mathfrak{S}_n)(f_\lambda)^2 \right) \tag{14}$$

where f^λ is the number of standard tableaux of shape λ.
Sketch of proof : *This bound is a straightforward counting of the reduction costs the r matrices of permutation for each irreducible representation of the symmetric group.*

Corollary 1. *Let G be a permutation group, subgroup of $\mathfrak{S}_n$, given by r generators. The complexity of Algorithm 1 computing the secondary invariants is in $O(r \cdot (n!)^{\frac{3}{2}})$.*
Sketch of proof : *We apply Formula (6) and a rough bound of the maximum of the f^λ.*

Acknowledgments. This research was driven by computer exploration using the open-source mathematical software `Sage` [18]. In particular, we perused its algebraic combinatorics features developed by the `Sage-Combinat` community [6], as well as its group theoretical features provided by `GAP` [9].

References

1. Abdeljaouad, I.: Théorie des Invariants et Applications à la Théorie de Galois effective. Ph.D. thesis, Université Paris 6 (2000)
2. Bergeron, F.: Algebraic combinatorics and coinvariant spaces. CMS Treatises in Mathematics (2009)
3. Borie, N.: Calcul des invariants des groupes de permutations par transformée de Fourier. Ph.D. thesis, Laboratoire de Mathématiques, Université Paris Sud (2011)
4. Colin, A.: Solving a system of algebraic equations with symmetries. J. Pure Appl. Algebra **117/118**, 195–215 (1997). algorithms for algebra (Eindhoven, 1996)
5. Colin, A.: Théorie des invariants effective; Applications à la théorie de Galois et à la résolution de systèmes algébriques; Implantation en AXIOM. Ph.D. thesis, École polytechnique (1997)

6. Sage-Combinat community, T.: Sage-Combinat: enhancing Sage as a toolbox for computer exploration in algebraic combinatorics (2008)
7. Derksen, H., Kemper, G.: Computational invariant theory. Springer-Verlag, Berlin (2002)
8. Faugère, J., Rahmany, S.: Solving systems of polynomial equations with symmetries using SAGBI-Gröbner bases. In: Proceedings of the 2009 international symposium on Symbolic and algebraic computation, pp. 151–158 (2009)
9. The GAP Group: Lehrstuhl D für Mathematik, RWTH Aachen, Germany and SMCS, U. St. Andrews, Scotland: GAP - Groups, Algorithms, and Programming (1997)
10. Garsia, A., Stanton, D.: Group actions on stanley-reisner rings and invariants of permutation groups. Advances in Mathematics **51**(2), 107–201 (1984)
11. Gatermann, K.: Symbolic solution of polynomial equation systems with symmetry. Konrad-Zuse-Zentrum für Informationstechnik, Berlin (1990)
12. Geissler, K., Klüners, J.: Galois group computation for rational polynomials. J. Symbolic Comput. **30**(6), 653–674 (2000). algorithmic methods in Galois theory
13. Kemper, G.: The *invar* package for calculating rings of invariants. IWR Preprint 93–94, University of Heidelberg (1993)
14. King, S.: Fast Computation of Secondary Invariants (2007). Arxiv preprint math/0701270
15. Lascoux, A., Schützenberger, M.P.: Polynômes de Schubert. C. R. Acad. Sci. Paris Sér. I Math. **294**(13), 447–450 (1982)
16. Pouzet, M., Thiéry, N.M.: Invariants algébriques de graphes et reconstruction. C. R. Acad. Sci. Paris Sér. I Math. **333**(9), 821–826 (2001)
17. Stanley, R.P.: Invariants of finite groups and their applications to combinatorics. Bull. Amer. Math. Soc. (N.S.) **1**(3), 475–511 (1979)
18. Stein, W., et al.: Sage Mathematics Software (Version 3.3). The Sage Development Team (2009). http://www.sagemath.org
19. Sturmfels, B.: Algorithms in invariant theory. Springer-Verlag, Vienna (1993)
20. Terasoma, T., Yamada, H.: Higher Specht polynomials for the symmetric group. Proc. Japan Acad. Ser. A Math. Sci. **69**(2), 41–44 (1993)
21. Thiéry, N.M.: Computing minimal generating sets of invariant rings of permutation groups with SAGBI-Gröbner basis. In: Discrete models (Paris, 2001), pp. 315–328 (electronic) (2001)

On Differential Uniformity of Maps
that May Hide an Algebraic Trapdoor

Marco Calderini$^{(\boxtimes)}$ and Massimiliano Sala

Department of Mathematics, University of Trento, Trento, Italy
marco.calderini@unitn.it, maxsalacodes@gmail.com

Abstract. We investigate some differential properties for permutations in the affine group, of a vector space V over the binary field, with respect to a new group operation $\circ$, inducing an alternative vector space structure on V.

Keywords: Trapdoors · Differential uniformity · Block ciphers · Boolean functions

1   Introduction

Most modern block ciphers are built using components whose cryptographic strength is evaluated in terms of the resistance offered to attacks on the whole cipher. For example, differential properties of Boolean functions are studied for the S-Boxes to thwart differential cryptanalysis ([3,10]).

Little is known on similar properties to avoid trapdoors in the design of the block cipher. In [7] the authors investigate the minimal properties for the S-Boxes (and the mixing layer) of an AES-like cipher (more precisely, a translation-based cipher, or tb cipher) to thwart the trapdoor coming from the imprimitivity action, first noted in [11] .

In [9], Li observed that if V is a finite vector space over a finite field $\mathbb{F}_p$, the symmetric group $\mathrm{Sym}(V)$ will contain many isomorphic copies of the affine group $\mathrm{AGL}(V)$, which are its conjugates in $\mathrm{Sym}(V)$. So there are several structures $(V, \circ)$ of a $\mathbb{F}_p$-vector space on the set V , where $(V, \circ)$ is the abelian additive group of the vector space. Each of these structure will yield in general a different copy $\mathrm{AGL}(V, \circ)$ of the affine group within $\mathrm{Sym}(V)$. So, a trapdoor coming from an alternative vector space structure, which we call *hidden sum*, can be embedded in a cipher, whenever the permutation group generated by the round functions of the cipher is contained in a conjugate of $\mathrm{AGL}(V)$. In [6] the authors provide conditions on the S-Boxes of a tb cipher that avoid attacks coming from hidden sums. This result has been generalized to tb ciphers over any field in [2]. Also, in [1], the authors studied such trapdoors, characterizing a new class of vectorial Boolean functions, which they call *anti-crooked*, able to avoid any hidden sum.

In the yet unpublished Ph.D thesis [4] the author investigated some properties of affine groups, of a vector space over the binary field, with respect to a

© Springer International Publishing Switzerland 2015
A. Maletti (Ed.): CAI 2015, LNCS 9270, pp. 70–78, 2015.
DOI: 10.1007/978-3-319-23021-4_7

hidden sum $\circ$. In particular, he focused on affine groups which contain the translation group with respect to the usual sum $+$, and affine groups whom translation group is contained in AGL(V). In this paper we study the differential properties of maps which are affine w.r.t. a hidden sum. Our results are presented in Section 3, while in Section 2 we provide some preliminaries from previous works. Our main result, Theorem 3, concludes Section 3. Section 4 concludes this paper with the sketch of an actual attack to a cipher in which a hidden sum trapdoor is embedded.

2   Preliminaries

Here we give some notation and some known results that we are going to use along the paper. In the following, if not specified, V will be an n-dimensional vector space over $\mathbb{F}_2$.

With the symbol $+$ we refer to the usual sum over the vector space V, and we denote by T_+, AGL($V, +$) and GL($V, +$), respectively, the translation, affine and linear groups w.r.t. $+$.

We recall that a p-elementary group G acting on a set Ω is a group of permutations on Ω such that for all g in G we have $g^p = Id_\Omega$.

A group G is called regular if for all a and b in Ω there exists a unique g in G such that $g(a) = b$.

Remark 1. An elementary group acting on a vector space $V = \mathbb{F}_p^n$ is obviously a p-elementary group. The translation group of V is an elementary abelian regular group. Vice versa, we claim that if T is an elementary abelian regular group, there exists a vector space structure $(V, \circ)$ such that T is the related translation group. In fact, from the regularity of T we have $T = \{\tau_a \mid a \in V\}$ where τ_a is the unique map in T such that $0 \mapsto a$. Then, defining the sum $x \circ a := \tau_a(x)$, it is easy to check that $(V, \circ)$ is a commutative group, and so we can consider the group operation as a sum, making it an additive group without loss of generality. Moreover, let the multiplication of a vector by an element of $\mathbb{F}_p$ defined by

$$sv := \underbrace{v \circ \cdots \circ v}_{s}, \text{ for all } s \in \mathbb{F}_p,$$

then it is easy to check that for all $s, t \in \mathbb{F}_p$, and $v, w \in V$

$$s(v \circ w) = sv \circ sw,$$

$$(s + t)v = sv \circ tv,$$

$$(st)v = s(tv)$$

and being T p-elementary $pv = 0$. Thus $(V, \circ)$ is a vector space over $\mathbb{F}_p$. Observe that $(V, \circ)$ and $(V, +)$ are isomorphic vector space (since $|V| < \infty$).

For abelian regular subgroups of the affine group in [5] the authors give a description of these in terms of commutative associative algebras that one can impose on the vector space $(V, +)$ or, in other words, of products that can be defined on V and distribute the sum $+$. We report the principal result shown in [5]. Recall that a (Jacobson) radical ring is a ring $(V, +, \cdot)$ in which every element is invertible with respect to the circle operation $x \circ y = x + y + x \cdot y$, so that $(V, \circ)$ is a group. The circle operation may induce a vector space structure on V or not.

Theorem 1. *Let $\mathbb{F}$ be an arbitrary field, and $(V, +)$ a vector space of arbitrary dimension over $\mathbb{F}$.*

There is a one-to-one correspondence between

1) *abelian regular subgroups T of $\mathrm{AGL}(V, +)$, and*
2) *commutative, associative $\mathbb{F}$-algebra structures $(V, +, \cdot)$ that one can impose on the vector space structure $(V, +)$, such that the resulting ring is radical.*

In this correspondence, isomorphism classes of $\mathbb{F}$-algebras correspond to conjugacy classes under the action of $\mathrm{GL}(V, +)$ of abelian regular subgroups of $\mathrm{AGL}(V, +)$.

We recall that an exterior algebra over an $\mathbb{F}$-vector space V is the $\mathbb{F}$-algebra whose product is the wedge product $\wedge$ having the following properties:

1) $x \wedge x = 0$ for all $x \in V$,
2) $x \wedge y = -y \wedge x$.

The elements of the exterior algebra over V are linear combinations of monomials such as $u, v \wedge w, x \wedge y \wedge z$, etc., where u, v, w, x, y, and z are vectors of V.

Remark 2. From the theorem above we can note that in characteristic 2, algebras corresponding to elementary abelian regular subgroups of $\mathrm{AGL}(V, +)$ are exterior algebras or a quotient thereof.

We will denote by σ_a the translation in T_+ such that $x \mapsto x + a$. We will use $T_\circ$ and $\mathrm{AGL}(V, \circ)$ to denote the translation and affine group corresponding to a hidden sum $\circ$, that is when $(V, \circ)$ is a vector space and so $T_\circ$ is elementary abelian and regular.

As noted in the remark above, since $T_\circ$ is regular, for each $a \in V$ there is a unique map $\tau_a \in T_\circ$ such that $0 \mapsto a$. Thus

$$T_\circ = \{\tau_a \mid a \in V\}.$$

The relation between $T_\circ$ and $\mathrm{AGL}(V, \circ)$ is that $\mathrm{AGL}(V, \circ)$ is the normalizer of $T_\circ$ in $\mathrm{Sym}(V)$, that is $\mathrm{AGL}(V, \circ)$ is the largest subgroup of $\mathrm{Sym}(V)$ containing $T_\circ$ such that $T_\circ$ is normal in it. Indeed, $\mathrm{AGL}(V, +)$ is the normalizer of T_+ and they are, respectively, the isomorphic images of $\mathrm{AGL}(V, \circ)$ and $T_\circ$. With 1_V we will denote the identity map of V.

Remark 3. If $T_{\circ} \subseteq \mathrm{AGL}(V, +)$, then $\tau_a = \sigma_a \kappa$ for some $\kappa \in \mathrm{GL}(V, +)$, since $\mathrm{AGL}(V, +) = \mathrm{GL}(V, +) \ltimes T_+$. We will denote by κ_a the linear map κ corresponding to τ_a.

Let $T \subseteq \mathrm{AGL}(V, +)$ and define the set

$$U(T) = \{a \mid \tau = \sigma_a, \tau \in T\}.$$

It is easy to check that $U(T)$ is a subspace of V, whenever T is a subgroup. If $T = T_{\circ}$ for some operation $\circ$, then $U(T_{\circ})$ is not empty for the following lemma.

Lemma 1 ([5]). *Let T_+ be the group of translation in $\mathrm{AGL}(V, +)$ and let $T \subseteq \mathrm{AGL}(V, +)$ be a regular subgroup. Then, if V is finite $T_+ \cap T$ is nontrivial.*

$U(T_{\circ})$ is important in the context of our theory and its dimension gives fundamental information on the corresponding hidden sum.

3   On the Differential Uniformity of a ∘-affine Map

Any round function of a translation-based block cipher (Definition 3.1 [7]) is composed by a parallel s-Box γ, a mixing layer λ and a translation σ_k by the round key. The map γ must be as non-linear as possible to create confusion in the message. An important notion of "non-linearity" of Boolean functions is the differential uniformity.

In this section we establish a lower bound on the differential uniformity of the maps lying in some $\mathrm{AGL}(V, \circ)$. We will consider the two cases of affine group $\mathrm{AGL}(V, \circ)$ such that $T_{\circ} \subseteq \mathrm{AGL}(V, +)$ and/or $T_+ \subseteq \mathrm{AGL}(V, \circ)$. In both cases in the following proofs we can consider w.l.o.g. maps f such that $f(0) = 0$. In fact in the first case we can compose f with $\tau_{f(0)}$ that maps $f(0)$ to 0 and in the second case we compose with $\sigma_{f(0)}$, in both cases we compose with an affine map.

We recall the definition of differential uniformity.

Definition 1. *Let $m, n \geq 1$. Let $f : \mathbb{F}_2^m \to \mathbb{F}_2^n$, for any $a \in \mathbb{F}_2^m$ and $b \in \mathbb{F}_2^n$ we define*

$$\delta_f(a, b) = |\{x \in \mathbb{F}_2^m \mid f(x + a) + f(x) = b\}|.$$

The differential uniformity of f is

$$\delta(f) = \max_{\substack{a \in \mathbb{F}_2^m, \, b \in \mathbb{F}_2^n \\ a \neq 0}} \delta_f(a, b).$$

f is said δ-differential uniform if $\delta = \delta(f)$.

We are ready for our first result.

Lemma 2. *Let $T_\circ \subseteq \mathrm{AGL}(V,+)$ and $\dim(U(T_\circ)) = k$. Then $f \in \mathrm{AGL}(V,\circ)$ is at least 2^k differentially uniform.*

Proof. Let $a \in U(T_\circ)$, then

$$f(x+a) + f(x) = f(x \circ a) + f(x) = (f(x) \circ f(a)) + f(x).$$

So, for all $f(x) \in U(T_\circ)$ we have

$$(f(x) \circ f(a)) + f(x) = (f(x) + f(a)) + f(x) = f(a),$$

that implies $|\{x \mid f(x+a) + f(x) = f(a)\}| \geq 2^k$.

When $T_+ \subseteq \mathrm{AGL}(V,\circ)$, we can define $U_\circ(T_+) = \{a \mid \sigma_a \in T_+ \cap T_\circ\}$ and it is a vector subspace of $(V,\circ)$. Then we obtain, analogously, the following lemma.

Lemma 3. *Let $T_+ \subseteq \mathrm{AGL}(V,\circ)$ and $\dim(U_\circ(T_+)) = k$, as a subspace of $(V,\circ)$. Then $f \in \mathrm{AGL}(V,\circ)$ is at least 2^k differentially uniform.*

Recalling that given a ring R, $r \in R$ is called nilpotent if there exists an integer n such that $r^n = 0$, while $r \in R$ is called unipotent if and only if $r - 1$ is nilpotent, we have the following:

Lemma 4. *Let $T_\circ \subseteq \mathrm{AGL}(V,+)$. Then for each $a \in V$, κ_a has order 2 and it is unipotent.*

Proof. We know that τ_a has order 2, because $T_\circ$ is elementary. Then, $\tau_a^2 = 1_V$ implies $\tau_a(a) = 0$, and in particular $\kappa_a(a) = a$. So

$$x = \tau_a^2(x) = \kappa_a(\kappa_a(x) + a) + a = \kappa_a^2(x) + a + a = \kappa_a^2(x) \quad \text{for all } x \in V.$$

That implies $(\kappa_a - 1_V)^2 = \kappa_a^2 - 1_V = 0$.

Remark 4. The lemma above can be easily generalized to any characteristic p, in this case the order of κ_a would be p.

Remark 5. It is well known that a square matrix is unipotent if and only if its characteristic polynomial $P(t)$ is a power of $t - 1$, i.e. it has a unique eigenvalue equals to 1.

We recall the following definition.

Definition 2. *Let A be an $n \times n$ matrix over a field $\mathbb{F}$, with $\lambda \in \mathbb{F}$ along the main diagonal and 1 along the diagonal above it, that is*

$$A = \begin{bmatrix} \lambda & 1 & \cdots & 0 \\ 0 & \lambda & 1 \cdots & 0 \\ \vdots & & & \vdots \\ 0 & \cdots & & \lambda \end{bmatrix}.$$

Then A is called the $n \times n$ elementary Jordan matrix or Jordan block of size n.

Definition 3. *A matrix A defined over a field $\mathbb{F}$ is said to be in Jordan canonical form if A is block-diagonal where each block is a Jordan block defined over $\mathbb{F}$.*

The following theorem is well-known (see for instance [8]).

Theorem 2. *Let A be an $n \times n$ matrix over a field $\mathbb{F}$ such that any eigenvalue of A is contained in $\mathbb{F}$, then there exists a matrix J defined over $\mathbb{F}$, which is in Jordan canonical form and similar to A.*

Lemma 5. *Let $T_\circ \subseteq \mathrm{AGL}(V, +)$. Then for each $a \in V$, κ_a fixes at least $2^{\lfloor \frac{n-1}{2} \rfloor + 1}$ elements of V.*

Proof. From Lemma 4, κ_a has a unique eigenvalue equals to $1 \in \mathbb{F}_2$, then from Theorem 2 there exists a matrix over $\mathbb{F}_2$ in the Jordan form similar to κ_a. Thus, $\kappa_a = AJA^{-1}$, for some $A, J \in GL(V, +)$ with

$$
J = \begin{bmatrix} 1 & \alpha_1 & \cdots & & 0 \\ 0 & 1 & \alpha_2 & \cdots & 0 \\ \vdots & & & & \vdots \\ 0 & \cdots & & 1 & \alpha_{n-1} \\ 0 & \cdots & & & 1 \end{bmatrix} \quad \text{and} \quad J^2 = \begin{bmatrix} 1 & 0 & \alpha_1\alpha_2 & \cdots & & 0 \\ 0 & 1 & 0 & \alpha_2\alpha_3 & \cdots & 0 \\ \vdots & & & & & \vdots \\ 0 & \cdots & & 1 & 0 & \alpha_{n-2}\alpha_{n-1} \\ 0 & \cdots & & & 1 & 0 \\ 0 & \cdots & & & & 1 \end{bmatrix}.
$$

where $\alpha_i \in \mathbb{F}_2$ for $1 \leq i \leq n-1$.

From the fact that J is conjugated to κ_a we have $J^2 = 1_V$, and that implies $\alpha_i \alpha_{i+1} = 0$ for all $1 \leq i \leq n-2$.

Note that if $\alpha_i = 1$ then α_{i-1} and α_{i+1} have to be equal to 0. Thus we have that when n is even at most $\frac{n}{2}$ α_i's can be equal to 1. Then at least $\frac{n}{2}$ elements of the canonical basis are fixed by J. When n is odd we have at most $\frac{n-1}{2}$ α_i's equal to 1 and then at least $\frac{n-1}{2} + 1$ elements of the canonical basis are fixed by J. Our claim follows from the fact that κ_a is conjugated to J.

In terms of algebras we have the following corollary.

Corollary 1. *Let $T_\circ \subseteq \mathrm{AGL}(V, +)$, and let $(V, +, \cdot)$ be the associated algebra of Theorem 1. Then for each $a \in V$, $a \cdot x$ is equal to 0 for at least $2^{\lfloor \frac{n-1}{2} \rfloor + 1}$ elements x of V.*

Remark 6. The bound on the number of elements fixed by κ_a given in Lemma 5 is tight. In fact let $(V, +, \cdot)$ be the exterior algebra over a vector space of dimension three, spanned by e_1, e_2, e_3. That is, V has basis

$$
e_1, e_2, e_3, e_1 \wedge e_2, e_1 \wedge e_3, e_2 \wedge e_3, e_1 \wedge e_2 \wedge e_3.
$$

We have that $e_1 \cdot x = 0$ for all $x \in E = \langle e_1, e_1 \wedge e_2, e_1 \wedge e_3, e_1 \wedge e_2 \wedge e_3 \rangle$. So, for all $x \in E$

$$
x \circ e_1 = x + e_1 + x \cdot e_1 = x + e_1.
$$

Vice versa if $x \circ e_1 = x + e_1$ then $x \in E$. The size of E is 2^4.

Lemma 6. *Let $T_\circ \subseteq AGL(V, +)$. Then $f \in AGL(V, \circ)$ is at least $2^{\lfloor \frac{n-1}{2} \rfloor + 1}$ differentially uniform.*

Proof. From Lemma 1 there exists $a \in U(T_\circ)$ different from zero. So

$$f(x + a) + f(x) = f(x \circ a) + f(x) = (f(x) \circ f(a)) + f(x) =$$

$$(f(x) + f(a) + f(a) \cdot f(x)) + f(x)$$

Now, from Corollary 1 we have that $f(a) \cdot f(x) = 0$ for at least $2^{\lfloor \frac{n-1}{2} \rfloor + 1}$ elements of V.
This implies $|\{x \mid f(x + a) + f(x) = f(a)\}| \geq 2^{\lfloor \frac{n-1}{2} \rfloor + 1}$.

Lemma 7. *Let $T_+ \subseteq AGL(V, \circ)$. Then $f \in AGL(V, \circ)$ is at least $2^{\lfloor \frac{n-1}{2} \rfloor + 1}$ differentially uniform.*

Proof. Note that Theorem 1, Lemma 1 and Corollary 1 hold also inverting the operation $\circ$ and $+$. Then, there exists $a \in V$ different from zero such that $x + a = x \circ a$ for all $x \in V$. Considering the algebra $(V, \circ, \cdot)$ such that $x + y = x \circ y \circ x \cdot y$ for all $x, y \in V$, we have

$$f(x + a) + f(x) = f(x \circ a) + f(x) = (f(x) \circ f(a)) + f(x) =$$

$$(f(x) \circ f(a)) \circ f(x) \circ f(x) \cdot (f(x) \circ f(a)) =$$

$$f(x) \circ f(a) \circ f(x) \circ f(x) \cdot f(x) \circ f(x) \cdot f(a).$$

From Remark 2, we have $y^2 = 0$ for all $y \in V$, and from Corollary 1 $f(x) \cdot f(a) = 0$ for at least $2^{\lfloor \frac{n-1}{2} \rfloor + 1}$ elements. Thus

$$|\{x \mid f(x + a) + f(x) = f(a)\}| \geq 2^{\lfloor \frac{n-1}{2} \rfloor + 1}.$$

Summarizing our results in this section, especially Lemma 2, 3, 6, 7, we obtain our theorem on the claimed differentiability.

Theorem 3. *Let $T_\circ \subseteq AGL(V, +)$ ($T_+ \subseteq AGL(V, \circ)$, respectively). Let $f \in AGL(V, \circ)$. Then $\delta(f) \geq 2^m$, where*

- $m = \max\{\lfloor \frac{n-1}{2} \rfloor + 1, \dim(U(T_\circ))\}$
- *($m = \max\{\lfloor \frac{n-1}{2} \rfloor + 1, \dim(U_\circ(T_+))\}$, respectively).*

By a computer check we obtain the following fact.

Fact 1. *Let $V = \mathbb{F}_2^n$ with $n = 3, 4, 5$. If $T_+ \subseteq AGL(V, \circ)$, let $f \in AGL(V, \circ)$. Then $\delta(f) \geq 2^{n-1}$.*

Remark 7. For $n = 7, 8$ there exist examples of functions that are affine w.r.t. a hidden sum $\circ$ satisfying $T_+ \subseteq AGL(V, \circ)$ and $\delta(f) = 2^{n-2}$. The existence of these permutations and Fact 1 suggest that probably there may exist bounds which are sharper than those in Theorem 3.

Remark 8. Note that if we consider $f \in Sym(\mathbb{F}_2^4)$ with $\delta(f) = 4$ then the parallel map (f, f) acting on $\mathbb{F}_2^8$ is 2^6 differentially uniform. Thus the differential uniformity may not guarantee, alone, security from a hidden sum trapdoor!

4   A Block Cipher With a Hidden Sum

In [4] (but see also [1]) the author constructs a toy block cipher with messages in $V = (\mathbb{F}_2)^6$ that involves two 3-bit invertible S-Boxes which are 4-differentially uniform. That cipher is such that the group generated by the parallel S-Box γ, the mixing layer λ and the translation group T_+ is contained in an affine group $\mathrm{AGL}(V, \circ)$ for an operation $\circ$, i.e.

$$\langle \lambda\gamma, T_+ \rangle \subseteq \mathrm{AGL}(V, \circ).$$

Moreover the authors report an algorithm that, in linear time, for all vector $v \in V$, it returns the vector $[v] = [\alpha_1, \ldots, \alpha_6]$ such that $v = \alpha_1 v_1 \circ \cdots \circ \alpha_6 v_6$, where $v_1, \ldots, v_6$ is a fixed basis of $(V, \circ)$.

Let $\varphi = \varphi_k$ be the encryption function, with a given unknown session key k. Assuming that an attacker can call the encryption oracle then the attack based on the hidden sum is the following:

1) compute $[\varphi(0)], [\varphi(v_1)], \ldots, [\varphi(v_6)]$,
2) construct the affinity $M + t$ given by the matrix M with rows $M_i = [\varphi(v_i)] + [\varphi(0)]$ and translation vector $t = [\varphi(0)]$.

We will have

$$[\varphi(v)] = [v] \cdot M + [t], \quad [\varphi^{-1}(v)] = ([v] + [t]) \cdot M^{-1},$$

for all $v \in V$, where the product row by column is the standard scalar product. The knowledge of M, t and M^{-1} provides a global deduction (reconstruction), since it becomes trivial to encrypt and decrypt. The attacks requires only 7 encryption that is much faster than brute-force searching in the key-space, considering that the key-space is $(\mathbb{F}_2)^6$.

Acknowledgments. Most of these results are present in the first author's Ph.D. thesis [4] and so he would like to thank his supervisor, the second author. The authors would like to thank Riccardo Aragona and the anonymous reviewers for their helpful comments.

References

1. Aragona, R., Calderini, M., Sala, M.: The role of Boolean functions in hiding sums as trapdoors for some block ciphers (2014). arXiv preprint arXiv:1411.7681
2. Aragona, R., Caranti, A., Dalla Volta, F., Sala, M.: On the group generated by the round functions of translation based ciphers over arbitrary finite fields. Finite Fields and Their Applications **25**, 293–305 (2014)
3. Biham, E., Shamir, A.: Differential cryptanalysis of DES-like cryptosystems. J. Cryptol. **4**(1), 3–72 (1991)
4. Calderini, M.: On Boolean functions, symmetric cryptography and algebraic coding theory, Ph.D. thesis, University of Trento (2015)

5. Caranti, A., Dalla Volta, F., Sala, M.: Abelian regular subgroups of the affine group and radical rings. Publ. Math. Debrecen **69**(3), 297–308 (2006)

6. Caranti, A., Dalla Volta, F., Sala, M.: An application of the ONan-Scott theorem to the group generated by the round functions of an AES-like cipher. Designs, Codes and Cryptography **52**(3), 293–301 (2009)

7. Caranti, A., Dalla Volta, F., Sala, M.: On some block ciphers and imprimitive groups. AAECC **20**(5–6), 229–350 (2009)

8. Lang, S.: Linear Algebra. Springer Undergraduate Texts in Mathematics and Technology. Springer (1987)

9. Li, C.H.: The finite primitive permutation groups containing an abelian regular subgroup. Proceedings of the London Mathematical Society **87**(03), 725–747 (2003)

10. Nyberg, K.: Differentially uniform mappings for cryptography. In: Helleseth, T. (ed.) EUROCRYPT 1993. LNCS, vol. 765, pp. 55–64. Springer, Heidelberg (1994)

11. Paterson, K.G.: Imprimitive permutation groups and trapdoors in iterated block ciphers. In: Knudsen, L.R. (ed.) FSE 1999. LNCS, vol. 1636, pp. 201–214. Springer, Heidelberg (1999)

On the Lower Block Triangular Nature of the Incidence Matrices to Compute the Algebraic Immunity of Boolean Functions

Deepak Kumar Dalai$^{(\boxtimes)}$

School of Mathematical Sciences, NISER, Bhubaneswar 751005, India
deepak@niser.ac.in

Abstract. The incidence matrix between two sets of vectors in $\mathbb{F}_2$ has a great importance in different areas of mathematics and sciences. The rank of these matrices are very useful while computing the algebraic immunity(AI) of Boolean functions in cryptography literature [3,7]. With a proper ordering of monomial (exponent) vectors and support vectors, some interesting algebraic structures in the incidence matrices can be observed. We have exploited the lower-block triangular structure of these matrices to find their rank. This structure is used for faster computation of the AI and the low degree annihilators of an n-variable Boolean functions than the known algorithms. On the basis of experiments on at least 20 variable Boolean functions, we conjecture about the characterization of power functions of algebraic immunity 1, could verify the result on the AI of n-variable inverse S-box presented in [6](i.e., $\lceil 2\sqrt{n} \rceil - 2$), and presented some results on the AI of some important power S-boxes.

Keywords: Cryptography · Boolean function · Power function · Algebraic immunity

1 Notation

In this section, we introduce the basic notations and definitions which are required to read the later part of the article.

V_n: The n dimensional vector space over the two element field $\mathbb{F}_2 = \{0, 1\}$.
$\mathsf{wt}(v)$: The weight of a vector $v = (v_1, v_2, \ldots, v_n) \in V_n$ is $\mathsf{wt}(v) = |\{v_i : v_i = 1\}|$.
$V_{n,d}$: The set of vectors in V_n of weight d or less i.e., $V_{n,d} = \{v \in V_n : \mathsf{wt}(v) \le d\}$.
$u \subseteq v$: For $u = (u_1, u_2, \ldots, u_n), v = (v_1, v_2, \ldots, v_n) \in V_n$, we denote $u \subseteq v$ if $u_i = 1$ implies $v_i = 1$ for $1 \le i \le n$.
$+, \sum$: The addition operators on $\mathbb{F}_2$ or, on reals $\mathbb{R}$, which is context based.
$\mathsf{int}(u)$: The integer value of the binary string representation of the vector $u \in V_n$.

Ordering of vectors: If $u, v \in V_n$, then
1. Lexicographic ordering: $u < v$ if $\mathsf{int}(u) < \mathsf{int}(v)$.
2. Weighted ordering: $u <_w v$ if $(\mathsf{wt}(u) < \mathsf{wt}(v))$ or, $(\mathsf{wt}(u) = \mathsf{wt}(v)$ and $\mathsf{int}(u) < \mathsf{int}(v))$.

© Springer International Publishing Switzerland 2015
A. Maletti (Ed.): CAI 2015, LNCS 9270, pp. 79–89, 2015.
DOI: 10.1007/978-3-319-23021-4_8

Incidence matrix (M_V^X): For $v, x \in V_n$, x is incident on v if $x \subseteq v$. We denote $v^x = 1$ if $x \subseteq v$ and 0 otherwise. For given two ordered sets of vectors V, X, the incidence matrix M_V^X of V on X is defined as $M_V^X[i,j] = v_i^{x_j}$, where v_i and x_j are i-th and j-th element in V and X respectively. We call X as exponent vector set and V as support vector set.

Incidence matrix (M_V^d): If the exponent vector set $X = V_{n,d}$, then the incidence matrix M_V^X is denoted as M_V^d.

Boolean function: A function $f : V_n \mapsto \mathbb{F}_2$ is called n variable Boolean function. The set of all Boolean functions on n-variable is denoted as B_n. The polynomial form of a Boolean function can be represented as an element of binary quotient ring on n variables $\mathbb{F}_2[x_1, \ldots, x_n]/\langle x_1^2 - x_1, \ldots, x_n^2 - x_n \rangle$ and this form is called the *algebraic normal form* (ANF) of the Boolean function. The degree of $f \in B_n$ (i.e., $\deg(f)$) is the algebraic degree. We also denote $B_{n,d} = \{f \in B_n : \deg(f) \leq d\}$ and $m_{n,d}$ is the set of all monomials of degree d or less. The evaluations of f at each vector in V_n with an order is known as the *truth table* representation of f and the representation can be viewed as a 2^n-tuple binary vector. The support set and the weight of $f \in B_n$ is defined as $S(f) = \{v \in V_n : f(v) = 1\}$ and $\mathsf{wt}(f) = |S(f)|$ respectively.

Algebraic immunity (AI): Given $f \in B_n$, a nonzero $g \in B_n$ is called an annihilator of f if $f.g = 0$, i.e., $f(v)g(v) = 0$ for all $v \in V_n$. The set of all annihilators of $f \in B_n$ is denoted by $An(f)$. The algebraic immunity of $f \in B_n$ is defined as $\mathsf{AI}(f) = \min\{\deg(g) : g \in An(f) \cup An(1 + f)\}$.

$\mathsf{wt}(M), \mathsf{den}(M)$: The weight and density of an $m \times n$ binary matrix M are defined as $\mathsf{wt}(M) = |\{M[i,j] : M[i,j] = 1\}|$ and $\mathsf{den}(M) = \frac{\mathsf{wt}(M)}{mn}$ respectively.

2   Introduction

The incidence matrix M_V^X is an interesting tool in the study of several branches in mathematics and computer sciences like combinatorics, coding theory, cryptography and polynomial interpolation. The incidence matrix M_V^d has an important role in the study of algebraic cryptanalysis. The problem to find the rank of this matrix is equivalent to compute the AI of a Boolean function [7]. Some algorithms are available in [4,5,7] to find the rank of M_V^d and the solution of the system of equations $M_V^d \gamma = 0$ to find the annihilators of degree d of the Boolean function of support set V.

From the point of view of algebraic cryptanalysis, $f \in B_n$ should not be used to design a cryptosystem if $\mathsf{AI}(f)$ is low [1,7]. It is known that for any $f \in B_n$, $\mathsf{AI}(f) \leq \lceil \frac{n}{2} \rceil$. Thus, the target of a good design is to use a $f \in B_n$ such that neither f nor $1 + f$ has an annihilator of degree much less than $\lceil \frac{n}{2} \rceil$.

If $g \in B_n$ is an annihilator of $f \in B_n$ then $g(v) = 0$ for $v \in S(f)$. To find a d or lesser degree annihilator $g \in B_{n,d}$, one has to solve the system of linear equations

$$\sum_{\alpha \in V_n, \mathsf{wt}(\alpha) \leq d} a_\alpha v^\alpha = 0 \quad \text{for } v \in S(f) \quad \text{i.e.,} \quad \sum_{\alpha \in V_n, \mathsf{wt}(\alpha) \leq d, \alpha \subseteq v} a_\alpha = 0 \text{ for } v \in S(f).$$

That is, $M_{S(f)}^d \gamma = 0$. $\qquad (1)$

where the transpose of γ is the unknown row vector (a_α). If $\mathrm{rank}(M_{S(f)}^d) < |m_{n,d}| = \sum_{i=0}^d \binom{n}{i}$ then f has a d or lesser degree annihilator.

For $f \in B_n$, the incidence matrix $M_{S(f)}^d$ is a particular case of M_V^X, whose rank tells about the $\mathsf{AI}(f)$. In this article, we study the rank of M_V^X, with special attention on $M_{S(f)}^d$. Some structures of M_V^X, which are not seen in a random binary matrix are addressed in [2]. Thus, the system of equations in Equation 1 can be solved faster as compared to solving an arbitrary system of equations of same order if the algebraic structures in M_V^X are carefully exploited. For example, in [4], some structures have been exploited to make it constant time faster in average case.

In Section 3, we have proposed a technique on the ordering of vectors in X and V which makes the matrix M_V^X and $M_{S(f)}^d$ a lower block triangular. The Section 3.2 and 3.3 contain the main results of this article to reduce the computation time. Experimental results of some important exponent S-boxes are presented in Section 4. On the basis of experiments, we conjecture about the complete characterization of power functions of algebraic immunity 1. We too verified the result on the AI of inverse power function in [6] till 20 variable Boolean functions which was conjectured in [6]. Some experimental results on some important power functions are too presented in this section.

3   Lower-Block Triangular Nature of M_V^X

An $n \times m$ matrix M is a lower-block triangular if its form is as

$$M = \begin{pmatrix} M_{11} & M_{12} & \dots & M_{1l} \\ M_{21} & M_{22} & \dots & M_{2l} \\ \dots & \dots & \ddots & \dots \\ M_{l1} & M_{l2} & \dots & M_{ll} \end{pmatrix} \qquad (2)$$

where M_{ij} are $n_i \times m_j$ sub-matrices for $1 \leq i, j \leq l$ with $\sum_{i=0}^l n_i = n$ and $\sum_{j=0}^l m_j = m$ and $M_{i,j}$ are zero sub-matrices for $j > i$.

3.1   Using the Ordering $<_w$

Consider two ordered sets of vectors $V, X \subseteq V_n$ with the ordering $<_w$. Let $V^0, V^1, \dots, V^n$ and $X^0, X^1, \dots, X^n$ be the disjoint partitions of V and X such that $V^i = \{v \in V : \mathsf{wt}(v) = i\}$ and $X^i = \{x \in X : \mathsf{wt}(x) = i\}, 0 \leq i \leq n$ respectively. If $v \in V^i$, $x \in X^j$ and $i < j$, it is clear that $v <_w x$ and $x \not\subseteq v$. Hence, from the definition of incidence, we have the following theorem.

Theorem 1. *The incidence matrix M_V^X is a lower block triangular matrix with $M_{ij} = M_{V^i}^{X^j}$ on the ordering $<_w$ of elements of V and X.*

Since M_V^X is lower block triangular, block wise Gaussian row elimination can be performed to find its rank. Consider that V and X are chosen randomly such that $|V| = |X| = 2^{n-1}$. Here, $|X^i|$ and $|V^i|$ are approximately $\frac{1}{2}\binom{n}{i}$ for $0 \le i \le n$. The time complexity for ith block wise row elimination is $O(2^n \binom{n}{i}^2)$. Hence, the time complexity for finding the rank of M_V^X is $O(2^n \sum_{i=0}^n \binom{n}{i}^2) = O(2^n \binom{2n}{n})$.

For the case of $M_{S(f)}^d$, $X = V_{n,d}$ and $V = S(f)$. So, $|X^i| = \binom{n}{i}$ for $0 \le i \le d$ and $|X^i| = 0$ for $d + 1 \le i \le n$. If $f \in B_n$ is a randomly chosen Boolean function, then $|V^i| \approx \frac{1}{2}\binom{n}{i}$, for $0 \le i \le n$. During each block wise row operation of matrix $M_{S(f)}^d$ from down to top, all columns in the block should be eliminated to have the rank equal to the number of columns. So, the same number of rows are eliminated and rest of the rows augmented to the next block of rows. For $0 \le j < n - d$, no computation is needed for the jth block wise row elimination as $|X^{n-j}| = 0$. For $n - d \le j \le n$, the number of rows in jth block operation is

$$r_j = |V^{n-j}| + \left(\sum_{i=0}^{j-1} |V^{n-i}| - \sum_{i=n-d}^{j-1} |X^{n-i}|\right)$$

$$= \sum_{i=0}^{j} |V^{n-i}| - \sum_{i=n-d}^{j-1} |X^{n-i}| \approx \frac{1}{2}\sum_{i=0}^{j}\binom{n}{i} - \sum_{i=n-d}^{j-1}\binom{n}{i}.$$

For $d < \frac{n}{2}$,

$$r_j \approx \frac{1}{2}\left(\binom{n}{j} + \sum_{i=n-d}^{j-1}\binom{n}{i} + \sum_{i=d+1}^{n-(d+1)}\binom{n}{i} + \sum_{i=n-(j-1)}^{d}\binom{n}{i} + \sum_{i=0}^{n-j}\binom{n}{i}\right) - \sum_{i=n-d}^{j-1}\binom{n}{i}$$

$$= \frac{1}{2}\left(\binom{n}{j} + \sum_{i=d+1}^{n-(d+1)}\binom{n}{i} + \sum_{i=0}^{n-j}\binom{n}{i}\right) = O(2^n).$$

During the jth block wise operation, the sub matrix has r_j many rows and $\sum_{i=0}^{n-j}\binom{n}{i}$ many columns and from there $\binom{n}{n-j}$ many columns (and as many rows) to be eliminated. The time complexity in the jth block wise row elimination is $O(r_j\binom{n}{n-j}(\sum_{i=0}^{n-j}\binom{n}{i})) = O(r_j\binom{n}{j}^2)$ and hence, the time complexity for finding the rank of $M_{S(f)}^d$ is $O(\sum_{j=n-d}^n (r_j\binom{n}{j}^2)) = O(2^n \sum_{j=n-d}^n \binom{n}{j}^2) = O(2^n \sum_{j=0}^d \binom{n}{j}^2)$.

Moreover, as discussed in [2, Section 3.2], each sub-matrix is sparser by $O(2^d)$ than a random matrix, which can further be exploited to speed up the process by $O(2^d)$. Moreover, there is advantage in space complexity as only the sub-matrix of size $r_j \times \binom{n}{j} = O(2^n \binom{n}{j})$ is needed during the jth block operation in stead of the whole $2^{n-1} \times 2^{n-1}$ matrix.

3.2 Using the Ordering $<$

Consider two ordered subsets V, X of V_n with the ordering $<$. Here onwards, we mean the notation $K = 2^k - 1$ and $N = 2^n - 1$. Let $V^0, V^1, \ldots, V^K$, and $X^0, X^1, \ldots, X^K$, $k \leq n$, be disjoint subsets of V and X, partitioned on the value of left most k coordinates of the vectors in V and X respectively. The superscript i of V^i and X^i denotes the integer value of left most k-coordinates of vectors in V and X. If $v \in V^i$, $x \in X^j$ and $i < j$, then $v < x$ and that implies $x \not\subseteq v$. Let denote $\mathsf{vect}(i)$ is the vector form of binary representation of i. Hence, we have the following lemma.

Lemma 1. *The incidence matrix $M_{V^i}^{X^j}$ is a zero matrix if $\mathsf{vect}(j) \not\subseteq \mathsf{vect}(i)$ for $0 \leq i, j \leq K$.*

Since $\mathsf{vect}(j) \not\subseteq \mathsf{vect}(i)$ for $j > i$, $M_{V^i}^{X^j}$ is zero matrix for $j > i$ and we have the following theorem.

Theorem 2. *The incidence matrix M_V^X is a lower block triangular matrix with $M_{ij} = M_{V^i}^{X^j}$ on the ordering $<$ of elements of V and X.*

Since M_V^X is lower block triangular, block wise Gaussian row elimination from down to top can be implemented for reducing the computation time. Hence we have the following results on the rank of M_V^X.

Corollary 1. $rank(M_V^X) < |X|$ *iff* $rank(M_{\overline{V}}^{\overline{X}}) < |\overline{X}|$ *where* $\overline{V} = \cup_{i=0}^{p} V^{K-i}$ *and* $\overline{X} = \cup_{i=0}^{p} X^{K-i}$ *for some* $0 \leq p \leq K$.

Corollary 2. *If* $\sum_{i=0}^{p} |V^{K-i}| < \sum_{i=0}^{p} |X^{K-i}|$ *for some* $0 \leq p \leq K$, *then* $rank(M_V^X) < |X|$. *Therefore, if* $|V| = |X|$ *and* $\sum_{i=0}^{p} |V^i| > \sum_{i=0}^{p} |X^i|$ *for some* $0 \leq p \leq K$, *then* $rank(M_V^X) < |X|$.

Corollary 2 classifies some Boolean functions of having low AI. It can be used in better way by finding a possible permutation on the variables $x_1, x_2, \ldots, x_n$, such that $\sum_{i=0}^{p} |V^{K-i}| < \sum_{i=0}^{p} |X^{K-i}|$ for a some p.

Corollary 3. *If* $rank(M_V^X) = |X|$ *then for every permutation on variables* $x_1, x_2, \ldots, x_n$ *and* k, p, $0 \leq k \leq n, 0 \leq p < 2^k$, $\sum_{i=0}^{p} |V^{K-i}| \geq \sum_{i=0}^{p} |X^{K-i}|$.

Example 1. Let $X = \{1, 2, 3, 4, 8, 9, 10, 14\}$ and $V = \{0, 3, 4, 5, 7, 9, 12, 15\}$ be two subsets of V_4. Here, the vectors are shown in their integer form. If we fix the left most two coordinates, then $X^0 = \{1, 2, 3\}, X^1 = \{4\}, X^2 = \{8, 9, 10\}, X^3 = \{14\}$ and $V^0 = \{0, 3\}, V^1 = \{4, 5, 7\}, V^2 = \{9\}, V^3 = \{12, 15\}$. Here, $|V^0| + |V^1| = 5$ and $|X^0| + |X^1| = 4$. Hence, following the corollary 2, we have $rank(M_V^X) < |X|$. To find the exact value of $rank(M_V^X)$ the block wise row reduction of M_V^X can be done as following. The block of rows enclosed by double lines are to be reduced.

$$
M_V^X =
\begin{pmatrix}
0\,0\,0 & 0\,0\,0\,0 \\
1\,1\,1 & 0\,0\,0\,0 \\
0\,0\,0 & 1\,0\,0\,0 \\
1\,0\,0 & 1\,0\,0\,0 \\
1\,1\,1 & 1\,0\,0\,0 \\
1\,0\,0 & 0\,1\,1\,0\,0 \\
0\,0\,0 & 1\,1\,0\,0 \\
1\,1\,1 & 1\,1\,1\,1\,1
\end{pmatrix}
\rightarrow
\begin{pmatrix}
0\,0\,0 & 0\,0\,0\,0 \\
1\,1\,1 & 0\,0\,0\,0 \\
0\,0\,0 & 1\,0\,0\,0 \\
1\,0\,0 & 1\,0\,0\,0 \\
1\,1\,1 & 1\,0\,0\,0 \\
1\,0\,0 & 0\,1\,1\,0\,0 \\
0\,0\,0 & 1\,1\,0\,0 \\
1\,1\,1 & 1\,1\,1\,1\,1
\end{pmatrix}
\rightarrow
\begin{pmatrix}
0\,0\,0 & 0\,0\,0\,0 \\
1\,1\,1 & 0\,0\,0\,0 \\
0\,0\,0 & 1\,0\,0\,0 \\
1\,0\,0 & 1\,0\,0\,0 \\
1\,1\,1 & 1\,0\,0\,0 \\
0\,0\,0 & 1\,1\,0\,0\,0 \\
1\,0\,0 & 0\,1\,1\,0\,0 \\
1\,1\,1 & 1\,1\,1\,1\,1
\end{pmatrix}
\rightarrow
\begin{pmatrix}
0\,0\,0 & 0\,0\,0\,0 \\
1\,1\,1 & 0\,0\,0\,0 \\
1\,1\,1 & 0\,0\,0\,0 \\
0\,1\,1 & 0\,0\,0\,0 \\
1\,1\,1 & 1\,0\,0\,0 \\
0\,0\,0 & 1\,1\,0\,0\,0 \\
1\,0\,0 & 0\,1\,1\,0\,0 \\
1\,1\,1 & 1\,1\,1\,1\,1
\end{pmatrix}
\rightarrow
\begin{pmatrix}
0\,0\,0 & 0\,0\,0\,0 \\
0\,0\,0 & 0\,0\,0\,0 \\
1\,0\,0 & 0\,0\,0\,0 \\
0\,1\,1 & 0\,0\,0\,0 \\
1\,1\,1 & 1\,0\,0\,0 \\
0\,0\,0 & 1\,1\,0\,0\,0 \\
1\,0\,0 & 0\,1\,1\,0\,0 \\
1\,1\,1 & 1\,1\,1\,1\,1
\end{pmatrix}
$$

Here $rank(M_V^X) = 6$ i.e., there are two free monomials corresponding to the vectors 2 and 10 in X i.e., x_2 and $x_2 x_4$. So, there are 2 linearly independent annihilators on the monomials of exponent vectors from X of the Boolean function having support set V.

Now consider that V and X are chosen randomly such that $|V| = |X| = \eta$. Fixing k variables, there are 2^k blocks of rows of size approximately $\frac{\eta}{2^k}$. The time complexity for row elimination of each block is $O(\eta \times (\frac{\eta}{2^k})^2) = O(\eta^3 2^{-2k})$. Hence, the time complexity for finding the rank of M_V^X is $O(2^k \times \eta^3 2^{-2k}) = O(\eta^3 2^{-k})$. If $|V| = |X| = 2^{n-1}$, the time complexity for finding the rank of M_V^X is $O(2^{3n-k})$. If one fixes all n variables, the theoretical time complexity becomes $O(2^{2n})$, i.e., quadratic time complexity on number of monomials. Moreover, the space complexity for the computation is $O(2^n)$ (i.e., linear) as only one block of rows is needed during the computation. Hence, we have the following theorem.

Theorem 3. *For a randomly chosen subsets V and X of V_n such that $|V| = |X| = 2^{n-1}$, the expected time complexity and space complexity to compute the rank of the $2^{n-1} \times 2^{n-1}$ matrix M_V^X is $O(2^{2n})$ and $O(2^n)$ i.e., quadratic time complexity and linear space complexity on the $|X|$ respectively.*

Now we shall discuss about the rank of $M_{S(f)}^d$, which is needed to compute $\mathsf{AI}(f)$ for $f \in B_n$. In this case, $X = V_{n,d}$ and $V = S(f)$. Since the exponent set X is not a random set, the time and space complexity is not expected as the described one in Theorem 3. For $0 \leq k \leq n$, we have $|X^i| = |V_{n,d}^i| = b_i = \sum_{j=0}^{d-\mathrm{wt}(i)} \binom{n-k}{j}$, $0 \leq i < 2^k$. If $f \in B_n$ is randomly chosen, then we have $|V^i| \approx 2^{n-k-1}$, $0 \leq i < 2^k$. In each block wise row operation (from down to top) of matrix $M_{S(f)}^d$, every time all columns in the block need to be eliminated. So, the same number of rows are also eliminated and rest of the rows are augmented to the next block of rows. Hence, during the j-th block wise row operation, for $0 \leq j \leq K$, the number of rows is

$$
r_j = |V^{K-j}| + \sum_{i=0}^{j-1}(|V^{K-i}| - b_{K-i})
$$

$$
= \sum_{i=0}^{j} |V^{K-i}| - \sum_{i=0}^{j-1} b_{K-i} \approx (j+1)2^{n-k-1} - \sum_{i=0}^{j-1} b_{K-i}.
$$

At the j-th block operation, the sub-matrix contains r_j rows, $c_j = \sum_{i=0}^{K-j} b_i$ columns and b_{K-j} columns from these c_j columns to be eliminated. So, the time complexity for the jth block row elimination is $O(r_j c_j b_{K-j})$ and hence, time complexity to find the rank of $M^d_{S(f)}$ is $O(\sum^K_{j=0} r_j c_j b_{K-j})$.

If $k = n$, then the time to compute the rank of $M^d_{S(f)}$ is $O(\sum^N_{j=0} r_j c_j b_{N-j})$. In this case $b_i = \sum\limits^{d-\text{wt}(i)}_{i=0} \binom{0}{i} = \begin{cases} 1 \text{ if } \text{wt}(i) \leq d \\ 0 \text{ if } \text{wt}(i) > d, \end{cases}$ i.e.,

$$b_{N-j} = \begin{cases} 1 \text{ if } \text{wt}(j) \geq n-d \\ 0 \text{ if } \text{wt}(i) < n-d. \end{cases}$$

So,

$$c_j = \sum^{N-j}_{i=0} b_i = \sum_{\substack{0 \leq i \leq N_j \\ \text{wt}(i) \leq d}} 1 = \sum^d_{i=0} \binom{n}{i} - \sum_{\substack{0 \leq i \leq j-1 \\ \text{wt}(i) \geq n-d}} 1$$

and

$$r_j \approx \frac{j+1}{2} - \sum^{j-1}_{i=0} b_{N-i} = \frac{j+1}{2} - \sum_{\substack{0 \leq i \leq j-1 \\ \text{wt}(i) \geq n-d}} 1.$$

When $\text{wt}(j) < n - d$ i.e., $b_{N-j} = 0$, there is no column to eliminate and hence no operation is done. When $\text{wt}(j) \geq n-d$, i.e., $b_{N-j} = 1$, there is only one column to eliminate. So, the time complexity for j-th block operation is $O(r_j c_j)$. Therefore, the time complexity to find the rank of $M^d_{S(f)}$ is $O(\sum_{\substack{0 \leq j \leq N \\ \text{wt}(j) \geq n-d}} r_j c_j)$.

Simplifying it, we have

$$\sum_{\substack{0 \leq j \leq N \\ \text{wt}(j) \geq n-d}} r_j c_j = \sum_{\substack{0 \leq j \leq N \\ \text{wt}(j) \geq n-d}} (\frac{j+1}{2} - \sum_{\substack{0 \leq i \leq j-1 \\ \text{wt}(i) \geq n-d}} 1)(\sum^d_{i=0} \binom{n}{i} - \sum_{\substack{0 \leq i \leq j-1 \\ \text{wt}(i) \geq n-d}} 1)$$

$$\leq \sum_{\substack{0 \leq j \leq N \\ \text{wt}(j) \geq n-d}} (\frac{j+1}{2} - \sum_{\substack{0 \leq i \leq j-1 \\ \text{wt}(i) \geq n-d}} 1)(\sum^d_{i=0} \binom{n}{i}).$$

Now, we will find the value of the summation $\sum_{\substack{0 \leq j \leq N \\ \text{wt}(j) \geq n-d}} j$. If j is in the summation, then j has $\text{wt}(j)$ many non-zero positions in the binary expansion of j and each non-zero position k contributes the value 2^k to the summation. In the summation, each position occurs $\frac{1}{n} \sum^n_{i=n-d} i \binom{n}{i} = \sum^n_{i=n-d} \binom{n-1}{i-1}$ many times. So, for $0 \leq k < n$, k-th position contributes the value $2^k \sum^n_{i=n-d} \binom{n-1}{i-1}$ to the summation. Hence, $\sum_{\substack{0 \leq j \leq N \\ \text{wt}(j) \geq n-d}} j = \sum^n_{i=n-d} \binom{n-1}{i-1} \sum^{n-1}_{k=0} 2^k = \sum^n_{i=n-d} \binom{n-1}{i-1} N.$

So,

$$\sum_{\substack{0 \leq j \leq N \\ \text{wt}(j) \geq n-d}} \frac{j+1}{2} = \frac{1}{2}(\sum_{\substack{0 \leq j \leq N \\ \text{wt}(j) \geq n-d}} j + \sum^n_{i=n-d} \binom{n}{i}) = \frac{1}{2}(\sum^n_{i=n-d} \binom{n-1}{i-1} N + \sum^n_{i=n-d} \binom{n}{i})$$

Now, in the summation $\sum_{\substack{0 \leq j \leq N \\ \text{wt}(j) \geq n-d}} \sum_{\substack{0 \leq i \leq j-1 \\ \text{wt}(i) \geq n-d}} 1$, an integer i with $\text{wt}(i) \geq n-d$, is counted l times, where $l = |\{j : i < j \leq N, \text{wt}(j) \geq n-d\}|$. Let $i_1 < i_2 < \cdots < N$ are integers with weight at least $n-d$, then i_1 is counted

$\sum_{i=n-d}^n \binom{n}{i} - 1$ times, i_2 is counted $\sum_{i=n-d}^n \binom{n}{i} - 2$ times and so on.

So, $\sum_{\substack{0 \le j \le N \\ \text{wt}(j) \ge n-d}} \sum_{\substack{0 \le i \le j-1 \\ \text{wt}(i) \ge n-d}} 1 = (\sum_{i=n-d}^n \binom{n}{i} - 1) + (\sum_{i=n-d}^n \binom{n}{i} - 2) + \cdots + 0$

$= \frac{1}{2} \sum_{i=n-d}^n \binom{n}{i} (\sum_{i=n-d}^n \binom{n}{i} - 1)$.

     Hence, $\sum_{\substack{0 \le j \le N \\ \text{wt}(j) \ge n-d}} r_j c_j \le (2^n \sum_{i=n-d}^n \binom{n}{i} - (\sum_{i=n-d}^n \binom{n}{i})^2) \sum_{i=0}^d \binom{n}{i}$

$= (\sum_{i=0}^d \binom{n}{i})^2 \sum_{i=d+1}^n \binom{n}{i}$.

Theorem 4. *For a randomly chosen Boolean function $f \in B_n$, the expected time complexity and space complexity to compute the rank of the matrix $M_{S(f)}^d$*

is $O((\sum_{i=0}^d \binom{n}{i})^2 \sum_{i=d+1}^n \binom{n}{i})$ *and* $O(\max_{0 \le j \le N} r_j c_j)$ *respectively.*

Since the simplification of the above expression is not very easy, the time complexity bound given in the Theorem 4 is not a tight upper bound. Hence the theoretical time complexity mentioned in Theorem 4 is not a significant improvement over other algorithms. However, in practice, it is very fast and can be used to compute for $n = 20$. Moreover, exploiting the sparseness of the sub-matrices, the computation speed can further be improved.

3.3   Ordering $<$ and Dalai-Maitra Algorithm [4]

As we discussed in above, to find AI of $f \in B_n$, one needs to compute the rank of $M_{S(f)}^d$. The involutory property of $M_{V_{n,d}}^{V_{n,d}}$ (i.e., $(M_{V_{n,d}}^{V_{n,d}})^2 = I$) is exploited to reduce the size of incidence matrix $M_{S(f)}^d$ to compute its rank in Dalai-Maitra algorithm [4]. Instead of computing the rank of $M_{S(f)}^d$ of order $|V_{n,d}| \times |S(f)|$, it is proposed to compute the rank of a smaller matrix I_f^d of order $|S(f) \setminus V_{n,d}| \times |V_{n,d} \setminus S(f)|$. Given a $f \in B_n$ and $d \le n$ the matrix I_f^d is defined as

$$I_f^d[v, x] = \begin{cases} \sum_{i=0}^{d-\text{wt}(x)} \binom{\text{wt}(v)-\text{wt}(x)}{i} \bmod 2 & \text{if } x \subseteq v \\ 0 & \text{if } x \not\subseteq v, \end{cases}$$

where $v \in Y = S(f) \setminus V_{n,d}$ and $x \in Z = V_{n,d} \setminus S(f)$.

Theorem 5. *[4] The matrix $M_{S(f)}^d$ is of full rank (i.e., $|V_{n,d}|$) iff the matrix I_f^d is of full rank (i.e., $|Z|$).*

We can see that the order of matrix I_f^d is reduced by half in average in both the number of rows and columns. To find AI(f), finding rank of $M_{S(f)}^d$ can speed up the process approximately by 8 times. We further speed up the process by observing the lower block triangular nature of $M_{S(f)}^d$ by proper ordering of the vectors in Y and Z.

     Let the vectors in Y and Z be ordered by $<$. For $0 \le k \le n$, let $Y^0, \ldots, Y^{2^k-1}$ and $Z^0, \ldots, Z^{2^k-1}$ be the partitions of Y and Z on their left most k coordinates of vectors in Y and Z respectively. Let denote $I_f^d[Y^i, Z^j]$ be the sub-matrix in I_f^d corresponding to the vector subsets Y^i and Z^j.

Lemma 2. *The sub-matrix $I_f^d[Y^i, Z^j]$ is a zero matrix if $\mathsf{vect}(j) \not\subseteq \mathsf{vect}(i)$ for $0 \leq i, j \leq K$.*

Since $\mathsf{vect}(j) \not\subseteq \mathsf{vect}(i)$ for $j > i$, $I_f^d[Y^i, Z^j]$ is zero matrix for $j > i$ and we have the following theorem.

Theorem 6. *The matrix I_f^d is a lower block triangular matrix with submatrices $I_{f_{ij}}^d = I_f^d[Y^i, Z^j]$, $0 \leq i, j \leq 2^k - 1$ on the ordering $<$ of elements of Y and Z.*

Comparing the partitions in matrix $M_{S(f)}^d$ in subsection 3.2, here we have $|Y^i| \approx \frac{|V^i|}{2}$ and $|Z^i| \approx \frac{|X^i|}{2}$. Therefore, the computation in this technique is expected to be 8 times faster than the technique described in the earlier subsection. Therefore, the technique presented here is so far the best technique to evaluate AI of a Boolean function. It is possible to find AI of a Boolean function of 20 variables or, a few more variables with less memory.

4 Experiments on the AI of Power Functions

Since the vector space characteristic of finite field $\mathbb{F}_{2^n}$ can be viewed as $V_n = \mathbb{F}_2^n$, every function $F : \mathbb{F}_{2^n} \mapsto \mathbb{F}_{2^n}$ can be viewed as an ordered collection of n Boolean function. That is, $F(x) = (F_1(x), F_2(x), \cdots, F_n(x))$, where the Boolean functions F_is are called the co-ordinate Boolean functions of F. The nonzero linear combination of the co-ordinate functions, (i.e., $\sum_{i=1}^{n} a_i F_i, a_i \in \mathbb{F}_2$ but not all a_i are zero) are called component Boolean functions of F. The component functions of F can too be algebraically represented as $Tr(\lambda F)$ for non-zero constants $\lambda \in \mathbb{F}_{2^n}^*$.

Definition 1. *Let $F : \mathbb{F}_{2^n} \mapsto \mathbb{F}_{2^n}$ be a function. The algebraic immunity of F is $\mathsf{AI}(F) = \min_{(a_1,\ldots,a_n) \in V_n \setminus \{(0,\ldots,0)\}} \{\mathsf{AI}(\sum_{i=1}^{n} a_i F_i)\}$ i.e., the minimum of AI of the component functions of F.*

A function $F : \mathbb{F}_{2^n} \mapsto \mathbb{F}_{2^n}$ is called a power function if F is of the form $F(x) = x^d$ for $x \in \mathbb{F}_{2^n}$ and d is an integer. The degree of power function x^d is defined as the weight of the $\mathsf{vect}(d)$, which is the degree of each component function of x^d. In this section, we present some experimental results on the AI of power functions.

During the experiments, we observed a nice result for power functions of having algebraic immunity 1. It is known that $\mathsf{AI}(x^{d_1}) = \mathsf{AI}(x^{d_2})$ if d_1 and d_2 are in same 2-cyclotomic coset modulo $2^n - 1$ i.e., $d_2 = 2^i d_1 \bmod 2^n - 1$ for some integer i. The size of each 2-cyclotomic coset is a divisor of n. It is very clear that the AI of linear power functions, i.e., $\mathsf{AI}(x^{2^i \bmod 2^n - 1})$, is 1. We present a conjecture on the nonlinear power functions of algebraic immunity 1.

Conjecture 1. Let $n \geq 4$ and x^d be a power function from $\mathbb{F}_{2^n}$ to $\mathbb{F}_{2^n}$. Then $\mathsf{AI}(x^d) = 1$ iff one of the followings happens for d.

  i. $d \in \{1, 2, \ldots, 2^{n-1}\}$ i.e., x^d is a linear power function.
  ii. The size of 2-cyclotomic coset modulo $2^n - 1$ of d is a proper divisor of n.

Based on this conjecture, we have the following example and corollary.

Example 2. Let take $n = 6$. Here $\mathsf{AI}(x^d) = 1$ iff

1. $d \in \{1, 2, 4, 8, 16, 32\}$ (when x^d is linear) or,
2. $d \in \{9, 18, 36\} \cup \{21, 42\} \cup \{27, 54, 45\}$ (when x^d is not linear).

Corollary 4. *If n is prime , then there is no non-linear power functions of algebraic immunity 1.*

Further, using the proposed technique, we computed AI of some cryptographic important power functions like inverse functions, Kasami exponents and Niho exponents up to 21 variables. The AI of n-variable inverse function, x^{-1}, is upper bounded by $\lceil 2\sqrt{n} \rceil - 2$, Kasami and Niho exponents are upper bounded by $\lceil 2\sqrt{n} \rceil$ [8]. Experimentally, we checked that the AI of the inverse function is exactly $\lceil 2\sqrt{n} \rceil - 2$ for $n \leq 21$ which is proved in [6].

A Kasami exponent $K : \mathbb{F}_{2^n} \mapsto \mathbb{F}_{2^n}$ is of the form $x^{2^{2k} - 2^k + 1}$ for $k \leq \frac{n}{2}$ and $\gcd(n, k) = 1$. The degree of Kasami exponent is $k + 1$. Therefore, $\mathsf{AI}(K) \leq min\{k+1, \lceil 2\sqrt{n} \rceil\}$. The following table presents the experimental result of $\mathsf{AI}(K)$ for the largest $k \leq \frac{n}{2}$ and $\gcd(n, k) = 1$.

n	k	$\deg(K)$	$\lceil 2\sqrt{n} \rceil$	$\mathsf{AI}(K)$	n	k	$\deg(K)$	$\lceil 2\sqrt{n} \rceil$	$\mathsf{AI}(K)$
10	3	4	7	4	14	5	6	8	6
11	5	6	7	5	15	7	8	8	7
12	5	6	7	5	16	7	8	8	7
13	6	7	8	6	17	8	9	9	8

For odd $n = 2s + 1$, a Niho exponent $N : \mathbb{F}_{2^n} \mapsto \mathbb{F}_{2^n}$ is of the form $x^{2^s + 2^{\frac{s}{2}} - 1}$ if s is even and $x^{2^{\frac{3s+1}{2}} + 2^s - 1}$ if s is odd. The degree of Niho exponent is $d = \frac{n+3}{4}$ if $n \equiv 1 \mod 4$ and $d = \frac{n+1}{2}$ if $n \equiv 3 \mod 4$. Therefore, $\mathsf{AI}(N) \leq min\{d, \lceil 2\sqrt{n} \rceil\}$. The following table presents the experimental results of $\mathsf{AI}(N)$.

n	$\deg(N)$	$\lceil \sqrt{n} \rceil$	$\mathsf{AI}(N)$	n	$\deg(N)$	$\lceil \sqrt{n} \rceil$	$\mathsf{AI}(N)$
9	3	7	3	15	8	8	7
11	6	7	5	17	5	9	5
13	4	8	4	19	10	9	9

Then we do experiments to find power functions of optimal AI (i.e., $\lceil \frac{n}{2} \rceil$) and we found that there are power functions of optimal AI but it becomes rarer as n increases. The experiment is tabulated below.

n	$m = |\{x^d : \mathsf{AI}(x^d) = \lceil \frac{n}{2} \rceil, 0 \leq d \leq 2^n - 2\}|$	$\frac{m}{2^n - 1}$	n	m	$\frac{m}{2^n - 1}$
3	3	≈ 0.4286	4	4	≈ 0.2667
5	15	≈ 0.4839	6	12	≈ 0.1905
7	21	≈ 0.1654	8	48	≈ 0.1882
9	45	≈ 0.0881	10	260	≈ 0.2542
11	154	≈ 0.0752	12	1236	≈ 0.3018

References

1. Courtois, N., Meier, W.: Algebraic attacks on stream ciphers with linear feedback. In: Biham, E. (ed.) EUROCRYPT 2003. LNCS, vol. 2656, pp. 345–359. Springer, Heidelberg (2003)
2. Dalai, D.K.: Computing the rank of incidence matrix and algebraic immunity of boolean functions. IACR Cryptology ePrint Archive, p. 273 (2013)
3. Dalai, D.K., Gupta, K.C., Maitra, S.: Results on algebraic immunity for cryptographically significant boolean functions. In: Canteaut, A., Viswanathan, K. (eds.) INDOCRYPT 2004. LNCS, vol. 3348, pp. 92–106. Springer, Heidelberg (2004)
4. Dalai, D.K., Maitra, S.: Reducing the number of homogeneous linear equations in finding annihilators. In: Gong, G., Helleseth, T., Song, H.-Y., Yang, K. (eds.) SETA 2006. LNCS, vol. 4086, pp. 376–390. Springer, Heidelberg (2006)
5. Didier, F.: Using Wiedemann's algorithm to compute the immunity against algebraic and fast algebraic attacks. In: Barua, R., Lange, T. (eds.) INDOCRYPT 2006. LNCS, vol. 4329, pp. 236–250. Springer, Heidelberg (2006)
6. Feng, X., Gong, G.: On algebraic immunity of trace inverse functions over finite fields with characteristic two. Cryptology ePrint Archive, Report 2013/585 (2013). http://eprint.iacr.org/
7. Meier, W., Pasalic, E., Carlet, C.: Algebraic attacks and decomposition of boolean functions. In: Cachin, C., Camenisch, J.L. (eds.) EUROCRYPT 2004. LNCS, vol. 3027, pp. 474–491. Springer, Heidelberg (2004)
8. Nawaz, Y., Gong, G., Gupta, K.C.: Upper bounds on algebraic immunity of boolean power functions. In: Robshaw, M. (ed.) FSE 2006. LNCS, vol. 4047, pp. 375–389. Springer, Heidelberg (2006)

Weighted Unranked Tree Automata over Tree Valuation Monoids and Their Characterization by Weighted Logics

Manfred Droste[1], Doreen Heusel[1], and Heiko Vogler[2]([✉])

[1] Institut für Informatik, Universität Leipzig, D-04109 Leipzig, Germany
{droste,dheusel}@informatik.uni-leipzig.de
[2] Institut für Theoretische Informatik,
Technische Universität Dresden, D-01062 Dresden, Germany
Heiko.Vogler@tu-dresden.de

Abstract. We introduce a new behavior of weighted unranked tree automata. We prove a characterization of this behavior by two fragments of weighted MSO logic and thereby provide a solution of an open equivalence problem of Droste and Vogler. The characterization works for valuation monoids as weight structures; they include all semirings and, in addition, enable us to cope with average.

1 Introduction

In 1967, Thatcher investigated the theory of pseudoterms (nowadays known as unranked trees) and pseudoautomata (or unranked tree automata), see [30]. Since then, this theory has been further developed, cf. e.g. [2,3,22,26,27] and Chapter 8 of [8], due to the development of the modern document language XML and the fact that (fully structured) XML-documents can be formalized as unranked trees. An automaton model for unranked trees with ordered data values was investigated in [29], and important closure properties of symbolic unranked tree transducers were given in [32,19]. In [15,21], weighted automata on unranked trees over semirings were investigated in order to be able to deal with quantitative queries. For further background on weighted tree automata we refer to [11,18].

Weighted logics over semirings represent another approach for the investigation of quantitative aspects. For words, a weighted MSO logic which is expressively equivalent to weighted word automata was developed in [9]. Several analogous formalisms followed for infinite words [13], ranked trees [14], infinite trees [28], trace languages [25], picture languages [17], texts [23], and nested words [24].

In [15] a logic counterpart for weighted unranked tree automata over semirings was established. More precisely, each unranked tree series which is definable in syntactically restricted MSO logic is recognizable [15, Thm.6.5], and

D. Heusel—Partially supported by DFG Graduiertenkolleg 1763 (QuantLA).

A. Maletti (Ed.): CAI 2015, LNCS 9270, pp. 90–102, 2015.
DOI: 10.1007/978-3-319-23021-4_9

every recognizable unranked tree series is MSO-definable [15, Thm.5.9] and, if the semiring is commutative, even syntactically restricted MSO-definable. But surprisingly, there is a recognizable unranked tree series over a non-commutative semiring which is not definable in syntactically restricted MSO logic. In [15] it is stated as an open problem to determine a weighted automata model expressively equivalent to syntactically restricted MSO logic. One goal of our paper is to solve this problem.

For this, we present a new class of weighted unranked tree automata. Syntactically they do not differ from the ones of [15]. They still consist of a state set and a family of weighted word automata. The latter are used to calculate the local weight at a position of a tree by letting the weighted word automaton run on the states at the children of the position. However, we will define the semantics (or: behavior) of weighted unranked tree automata in a different way. We do not use runs anymore, but we choose the technically more involved extended runs, which were already introduced in [15]. Additionally to the information of classical runs, extended runs also include runs of the weighted word automata called at positions of the input tree. In addition we change the way how the weight of such an extended run is calculated. In [15], the local weight of a position was defined by the weight of the run chosen for the word emerged of its children's labels. Here the local weight of a position equals the weight of the transition taken for this position in the run of the position's parent.

In this paper we consider tree valuation monoids as weight structures which were defined in [10] (cf. [4,5,6,7,12]). Tree valuation monoids are additive monoids equipped with a valuation function that assigns a value of this monoid to any tree with labels from the additive monoid. We will use the valuation function to calculate the weights of an extended run in a global way, i.e. given a run we apply the valuation function to all local weights which appear along the extended run. Tree valuation monoids are very general: each semiring, and each bounded (possibly non-distributive) lattice [20] is a tree valuation monoid. In addition, these structures enable us to cope with non-binary valuation functions like average or discounting. Thus our weighted unranked tree automata subsume the weighted unranked tree automata over commutative semirings of [15] and the weighted ranked tree automata over tree valuation monoids [10].

The main results of this paper are the following. We define a weighted MSO logic for unranked trees over product tree valuation monoids analogously to [12] and characterize the behavior of our weighted unranked tree automata by two different fragments of the logic, see Theorem 5.1. Thereby we solve the open equivalence problem of [15] in Corollary 5.7, and generalize the respective results of [15] about weighted unranked tree automata over commutative semirings and the respective results of [10].

2   Unranked Trees and (Product) Tree Valuation Monoids

Let $\mathbb{N} = \{1, 2, \ldots\}$ be the set of all natural numbers and $\mathbb{N}_0 = \mathbb{N} \cup \{0\}$. For a set X, the set X^* comprises all finite words over X. If $X_1, \ldots, X_n$ are sets and $x \in X_1 \times \ldots \times X_n$, then x_i equals the i-th component of x.

We will base unranked trees on tree domains. A *tree domain* $\mathcal{B}$ is a finite, non-empty subset of $\mathbb{N}^*$ such that for all $u \in \mathbb{N}^*$ and $i \in \mathbb{N}$, $u.i \in \mathcal{B}$ implies $u, u.1, \ldots, u.(i-1) \in \mathcal{B}$. An *unranked tree* over a set X (of labels) is a mapping $t : \mathcal{B} \to X$ such that $\mathrm{dom}(t) = \mathcal{B}$ is a tree domain. The elements of $\mathrm{dom}(t)$ are called *positions* of t and $t(u)$ is called *label* of t at $u \in \mathrm{dom}(t)$. We call $u \in \mathrm{dom}(t)$ a *leaf* of t if there is no i such that $u.i \in \mathrm{dom}(t)$. The *set of all leaves* of t is denoted by $\mathrm{dom}_{\mathrm{leaf}}(t)$. With $\mathrm{rk}_t(u) = \max\{i \in \mathbb{N} \mid u.i \in \mathrm{dom}(t)\}$ we denote the *rank* of position u. The image of t is $\mathrm{im}(t) = \{t(u) \mid u \in \mathrm{dom}(t)\}$. We denote the set of all unranked trees over X by U_X. A *tree language* is a subset of U_X. We view each $d \in X$ as unranked tree in U_X, also denoted by d, whose tree domain only consists of the position ε which is labeled by d.

Now we recall the notion of tree valuation monoids and product tree valuation monoids as defined in [10,12]. A *tree valuation monoid* (*tv-monoid* for short) is a quadruple $\mathbb{D} = (D, +, \mathrm{Val}, \mathbb{0})$ such that $(D, +, \mathbb{0})$ is a commutative monoid and $\mathrm{Val} : U_D \to D$ is a function, called *(tree) valuation function*, with $\mathrm{Val}(d) = d$ for every tree $d \in D$ and $\mathrm{Val}(t) = \mathbb{0}$ whenever $\mathbb{0} \in \mathrm{im}(t)$ for $t \in U_D$. A *product tree valuation monoid* (*ptv-monoid* for short) is a sextuple $\mathbb{D} = (D, +, \mathrm{Val}, \diamond, \mathbb{0}, \mathbb{1})$ which consists of a tv-monoid $(D, +, \mathrm{Val}, \mathbb{0})$, a constant $\mathbb{1} \in D$ with $\mathrm{Val}(t) = \mathbb{1}$ whenever $\mathrm{im}(t) = \{\mathbb{1}\}$ for $t \in U_D$, and an operation $\diamond \colon D^2 \to D$ with $\mathbb{0} \diamond d = d \diamond \mathbb{0} = \mathbb{0}$ and $\mathbb{1} \diamond d = d \diamond \mathbb{1} = d$ for all $d \in D$.

Example 2.1. $\mathbb{Q}_{\mathrm{max}} = (\mathbb{Q} \cup \{-\infty\}, \max, \mathrm{avg}, -\infty)$ with $\mathrm{avg}(t) = \frac{\sum_{u \in \mathrm{dom}(t)} t(u)}{|\mathrm{dom}(t)|}$ for all $t \in U_{\mathbb{Q} \cup \{-\infty\}}$ is a tv-monoid. The valuation function of this tv-monoid calculates the average of all weights of a tree. The idea for the average calculation was already suggested in [4,12] for words and in [10] for trees. From $\mathbb{Q}_{\mathrm{max}}$ we can obtain a ptv-monoid $\mathbb{Q}_{\mathrm{max}}^p$ by adding ∞ to the carrier set and setting $\diamond = \min$. We refer to [10] for further examples of (p)tv-monoids.

For the rest of this paper, let Σ be an alphabet, i.e., a finite, non-empty set, and $\mathbb{D}$ be a ptv-monoid.

3 Weighted Unranked Tree Automata

Here we introduce a new class of recognizable tree series. A tree series is recognizable if it can be recognized by a (classical) weighted unranked tree automaton over some tree valuation monoid using extended runs [15] for the definition of behavior. In the case of semirings, the semantics of a weighted unranked tree automaton based on runs and the semantics of this automaton based on extended runs are equivalent, cf. [15, Obs.6.8]. But for non-distributive structures, which are also considered here, this is not necessarily true. Besides, we define the weight of an extended run in a new way which is different from [15]. This will enable us to describe the behavior of weighted unranked tree automata by restricted weighted MSO formulas (see proof of Theorem 5.1).

A *weighted string automaton* (*WSA* for short) over Σ and $\mathbb{D}$ is a quadruple $\mathcal{A} = (P, I, \mu, F)$ where P is a non-empty, finite set of states, $I, F \subseteq P$ are the sets of initial and final states, respectively, and $\mu \colon P \times \Sigma \times P \to \mathbb{D}$.

A *run* of $\mathcal{A}$ on $w = w_1 \ldots w_n$ with $w_1, \ldots, w_n \in \Sigma$ and $n \geq 0$ is a sequence $\pi = (p_{i-1}, w_i, p_i)_{1 \leq i \leq n}$ if $n > 0$, and a state $\pi = p_0$ if $n = 0$ where $p_0, \ldots, p_n \in P$. The run π is *successful* if $p_0 \in I$ and $p_n \in F$. In order to define the weight $\text{wt}(\pi)$ of π using a tree valuation function Val, we define a tree t_π by letting $\text{dom}(t_\pi) = \{1^i \mid 0 \leq i < n\}$ and $t_\pi(1^i) = \mu(p_{i-1}, w_i, p_i)$ $(0 \leq i < n)$ if $n > 0$, and $t_\pi(\varepsilon) = \mathbb{0}$ if $n = 0$. Then let $\text{wt}(\pi) = \text{Val}(t_\pi)$. The *behavior* of $\mathcal{A}$ is the function $\|\mathcal{A}\| : \Sigma^* \to \mathbb{D}$ with $\|\mathcal{A}\|(w) = \sum_{\pi \text{ successful run on } w} \text{wt}(\pi)$ for $w \in \Sigma^*$. We call any mapping from Σ^* to $\mathbb{D}$ a *string series*. A string series S is called *recognizable* over $\mathbb{D}$ if there is a WSA $\mathcal{A}$ over Σ and $\mathbb{D}$ with $\|\mathcal{A}\| = S$.

A *weighted unranked tree automaton* (*WUTA* for short) over Σ and $\mathbb{D}$ is a triple $\mathcal{M} = (Q, \mathcal{A}, \gamma)$ where Q is a non-empty, finite set of states, $\mathcal{A} = (\mathcal{A}_{q,a} \mid q \in Q, a \in \Sigma)$ is a family of WSA over Q as alphabet and $\mathbb{D}$, and $\gamma : Q \to \mathbb{D}$ is a *root weight function*. Let $\mathcal{A}_{q,a} = (P_{q,a}, I_{q,a}, \mu_{q,a}, F_{q,a})$ for all $q \in Q$, $a \in \Sigma$. We assume the sets $P_{q,a}$ to be pairwise disjoint and let $P_{\mathcal{A}} = \bigcup_{q \in Q, a \in \Sigma} P_{q,a}$. Moreover, let $\mu_{\mathcal{A}}$ be the union of the transition functions $\mu_{q,a}$.

Intuitively, an extended run assigns a state $q \in Q$ to each position u of a given tree $t \in U_\Sigma$ and then consists of one run of $\mathcal{A}_{q,t(u)}$ on $q_1 \ldots q_{\text{rk}_t(u)}$ where q_i is the state assigned to the i-th child of u. Formally, an *extended run* of $\mathcal{M}$ on a tree t is a triple (q, s, l) such that

- $q \in Q$ is the *root state*;
- $s : \text{dom}(t) \setminus \{\varepsilon\} \to P_{\mathcal{A}} \times Q \times P_{\mathcal{A}}$ is a function such that $s(1) \ldots s(\text{rk}_t(\varepsilon))$ is a run of $\mathcal{A}_{q,t(\varepsilon)}$ and $s(u.1) \ldots s(u.\text{rk}_t(u))$ is a run of $\mathcal{A}_{s(u)_2, t(u)}$ for every $u \in \text{dom}(t) \setminus (\text{dom}_{\text{leaf}}(t) \cup \{\varepsilon\})$;
- $l : \text{dom}_{\text{leaf}}(t) \to P_{\mathcal{A}}$ is a function satisfying $l(\varepsilon) \in P_{q,t(\varepsilon)}$ if t only consists of the root, and if $u \neq \varepsilon$ is a leaf, then $l(u) \in P_{s(u)_2, t(u)}$.

An extended run is *successful* if $s(u.1) \ldots s(u.\text{rk}_t(u))$ is successful for all $u \in \text{dom}(t) \setminus \text{dom}_{\text{leaf}}(t)$ and if $l(u)$ is successful for all $u \in \text{dom}_{\text{leaf}}(t)$ (i.e., $l(u)$ is an initial and final state of $\mathcal{A}_{s(u)_2, t(u)}$ if $u \neq \varepsilon$ respectively of $\mathcal{A}_{q,t(\varepsilon)}$ if $u = \varepsilon$). We let $\text{succ}(\mathcal{M}, t)$ denote the set of all successful extended runs of $\mathcal{M}$ on t.

To define the weight of an extended run we proceed differently from [15] where the local weight of a position u was defined by the weight of the run chosen for the labels of the children of u. Here, we will define the local weight of u by the weight of the transition taken for u in the run of the parent of u. Each extended run (q, s, l) on t defines a tree $\mu(t, (q, s, l)) \in U_\mathbb{D}$ where $\text{dom}(\mu(t, (q, s, l))) = \text{dom}(t)$ and

$$\mu(t, (q, s, l))(u) = \begin{cases} \gamma(q) & \text{if } u = \varepsilon, \\ \mu_{\mathcal{A}}(s(u)) & \text{otherwise} \end{cases}$$

for all $u \in \text{dom}(t)$. We call $\mu(t, (q, s, l))(u)$ the *local weight* of u and $\text{Val}(\mu(t, (q, s, l)))$ the *weight of* (q, s, l) *on* t. The *behavior* of a WUTA $\mathcal{M}$ is the function $\|\mathcal{M}\| : U_\Sigma \to \mathbb{D}$ defined by

$$\|\mathcal{M}\|(t) = \sum_{(q,s,l) \in \text{succ}(\mathcal{M},t)} \text{Val}(\mu(t, (q, s, l)))$$

for all $t \in U_\Sigma$. Thus, if no successful extended run on t exists, we put $\|\mathcal{M}\|(t) = \mathbb{0}$.

Any mapping from U_Σ to $\mathbb{D}$ is called a *tree series*. A tree series $S: U_\Sigma \to \mathbb{D}$ is called *recognizable* over $\mathbb{D}$ if there is a WUTA $\mathcal{M}$ over Σ and $\mathbb{D}$ with $\|\mathcal{M}\| = S$.

Example 3.1. Let $\mathbb{Q}_{\max}$ be the tv-monoid from Example 2.1. We will consider a WUTA $\mathcal{M}$ which calculates the leaves-to-size ratio of a given input tree, where the size of a tree is the number of all positions of the tree. Let $\mathcal{M} = (\{c, n\}, \mathcal{A}, \gamma)$ over an arbitrary, but fixed alphabet Σ with $\gamma(c) = 1$, $\gamma(n) = 0$, and

- $\mathcal{A}_{n,a} = (\{i, f\}, \{i\}, \mu_{n,a}, \{f\})$ where $\mu_{n,a}(i, n, f) = \mu_{n,a}(f, n, f) = 0$, $\mu_{n,a}(i, c, f) = \mu_{n,a}(f, c, f) = 1$ and $\mu_{n,a}(f, q, i) = \mu_{n,a}(i, q, i) = -\infty$
- $\mathcal{A}_{c,a} = (\{p\}, \{p\}, \mu_{c,a}, \{p\})$ where $\mu_{c,a}(p, q, p) = -\infty$

for all $q \in \{c, n\}$ and $a \in \Sigma$; for notational convenience, here we have dropped the condition on pairwise disjointness of the state sets.

First, let us consider an example tree. For this, we choose $\Sigma = \{\alpha, \beta\}$ and tree $t_{ex} = $. Then (n, s, l) with $s = $ and $l = $

is an extended run on t_{ex}. Here an unlabeled position means that it is not in the domain of the represented function. Obviously (n, s, l) is successful, since the runs $s(1)s(2) = (i, c, f)(f, n, f)$ and $s(2.1) = (i, c, f)$ are successful in $\mathcal{A}_{n,\alpha}$ and $\mathcal{A}_{n,\beta}$, respectively, and the run p is successful in $\mathcal{A}_{c,\alpha}$ as well as in $\mathcal{A}_{c,\beta}$. The local weights of (n, s, l) are

$$\mu(t_{ex}, (n, s, l)) = $$ $$= $$

and thus the weight of (n, s, l) equals $\frac{1}{2}$.

Now let t be an arbitrary, but fixed tree. It is easy to see that for every successful extended run (q, s, l) on t, $l(u) = p$ for every leaf u of t. Assume that in addition (q, s, l) assigns the state n to each inner position of t. Let π_u be the unique run of $\mathcal{A}_{n,t(u)}$ for which t_{π_u} has no label equal to $-\infty$, thus, π_u leads directly from i to f and finally loops in f. If (q, s, l) consists for every inner position $u \neq \varepsilon$ of π_u, then (q, s, l) is the only successful extended run such that $\mu(t, (q, s, l))$ does not contain $-\infty$. Let π denote this unique extended run. For leaves u of t, $\mu(t, \pi)(u) = 1$ and for inner positions u', $\mu(t, \pi)(u') = 0$. Thus,

$$\|\mathcal{M}\|(t) = \mathrm{avg}(\mu(t, \pi)) = \frac{\sum_{u \in \mathrm{dom}(t)} \mu(t, \pi)(u)}{|\mathrm{dom}(t)|} = \frac{\text{``number of leaves of } t\text{''}}{\text{``size of } t\text{''}}.$$

Remark 3.2. The WUTA subsume the weighted ranked tree automata over tv-monoids of [10] as well as the weighted unranked tree automata over commutative semirings [15]. But there are tree series over non-commutative semirings which are recognizable by the weighted unranked tree automata of [15] but not by our WUTA. An example was given in the proof of [15, Thm.6.10].

Furthermore, it is easy to show that unranked tree automata over Σ [30,2,22,27] are equivalent to WUTA over Σ and the boolean semiring $\mathbb{B}$. Thus, for each WUTA over $\mathbb{B}$ there is an equivalent deterministic WUTA [30, Thm.1]. A WUTA over $\mathbb{B}$ is *deterministic* if for every $a \in \Sigma$ and $q_1, q_2 \in Q$, if $q_1 \neq q_2$, then there is no $w \in Q^*$ such that $\|\mathcal{A}_{q_1,a}\|(w) = \|\mathcal{A}_{q_2,a}\|(w) = \mathbb{1}$.

Next we will derive some properties of recognizable tree series. Let S_1, S_2 be two tree series and $d \in \mathbb{D}$. The *scalar product* $d \diamond S_1$, the *sum* $S_1 + S_2$ and the *(Hadamard) product* $S_1 \diamond S_2$ are defined pointwise by $(d \diamond S_1)(t) = d \diamond S_1(t)$, $(S_1 + S_2)(t) = S_1(t) + S_2(t)$ and $(S_1 \diamond S_2)(t) = S_1(t) \diamond S_2(t)$ for all $t \in U_\Sigma$. For a tree language $L \subseteq U_\Sigma$, the *characteristic function of* L, called $\mathbb{1}_L$, equals $\mathbb{1}$ for all $t \in L$ and $\mathbb{0}$ for all $t \in U_\Sigma \setminus L$. A tree series S is a *recognizable step function* if there are recognizable tree languages $L_1, \ldots, L_k$ forming a partition of U_Σ and values $d_1, \ldots, d_k \in \mathbb{D}$ such that $S = \sum_{i=1}^{k} d_i \diamond \mathbb{1}_{L_i}$.

Lemma 3.3. *([10], Lemma 5.9) The class of recognizable step functions over Σ and $\mathbb{D}$ is closed under the operations $+$ and the Hadamard product $\diamond$.*

The next theorem can be proved by applying standard automata constructions (assuming, in (2), the unranked tree automaton for L to be deterministic).

Theorem 3.4. *Let $\mathbb{D}$ be a ptv-monoid.*

1. *The class of recognizable tree series is closed under sum.*
2. *Let L be a recognizable tree language and S a recognizable tree series. Then $\mathbb{1}_L \diamond S$ (which equals $S \diamond \mathbb{1}_L$) is also recognizable.*

A ptv-monoid $\mathbb{D}$ is *regular* if for all $d \in \mathbb{D}$ and all alphabets Σ a WUTA $\mathcal{M}_d$ exists with $\|\mathcal{M}_d\|(t) = d$ for each $t \in U_\Sigma$. Using Theorem 3.4 one can easily show the following lemma.

Lemma 3.5. *Let $\mathbb{D}$ be a regular ptv-monoid. Each recognizable step function S over $\mathbb{D}$ is a recognizable tree series.*

Now we consider the closure under relabeling, similarly to [14,12]. Let Σ and Γ be two alphabets and $h : \Sigma \to 2^\Gamma$ be a mapping. Then h can be extended to a mapping $h' : U_\Sigma \to 2^{U_\Gamma}$ by letting $h'(t)$ be the set of all trees t' over Γ such that $\mathrm{dom}(t') = \mathrm{dom}(t)$ and $t'(u) \in h(t(u))$ for each position $u \in \mathrm{dom}(t)$. For every tree series S over $\mathbb{D}$ and Σ the tree series $h''(S)$ over $\mathbb{D}$ and Γ is defined by

$$h''(S)(t') = \sum_{t \in U_\Sigma \wedge t' \in h'(t)} S(t)$$

for all $t' \in U_\Gamma$. Clearly, the index set of the summation is finite. We denote h' and h'' also by h which we call a *relabeling*. The proof for the following lemma works by an automaton construction already applied in a similar way in [16,12].

Lemma 3.6. *Recognizable tree series are closed under relabeling.*

We will show that under suitable conditions the Hadamard product $\diamond$ preserves the recognizability of arbitrary tree series. For this, we recall some properties of ptv-monoids already defined in [10]. We call $\mathbb{D}$ *left-multiplicative* if $d \diamond \mathrm{Val}(t) = \mathrm{Val}(t')$ for all $d \in \mathbb{D}$, $t, t' \in U_{\mathbb{D}}$ with $\mathrm{dom}(t) = \mathrm{dom}(t')$, $t'(\varepsilon) = d \diamond t(\varepsilon)$, and $t'(u) = t(u)$ for every $u \in \mathrm{dom}(t) \backslash \{\varepsilon\}$. Furthermore, $\mathbb{D}$ is *left-Val-distributive* if $d \diamond \mathrm{Val}(t) = \mathrm{Val}(t')$ for all $d \in \mathbb{D}$, $t, t' \in U_D$ with $\mathrm{dom}(t) = \mathrm{dom}(t')$ and $t'(u) = d \diamond t(u)$ for every $u \in \mathrm{dom}(t)$. Two subsets $D_1, D_2 \subseteq \mathbb{D}$ *commute* if $d_1 \diamond d_2 = d_2 \diamond d_1$ for all $d_1 \in D_1$, $d_2 \in D_2$. We call $\mathbb{D}$ *conditionally commutative* if $\mathrm{Val}(t_1) \diamond \mathrm{Val}(t_2) = \mathrm{Val}(t)$ for all $t_1, t_2, t \in U_{\mathbb{D}}$ with $\mathrm{dom}(t_1) = \mathrm{dom}(t_2) = \mathrm{dom}(t)$, $\mathrm{im}(t_1)$ and $\mathrm{im}(t_2)$ commute and $t(u) = t_1(u) \diamond t_2(u)$ for all $u \in \mathrm{dom}(t)$. A ptv-monoid $\mathbb{D}$ is a *conditionally commutative tree valuation semiring (cctv-semiring)* if $(D, +, \diamond, \mathbb{0}, \mathbb{1})$ is a semiring and if $\mathbb{D}$ is conditionally commutative and, moreover, left-multiplicative or left-Val-distributive. For examples, we refer to [10].

Let $\mathcal{W}_{\mathcal{M}}$ comprises all the weights of automaton $\mathcal{M}$, i.e., all transition weights of any automaton $\mathcal{A}_{q,a}$ of $\mathcal{M}$ and all root weights of $\mathcal{M}$.

Theorem 3.7. *Let $\mathbb{D}$ be a cctv-semiring.*

1. *Let S_1 be a recognizable step function and S_2 a recognizable tree series. Then $S_1 \diamond S_2$ is also recognizable.*
2. *Let $\mathcal{M}_i = (Q_i, \mathcal{A}_i, \gamma_i)$ be a WUTA ($i \in \{1, 2\}$) such that $\mathcal{W}_{\mathcal{M}_1}$ and $\mathcal{W}_{\mathcal{M}_2}$ commute. Then $\|\mathcal{M}_1\| \diamond \|\mathcal{M}_2\|$ is recognizable.*

4   Weighted MSO Logic for Unranked Trees

We introduce a weighted MSO logic and its semantics for unranked trees over tv-monoids. As in [10], we follow [9] incorporating an idea of [1]. Let $\mathcal{V}_1$ and $\mathcal{V}_2$ be countable, infinite sets of first order and second order variables, respectively. The syntax of the weighted MSO logic over $\mathbb{D}$ is defined by the EBNF:

$$\beta ::= \mathrm{label}_a(x) \mid \mathrm{desc}(x, y) \mid x \leq y \mid x \sqsubseteq y \mid x \in X \mid \neg\beta \mid \beta \wedge \beta \mid \forall x.\beta \mid \forall X.\beta$$
$$\varphi ::= d \mid \beta \mid \varphi \vee \varphi \mid \varphi \wedge \varphi \mid \exists x.\varphi \mid \forall x.\varphi \mid \exists X.\varphi$$

where $d \in \mathbb{D}$, $a \in \Sigma$, $x, y \in \mathcal{V}_1$, and $X \in \mathcal{V}_2$. We call the formulas β *boolean formulas* and the formulas φ *weighted MSO formulas* (or *wMSO formulas*).

To define the semantics of the wMSO formulas, we follow the common approach for MSO logics using assignments and extended alphabets to deal with free variables, cf. [31]. The set $\mathrm{free}(\varphi)$ of free variables occurring in φ is defined as usual. A *sentence* is a formula without free variables. Let φ be a wMSO formula, $\mathcal{V}$ a finite set of variables with $\mathrm{free}(\varphi) \subseteq \mathcal{V}$, and $t \in U_{\Sigma}$. A $(\mathcal{V}, t)$-*assignment* is a mapping $\sigma : \mathcal{V} \to \mathrm{dom}(t) \cup 2^{\mathrm{dom}(t)}$ with $\sigma(x) \in \mathrm{dom}(t)$ for $x \in \mathcal{V}_1$ and $\sigma(X) \subseteq \mathrm{dom}(t)$ for $X \in \mathcal{V}_2$. As usual, we encode each $(\mathcal{V}, t)$-assignment by a tree over the extended alphabet $\Sigma_{\mathcal{V}} = \Sigma \times \{0, 1\}^{\mathcal{V}}$; we call a tree over $\Sigma_{\mathcal{V}}$ *valid* if it arises in this way. For details we refer to [9,15,10]. From now on we identify a pair (t, σ) and its encoding $s \in U_{\Sigma_{\mathcal{V}}}$. For $x \in \mathcal{V}_1$, the update $s[x \to u] \in U_{\Sigma_{\mathcal{V} \cup \{x\}}}$ for $u \in \mathrm{dom}(t)$ is defined by $s[x \to u] = (t, \sigma[x \to u]) = (t, \sigma')$ where

Table 1. The semantics of wMSO formulas

$$[\![\mathrm{label}_a(x)]\!]_{\mathcal{V}}(s) = \begin{cases} \mathbb{1} & \text{if } t(\sigma(x)) = a, \\ \mathbb{0} & \text{otherwise} \end{cases} \qquad [\![\mathrm{desc}(x,y)]\!]_{\mathcal{V}}(s) = \begin{cases} \mathbb{1} & \text{if } \exists i \in \mathbb{N} : \sigma(y) = \sigma(x).i, \\ \mathbb{0} & \text{otherwise} \end{cases}$$

$$[\![x \leq y]\!]_{\mathcal{V}}(s) = \begin{cases} \mathbb{1} & \text{if } \sigma(x) = \sigma(y) = \varepsilon \vee \exists u \in \mathrm{dom}(s) : \exists i, j \in \mathbb{N}, i \leq j : \\ & \sigma(x) = u.i, \sigma(y) = u.j, \\ \mathbb{0} & \text{otherwise} \end{cases}$$

$$[\![x \sqsubseteq y]\!]_{\mathcal{V}}(s) = \begin{cases} \mathbb{1} & \text{if } \sigma(x) \sqsubseteq_s \sigma(y), \\ \mathbb{0} & \text{otherwise} \end{cases} \qquad [\![x \in X]\!]_{\mathcal{V}}(s) = \begin{cases} \mathbb{1} & \text{if } \sigma(x) \in \sigma(X), \\ \mathbb{0} & \text{otherwise} \end{cases}$$

$$[\![\neg\beta]\!]_{\mathcal{V}}(s) = \begin{cases} \mathbb{1} & \text{if } [\![\beta]\!]_{\mathcal{V}}(s) = \mathbb{0}, \\ \mathbb{0} & \text{otherwise} \end{cases} \qquad [\![d]\!]_{\mathcal{V}}(s) = d$$

$$[\![\varphi \vee \psi]\!]_{\mathcal{V}}(s) = [\![\varphi]\!]_{\mathcal{V}}(s) + [\![\psi]\!]_{\mathcal{V}}(s) \qquad [\![\varphi \wedge \psi]\!]_{\mathcal{V}}(s) = [\![\varphi]\!]_{\mathcal{V}}(s) \diamond [\![\psi]\!]_{\mathcal{V}}(s)$$

$$[\![\exists x.\varphi]\!]_{\mathcal{V}}(s) = \sum_{u \in \mathrm{dom}(s)} [\![\varphi]\!]_{\mathcal{V} \cup \{x\}}(s[x \to u]) \quad [\![\exists X.\varphi]\!]_{\mathcal{V}}(s) = \sum_{I \subseteq \mathrm{dom}(s)} [\![\varphi]\!]_{\mathcal{V} \cup \{X\}}(s[X \to I])$$

$$[\![\forall X.\beta]\!]_{\mathcal{V}}(s) = \begin{cases} \mathbb{1} & \text{if } [\![\beta]\!]_{\mathcal{V} \cup \{X\}}(s[X \to I]) = \mathbb{1} \text{ for all } I \subseteq \mathrm{dom}(s), \\ \mathbb{0} & \text{otherwise} \end{cases}$$

$$[\![\forall x.\varphi]\!]_{\mathcal{V}}(s) = \mathrm{Val}(s_D) \text{ for } s_D \in U_D \text{ given by } \mathrm{dom}(s_D) = \mathrm{dom}(s) \text{ and}$$
$$s_D(u) = [\![\varphi]\!]_{\mathcal{V} \cup \{x\}}(s[x \to u]) \text{ for all } u \in \mathrm{dom}(s)$$

$\sigma'|_{\mathcal{V} \setminus \{x\}} = \sigma|_{\mathcal{V} \setminus \{x\}}$ and $\sigma'(x) = u$. The update $s[X \to I] \in U_{\Sigma_{\mathcal{V} \cup \{X\}}}$ for $X \in \mathcal{V}_2$ and $I \subseteq \mathrm{dom}(t)$ is defined similarly.

The *semantics of a wMSO formula* φ over a ptv-monoid $\mathbb{D}$ and an alphabet Σ is the tree series $[\![\varphi]\!]_{\mathcal{V}} : U_{\Sigma_{\mathcal{V}}} \to D$ which equals $\mathbb{0}$ for non-valid trees and which is defined inductively for each valid tree $s = (t, \sigma)$ as shown in Table 1. Here $\sqsubseteq_s$ is a linear ordering on the positions of s. For the rest of this paper this linear ordering will be the depth-first left-to-right traversal. Then the formula $x \leq y$ can be expressed with the help of $x \sqsubseteq y$. Subsequently, we write $[\![\varphi]\!]$ for $[\![\varphi]\!]_{\mathrm{free}(\varphi)}$. Any boolean wMSO formula β can be viewed as a classical MSO formula which defines the recognizable tree language $L_{\mathcal{V}}(\beta)$ and we can easily show that $[\![\beta]\!]_{\mathcal{V}} = \mathbb{1}_{L_{\mathcal{V}}(\beta)}$. Furthermore, we can prove by induction that $[\![\varphi]\!]_{\mathcal{V}}(t, \sigma) = [\![\varphi]\!](t, \sigma|_{\mathrm{free}(\varphi)})$ for every wMSO formula φ, $(t, \sigma) \in U_{\Sigma_{\mathcal{V}}}$, and set of variables $\mathcal{V}$ with $\mathrm{free}(\varphi) \subseteq \mathcal{V}$.

Example 4.1. Let $\mathbb{Q}^p_{\max}$ be the ptv-monoid from Example 2.1. The boolean formula $\mathrm{leaf}(x) = \forall y. \neg \mathrm{desc}(x, y)$ maps every $t \in U_{\Sigma}$ and assignment σ to ∞ if $\sigma(x)$ is a leaf and to $-\infty$ if $\sigma(x)$ is not a leaf. Analogously to [10], we can show that the formula $\varphi = \forall x.((\mathrm{leaf}(x) \wedge 1) \vee (\neg \mathrm{leaf}(x) \wedge 0))$ defines the leaves-to-size ratio for trees which was previously computed by the WUTA of Example 3.1.

Next we introduce some fragments of the weighted MSO logic which will be essential for our main result. A wMSO formula is an *almost boolean formula* if it consists only of conjunctions and disjunctions of boolean formulas and elements of D. We call a wMSO formula $\forall$-*restricted* if all its subformulas $\forall x.\varphi$ satisfy that

φ is almost boolean. Let const(φ) be the set of all $d \in D$ occurring in a formula φ. Similarly to [12,10] we call φ *strongly $\wedge$-restricted* if whenever φ contains a subformula $\varphi_1 \wedge \varphi_2$, then both φ_1 and φ_2 are almost boolean or φ_1 or φ_2 is boolean; and *commutatively $\wedge$-restricted* if whenever φ contains a subformula $\varphi_1 \wedge \varphi_2$, then φ_1 is almost boolean or const(φ_1) and const(φ_2) commute. Note that each strongly $\wedge$-restricted wMSO formula is commutatively $\wedge$-restricted. For examples of weighted logic formulas and a discussion on the above restrictions, we refer the reader to [9,14,15,12].

5   Weighted Tree Automata and Weighted MSO Logic

Here we characterize the class of behaviors of WUTA by the fragments of the weighted MSO logic.

Theorem 5.1. *Let $S : U_\Sigma \to \mathbb{D}$ be a tree series.*

1. *If $\mathbb{D}$ is regular, then S is recognizable iff $S = [\![\varphi]\!]$ for some $\forall$-restricted and strongly $\wedge$-restricted wMSO sentence φ.*
2. *If $\mathbb{D}$ is a cctv-semiring, then S is recognizable iff $S = [\![\varphi]\!]$ for some $\forall$-restricted and commutatively $\wedge$-restricted wMSO sentence φ.*

For ranked trees, examples were given in [10] showing that it is not possible to drop the constraints on $\mathbb{D}$ in statements (1) or (2). These examples could be easily extended to the unranked tree setting. It remains to prove Theorem 5.1. For this, the following proposition will be very useful; it can be proved as the corresponding result in [14] by using Theorem 3.4(2).

Proposition 5.2. *Let φ be a wMSO formula and $\mathcal{V}$ a finite set of variables with free$(\varphi) \subseteq \mathcal{V}$. Then $[\![\varphi]\!]$ is recognizable iff $[\![\varphi]\!]_\mathcal{V}$ is recognizable, and $[\![\varphi]\!]$ is a recognizable step function iff $[\![\varphi]\!]_\mathcal{V}$ is a recognizable step function.*

Analogously to [10] one can show:

Lemma 5.3. *If φ is an almost boolean formula, then $[\![\varphi]\!]$ is a recognizable step function. Conversely, if $S : U_\Sigma \to \mathbb{D}$ is a recognizable step function, then $S = [\![\varphi]\!]$ for some almost boolean sentence φ.*

Now we can show that our logic operators preserve the recognizability of the semantics of wMSO formulas by adapting the proofs for the corresponding Propositions 5.15-5.17 of [10].

Proposition 5.4. *Let φ and ψ be wMSO formulas over Σ and $\mathbb{D}$. If $[\![\varphi]\!]$ and $[\![\psi]\!]$ are recognizable, then $[\![\varphi \vee \psi]\!]$, $[\![\exists x.\varphi]\!]$, and $[\![\exists X.\varphi]\!]$ are recognizable. Furthermore, $[\![\varphi \wedge \psi]\!]$ and $[\![\psi \wedge \varphi]\!]$ are recognizable if $[\![\varphi]\!]$ is recognizable and ψ is boolean.*

Proposition 5.5. *Let φ be an almost boolean formula over $\mathbb{D}$ and Σ. Then $[\![\forall x.\varphi]\!]$ is recognizable.*

Proof. Let $\mathcal{W} = \text{free}(\varphi) \cup \{x\}$ and $\mathcal{V} = \text{free}(\forall x.\varphi) = \mathcal{W} \setminus \{x\}$. Since φ is almost boolean and by Lemma 5.3, $[\![\varphi]\!]_{\mathcal{W}} = \sum_{i=1}^{n} d_i \diamond \mathbb{1}_{L_i}$ for some partition $L_1, \ldots, L_n$ of all valid trees over $\Sigma_{\mathcal{W}}$ (for invalid trees s, $[\![\varphi]\!]_{\mathcal{W}}(s) = \mathbb{0}$). Let $\tilde{\Sigma} = \Sigma \times \{1, \ldots, n\}$. We extend every valid tree $(t, \sigma) \in U_{\Sigma_{\mathcal{W}}}$ to a tree (t, ν, σ) over $\tilde{\Sigma}_{\mathcal{V}}$ by the unique mapping $\nu : \text{dom}(t) \to \{1, \ldots, n\}$ that encodes to which L_i the update of (t, σ) and x belongs. Hence, $\nu(u) = i$ iff $(t, \sigma[x \to u]) \in L_i$ for all $u \in \text{dom}(t)$. Let $\tilde{L} \subseteq U_{\tilde{\Sigma}_{\mathcal{V}}}$ be the tree language of all such trees (t, ν, σ). In [15] it was already shown that $\tilde{L}$ is recognizable. Let $\mathcal{M} = (Q, \mathcal{B}, F)$ over $\tilde{\Sigma}_{\mathcal{V}}$ be a deterministic unranked tree automaton that recognizes $\tilde{L}$. We may assume that every subautomaton $\mathcal{B}_{q,\tilde{a}} = (Q_{q,\tilde{a}}, I_{q,\tilde{a}}, T_{q,\tilde{a}}, F_{q,\tilde{a}})$ (for $q \in Q$, $\tilde{a} \in \tilde{\Sigma}_{\mathcal{V}}$) of $\mathcal{M}$ is deterministic. Thus for every tree $\tilde{t} \in U_{\tilde{\Sigma}_{\mathcal{V}}}$ there is exactly one extended run π of $\mathcal{M}$ on $\tilde{t}$, and in addition there is exactly one run π_u of $\mathcal{B}_{\pi(u), \tilde{t}(u)}$ on $\pi(u.1) \ldots \pi(u.\text{rk}_{\tilde{t}}(u))$ for each $u \in \text{dom}(\tilde{t})$.

We wish to transform $\mathcal{M}$ into a WUTA $\mathcal{M}'$ over $\tilde{\Sigma}_{\mathcal{V}}$ such that for every tree $\tilde{t}$ the unique runs π and π_u ($u \in \text{dom}(\tilde{t})$) form an extended run $\tilde{\pi} = (q, s, l)$ with

$$\mu(\tilde{t}, \tilde{\pi})(u) = d_i \Leftrightarrow \|\mathcal{B}_{s(u)_2, \tilde{t}(u)}\|(s(u.1)_2 \ldots s(u.\text{rk}_{\tilde{t}}(u))_2) = \mathbb{1} \text{ and } \tilde{t}(u)_2 = i$$

for all $u \in \text{dom}(\tilde{t})$. Then $\mu(\tilde{t}, \tilde{\pi})(u) = [\![\varphi]\!]_{\mathcal{W}}(t, \sigma[x \to u])$ and $\text{Val}(\mu(\tilde{t}, \tilde{\pi})) = [\![\forall x \varphi]\!]_{\mathcal{V}}(t, \sigma)$ where $(t, \nu, \sigma) = \tilde{t}$. All other extended runs on $\tilde{t}$ shall get the weight $\mathbb{0}$. For this, we extend the states of $\mathcal{M}$ by values from $\{1, \ldots, n\}$. The value in the state assigned to a position u encodes $\tilde{t}(u)_2$. We let $\mathcal{A}_{(q,j),(a,i,f)}$ be a WUTA with an empty set of final states whenever $j \neq i$ to ensure that for a successful extended run a state with value i is assigned to a position with label (a, i, f). The automaton $\mathcal{A}_{(q,i),(a,i,f)}$ will be a modified version of $\mathcal{B}_{q,(a,i,f)}$; it is defined over the alphabet $Q \times \{1, \ldots, n\}$ such that there is a transition $(p_1, (q, i'), p_2)$ with weight $d_{i'}$ for every $i' \in \{1, \ldots, n\}$ iff $T_{q,(a,i,f)}(p_1, q, p_2) = \mathbb{1}$. Formally, $\mathcal{M}' = (Q', \mathcal{A}, \gamma)$ such that $Q' = Q \times \{1, \ldots, n\}$, $\gamma(q, i) = d_i$ if $F(q) = \mathbb{1}$ and $\gamma(q, i) = \mathbb{0}$ if $F(q) = \mathbb{0}$, and $\mathcal{A} = (\mathcal{A}_{q,a} \mid q \in Q', a \in \tilde{\Sigma}_{\mathcal{V}})$ where for $\tilde{a} = (a, i, f)$ we have $\mathcal{A}_{(q,i),\tilde{a}} = (Q_{q,\tilde{a}}, I_{q,\tilde{a}}, \mu_{(q,i),\tilde{a}}, F_{q,\tilde{a}})$ with

$$\mu_{(q,i),\tilde{a}}(p_1, (q', i'), p_2) = \begin{cases} d_{i'} & \text{if } T_{q,\tilde{a}}(p_1, q', p_2) = \mathbb{1}, \\ \mathbb{0} & \text{otherwise} \end{cases}$$

for $p_1, p_2 \in Q_{q,a}$, and $(q', i') \in Q'$; and $\mathcal{A}_{(q,j),(a,i,f)}$ has an empty set of final states if $i \neq j$.

Obviously, $\|\mathcal{M}'\|(\tilde{t}) = \text{Val}(\mu(\tilde{t}, \tilde{\pi})) = [\![\forall x.\varphi]\!]((t, \sigma))$ for all trees $\tilde{t} = (t, \nu, \sigma) \in U_{\tilde{\Sigma}_{\mathcal{V}}}$ where $\tilde{\pi}$ is the extended run arisen from π and the π_us. Let the relabeling $h : \tilde{\Sigma}_{\mathcal{V}} \to \Sigma_{\mathcal{V}}$ be defined by $h((a, i, f)) = (a, f)$. One can show that $h(\|\mathcal{M}\|)(s) = [\![\forall x.\varphi]\!](s)$ for all $s \in U_{\Sigma_{\mathcal{V}}}$. Hence, $[\![\forall x.\varphi]\!]$ is recognizable by Lemma 3.6. $\square$

Now we will prove our main result, Theorem 5.1.

Proof of Theorem 5.1. By Lemma 5.3 and Lemma 3.5, the semantics of almost boolean formulas over a regular ptv-monoid $\mathbb{D}$ is recognizable. For (1) the recognizability of the tree series $[\![\varphi]\!]$ for a formula φ is guaranteed by Propositions 5.4

and 5.5. For (2) we can proceed as in [10] and show by induction on the structure of φ that there is a WUTA recognizing $[\![\varphi]\!]$ whose weights are in the subsemiring generated by $\langle \mathrm{const}(\varphi) \cup \{0, 1\}, +, \diamond \rangle$.

For the converse, let $\mathcal{M}$ be a WUTA recognizing S. In [15, Thm.6.9] the behavior of $\mathcal{M}$ was described with a formula using two universal quantifiers which occur nested. Due to the current definition of the behavior, $\|\mathcal{M}\|$ can be expressed by a $\forall$-restricted and strongly $\wedge$-restricted wMSO sentence φ.   □

Remark 5.6. We can show that Theorem 5.1 generalizes the respective main theorem of [10] for ranked trees. For this, we use Remark 3.2, the transformation from the weighted MSO logic over ranked trees to the weighted MSO logic over unranked trees [15, Lemma7.3], and a reverse transformation for wMSO sentences without subformula of the form $x \sqsubseteq y$.

Theorems 6.5 and 6.10 of [15] show that weighted unranked tree automata over non-commutative semirings are more expressive than the restricted weighted MSO logic. Our slightly changed definition of the behavior of WUTA enables us to prove an equivalence result as follows. Let $\mathbb{K} = (K, +, \cdot, 0, 1)$ be a semiring. We associate $\mathbb{K}$ with the cctv-semiring $(K, +, \mathrm{Val}, \cdot, 0, 1)$ with $\mathrm{Val}(t) = \prod_{u \in \mathrm{dom}(t)} t(u)$ where we multiply according to a depth-first left-to-right traversal, i.e. for a position u we first collect the weights of its subtrees one by one from left to right and then we multiply with the weight of u itself. Now we obtain:

Corollary 5.7. *Let Σ be an alphabet, $(K, +, \cdot, 0, 1)$ a semiring, and S a tree series over Σ and $\mathbb{K}$. Then S is recognizable over $\mathbb{K}$ iff $S = [\![\varphi]\!]$ for a $\forall$-restricted and commutatively $\wedge$-restricted wMSO sentence φ.*

Hence, for commutative semirings, by Remark 3.2 and Corollary 5.7 we obtain the main equivalence results Theorem 6.5 and Theorem 6.9 of [15] as a consequence.

References

1. Bollig, B., Gastin, P.: Weighted versus probabilistic logics. In: Diekert, V., Nowotka, D. (eds.) DLT 2009. LNCS, vol. 5583, pp. 18–38. Springer, Heidelberg (2009)
2. Brüggemann-Klein, A., Murata, M., Wood, D.: Regular tree and regular hedge languages over unranked alphabets: version 1. Technical Report HKUST-TCSC-2001-0, The Honkong University of Sience and Technologie (2001)
3. Brüggemann-Klein, A., Wood, D.: Regular tree languages over non-ranked alphabets (1998). http://citeseerx.ist.psu.edu/viewdoc/summary?doi=10.1.1.50.5397
4. Chatterjee, K., Doyen, L., Henzinger, T.A.: Quantitative languages. In: Kaminski, M., Martini, S. (eds.) CSL 2008. LNCS, vol. 5213, pp. 385–400. Springer, Heidelberg (2008)
5. Chatterjee, K., Doyen, L., Henzinger, T.A.: Alternating weighted automata. In: Kutyłowski, M., Charatonik, W., Gębala, M. (eds.) FCT 2009. LNCS, vol. 5699, pp. 3–13. Springer, Heidelberg (2009)
6. Chatterjee, K., Doyen, L., Henzinger, T.A.: Probabilistic weighted automata. In: Bravetti, M., Zavattaro, G. (eds.) CONCUR 2009. LNCS, vol. 5710, pp. 244–258. Springer, Heidelberg (2009)

7. Chatterjee, K., Doyen, L., Henzinger, T.A.: Expressiveness and closure properties for quantitative languages. In: Proceedings of LICS 2009, pp. 199–208. IEEE Computer Society (2009)
8. Comon, H., Dauchet, M., Gilleron, R., Löding, C., Jacquemard, F., Lugiez, D., Tison, S., Tommasi, M.: Tree automata techniques and applications (2007). http://www.grappa.univ-lille3.fr/tata
9. Droste, M., Gastin, P.: Weighted automata and weighted logics. Theoretical Computer Science 380, 69–86 (2007)
10. Droste, M., Götze, D., Märcker, S., Meinecke, I.: Weighted tree automata over valuation monoids and their characterization by weighted logics. In: Kuich, W., Rahonis, G. (eds.) Algebraic Foundations in Computer Science. LNCS, vol. 7020, pp. 30–55. Springer, Heidelberg (2011)
11. Droste, M., Kuich, W., Vogler, H. (eds.): Handbook of Weighted Automata. EATCS Monographs on Theoretical Computer Science, Springer (2009)
12. Droste, M., Meinecke, I.: Weighted automata and weighted MSO logics for average- and longtime-behaviors. Information and Computation 220–221, 44–59 (2012)
13. Droste, M., Rahonis, G.: Weighted automata and weighted logics with discounting. Theory of Computing Systems 410(37), 3481–3494 (2009)
14. Droste, M., Vogler, H.: Weighted tree automata and weighted logics. Theoretical Computer Science 366, 228–247 (2006)
15. Droste, M., Vogler, H.: Weighted logics for unranked tree automata. Theory of Computing Systems 48, 23–47 (2011)
16. Droste, M., Vogler, H.: Kleene and Büchi theorems for weighted automata and multi-valued logics over arbitrary bounded lattices. Theoretical Computer Science 418, 14–36 (2012)
17. Fichtner, I.: Weighted picture automata and weighted logics. Theory of Computing Systems 48(1), 48–78 (2011)
18. Fülöp, Z., Vogler, H.: Weighted tree automata and tree transducers, chap. 9. In: Droste et al. [11] (2009)
19. Fülöp, Z., Vogler, H.: Forward and backward application of symbolic tree transducers. Acta Informatica 51, 297–325 (2014)
20. Grätzer, G.: General Lattice Theory, 2nd edn. Birkhäuser Verlag (January 2003)
21. Högberg, J., Maletti, A., Vogler, H.: Bisimulation minimisation of weighted automata on unranked trees. Fundamenta Informaticae 92, 103–130 (2009)
22. Libkin, L.: Logics for unranked trees: an overview. In: Caires, L., Italiano, G.F., Monteiro, L., Palamidessi, C., Yung, M. (eds.) ICALP 2005. LNCS, vol. 3580, pp. 35–50. Springer, Heidelberg (2005)
23. Mathissen, C.: Definable transductions and weighted logics for texts. Theoretical Computer Science 411, 631–659 (2010)
24. Mathissen, C.: Weighted logics for nested words and algebraic formal power series. Logical Methods in Computer Science 6 (2010)
25. Meinecke, I.: Weighted logics for traces. In: Grigoriev, D., Harrison, J., Hirsch, E.A. (eds.) CSR 2006. LNCS, vol. 3967, pp. 235–246. Springer, Heidelberg (2006)
26. Murata, M.: Forest-regular languages and tree-regular languages (1995). (unpublished manuscript)
27. Neven, F.: Automata, logic, and XML. In: Bradfield, J.C. (ed.) CSL 2002 and EACSL 2002. LNCS, vol. 2471, p. 2. Springer, Heidelberg (2002)
28. Rahonis, G.: Weighted Muller tree automata and weighted logics. Journal of Automata, Languages and Combinatorics 12, 455–483 (2007)
29. Tan, T.: Extending two-variable logic on data trees with order on data values and its automata. ACM Transactions on Computational Logic 15, 8:1–8:39 (2014)

30. Thatcher, J.W.: Characterizing derivation trees of context-free grammars through a generalization of finite automata theory. Journal of Computer and System Sciences **1**, 317–322 (1967)

31. Thomas, W.: Languages, automata, and logic. In: Rozenberg, G., Salomaa, A. (eds.) Handbook of Formal Languages, vol. A, pp. 389–455. Springer (1997)

32. Veanes, M., Bjørner, N.: Symbolic tree transducers. In: Clarke, E., Virbitskaite, I., Voronkov, A. (eds.) PSI 2011. LNCS, vol. 7162, pp. 377–393. Springer, Heidelberg (2012)

A New Partial Key Exposure Attack
on Multi-power RSA

Muhammed F. Esgin[1,2]([✉]), Mehmet S. Kiraz[1], and Osmanbey Uzunkol[1]

[1] TÜBİTAK BİLGEM UEKAE, Kocaeli, Turkey
{muhammed.esgin,mehmet.kiraz,osmanbey.uzunkol}@tubitak.gov.tr
[2] Graduate School of Natural and Applied Sciences,
İstanbul Şehir University, Istanbul, Turkey

Abstract. An important attack on multi-power RSA ($N = p^r q$) was introduced by Sarkar in 2014, by extending the small private exponent attack of Boneh and Durfee on classical RSA. In particular, he showed that N can be factored efficiently for $r = 2$ with private exponent d satisfying $d < N^{0.395}$. In this paper, we generalize this work by introducing a new partial key exposure attack for finding small roots of polynomials using Coppersmith's algorithm and Gröbner basis computation. Our attack works for all multi-power RSA exponents e (resp. d) when the exponent d (resp. e) has full size bit length. The attack requires prior knowledge of least significant bits (LSBs), and has the property that the required known part of LSB becomes smaller in the size of e. For practical validation of our attack, we demonstrate several computer algebra experiments.

Keywords: Multi-power RSA · Integer factorization · Partial key exposure · Coppersmith's method · Small roots of polynomials

1 Introduction

A natural way of speeding up the decryption/signing procedure of RSA based cryptographic schemes is to use a small private exponent d. However, Wiener [22] showed that classical RSA construction becomes insecure when $d < \frac{1}{3} N^{\frac{1}{4}}$. Later, this bound was further improved by Boneh and Durfee [2] to $N^{0.292}$ by using results of Coppersmith [6].

Kocher [15] initiated a new type of attack that obtains information about the bits of d using side-channel techniques in 1996. The idea is to exploit certain weaknesses of the actual implementation (e.g., execution time, power consumption, noise), which in turn reveals some bits of d. In general, the attacker gains information about either consecutive least significant bits (LSBs) or most significant bits (MSBs). Therefore, partial key exposure attacks mostly focus on these two rather specific cases.

Boneh, Durfee and Frankel [3] introduced the first algebraic partial key exposure attack using partial information of d. The attack finds the whole secret

© Springer International Publishing Switzerland 2015
A. Maletti (Ed.): CAI 2015, LNCS 9270, pp. 103–114, 2015.
DOI: 10.1007/978-3-319-23021-4_10

exponent d when sufficient partial knowledge of d is known. Coppersmith's algorithm for finding small roots of polynomials is used in such algebraic attacks [6,5,4]. This algorithm uses lattice reduction techniques to obtain efficient small roots of certain polynomials (in particular, the LLL algorithm [16]). Later, new partial key exposure attacks on classical RSA were described by Blömer and May in [1]. We refer to [9,14] for further partial key exposure attacks on standard RSA.

Notation: Let log denote the logarithm base 2 unless the base is given concretely. We use the following notation throughout this manuscript.

N	Multi-power RSA modulus
n	bitsize of N
p, q	prime factors of N
r	integer satisfying the relation $N = p^r q$
e	RSA public exponent
d	RSA private exponent
d_0	known part of d
α	$\log_N e$ (i.e., $e \approx N^\alpha$)
β	$\log_N d$ (i.e., $d \approx N^\beta$)
δ	$\log_N d_0$ (i.e., $d_0 \approx N^\delta$)

In this work, we focus on multi-power RSA (also referred as Takagi's RSA or prime power RSA) introduced by Takagi in [21]. One of the motivation of this variant is to speed up the RSA decryption/signing process. More concretely, $N = p^r q$ is chosen for two (distinct) primes of same bit length such that $r \geq 2$. Then, there are two different ways of generating public/private exponents. The first one imposes the condition $ed \equiv 1 \bmod (p-1)(q-1)$ while the other $ed \equiv 1 \bmod \phi(N)$, where $\phi(N) = p^{r-1}(p-1)(q-1)$. Decryption of a ciphertext c is computed more efficiently using simply a combination of Hensel lifting and Chinese Remainder Theorem modulo p^r and q (see [21] for details).

For the multi-power RSA variant when exponents are generated modulo $\phi(N)$, Takagi proved in [21] that if $d \leq N^{\frac{1}{2r+2}}$, then N can be factored. This was later improved by May in [18] to $d < \max\{N^{\frac{r}{(r+1)^2}}, N^{\frac{(r-1)^2}{(r+1)^2}}\}$. Recently, Sarkar [20] improved this bound even further for $r \leq 5$ and showed in particular that if $d < N^{0.395}$ and $r = 2$, then N can be factored efficiently. Thereafter, Lu et al. [17] improved Sarkar's result for $r \geq 4$. Their attack works when the unknown part $\tilde{d}$ of d (it may be all of d or an MSB/LSB part of it) satisfies $\tilde{d} < N^{\frac{r(r-1)}{(r+1)^2}}$.

In [13], a small private exponent attack is shown for the case when exponents are generated modulo $(p-1)(q-1)$. This attack shows that N can be factored if $d < N^{\frac{2-\sqrt{2}}{r+1}}$. Later, the idea of this work is used in [12] for partial key exposure attacks. For instance, for $r = 2$ and $e \approx N^{\frac{2}{3}}$, it is shown that N can be factored in any of the following conditions:

$$\gamma \leq \tfrac{7}{12} - \tfrac{1}{4}\sqrt{\tfrac{16+24\beta}{3} - \tfrac{39}{9}} \qquad \text{if MSBs or middle bits are known,}$$

$$\text{or} \quad \gamma \leq \tfrac{5}{9} - \tfrac{2}{3}\sqrt{\tfrac{2+3\beta}{3} - \tfrac{5}{9}} \qquad \text{if LSBs are known,}$$

where $d \approx N^\beta$ and the unknown part of d is approximately N^γ. Note that their attacks do not work when d is of full size modulo $(p-1)(q-1)$ (i.e., $d \approx N^{\frac{2}{3}}$).

Our Contribution. In this paper, we provide a new partial key exposure attack on multi-power RSA when the exponents are generated modulo $\phi(N)$. The attack basically uses partial knowledge of LSBs and works for all e (resp. d) when the exponent d (resp. e) has full size bit length.[1] More concretely, we prove the following theorem which generalizes Sarkar's result [20].

Theorem 1. *Let $r \geq 2$ be an integer and $N = p^r q$ be a multi-power RSA modulus, where p and q are distinct primes with the same bit size (i.e., $p, q \approx N^{\frac{1}{r+1}}$). Suppose that $ed \equiv 1 \bmod \phi(N)$ with $e \approx N^\alpha$ and $d \approx N^\beta$. Suppose further that an attacker obtains an LSB part d_0 of d, where $d_0 \geq N^\delta$ for some $\delta \in \mathbb{R}^{\geq 0}$. Then under Assumption 1, there exists an algorithm which finds the prime factors of N in polynomial time in $\log N$ provided that*

$$\rho(r, \beta, \alpha, \delta) < 0,$$

where ρ is a function of r, β, α and δ.

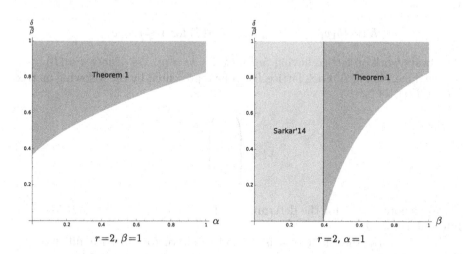

Fig. 1. The relation between the sizes of e (resp. d) and the fraction of the part of d required to be known.

---

[1] This rule is induced by the condition that $ed \equiv 1 \bmod \phi(N)$.

We show the improvement of our attack over Sarkar's result in Figure 1 for the case $r = 2$. Light grey area (indicated by "Sarkar'14") shows the attack region by [20] and darker grey areas are the applicable regions of our attack.

Organization of the Paper. In Section 2, we give preliminaries about lattices. In Section 3, we prove our main result, Theorem 1, extending the result of [20]. Section 4 demonstrates several experiments justifying our claims for the multi-power RSA moduli of length 1024 or 2048 bits. We conclude the paper in Section 5 and argue the improbability of using our attack for known MSBs by addressing an issue in [12].

2   Preliminaries

In this section, we give basic definitions and theorems about lattices. Let $v = (a_0, \cdots, a_s)$ be a vector in $\mathbb{R}^{s+1}$ for some $s \geq 0$. We use the Euclidean norm $||v||$ of v

$$||v|| := \sqrt{\sum_{i=0}^{s}(a_i)^2}.$$

For a multivariate polynomial f, the norm $||f||$ of f is the Euclidean norm of its coefficient vector. Let $v_1, \cdots, v_w \in \mathbb{R}^m$ be a set of $\mathbb{R}$-linearly independent vectors with $w, m \in \mathbb{N}^{>0}$ and $w \leq m$. Then, the lattice L generated by these vectors is

$$L := \{b_1 v_1 + \cdots + b_w v_w \ : \ b_i \in \mathbb{Z} \text{ for } 1 \leq i \leq w\}.$$

We always work on lattice having *full rank*, i.e. $w = m$. We denote $dim(L) := w$ for the dimension of L. Each lattice L can be represented by the following matrix $\mathcal{M} \in \mathrm{GL}(w, \mathbb{R})$:

$$\mathcal{M} = \begin{pmatrix} v_1 \\ \cdot \\ \cdot \\ \cdot \\ v_w \end{pmatrix}.$$

We denote $det(L)$ for the determinant of L. We have $det(L) = det(\mathcal{M})$ for a full rank lattice L.

In this work, the main goal is to find small vectors in such full lattices. Computational complexity of finding the smallest vector in a lattice increases exponentially in $dim(L)$. The reduction algorithm *LLL* introduced by Lenstra, Lenstra and Lovász [16] is generally used in practice to have an efficient lattice reduction technique for obtaining small enough basis vectors. The following theorem gives an upper bound on the norm of the reduced basis vectors output by the LLL algorithm.

Theorem 2. *Let L be a lattice with $dim(L) = w$ as above. The LLL algorithm produces a set of reduced basis vectors $\{R_1, \cdots, R_w\}$ such that*

$$\|R_i\| \leq 2^{\frac{w(w-1)}{4(w+1-i)}} det(L)^{\frac{1}{w+1-i}}.$$

The computational complexity of the LLL algorithm is polynomial in $dim(L)$ and in the maximal bitsize of an entry [19].

Coppersmith described methods for finding small roots of univariate and bivariate polynomials [4,5,6]. The methods can be extended to the polynomials having more variables, but the results become heuristic. Howgrave-Graham [11] reformulated these results and proved the following theorem:

Theorem 3 (Howgrave-Graham's Theorem, [11]). *Let $f(x_1, \cdots, x_s) \in \mathbb{Z}[x_1, \cdots, x_s]$ be a polynomial for $s \geq 1$. Assume that the number of monomials is less than or equal to w. If the following two conditions hold:*

1. *$M \in \mathbb{Z}^+$ and $f(x_1^0, \cdots, x_s^0) \equiv 0 \mod M$ for some $|x_1^0| < X_1, \cdots, |x_s^0| < X_s$,*
2. *$\|f(x_1 X_1, \cdots, x_s X_s)\| < \frac{M}{\sqrt{w}}$,*

then $(x_1^0, \cdots, x_s^0)$ is a root of f over $\mathbb{Z}$.

After finding multivariate polynomials carrying a common root over integers, we need to extract this root using Gröbner basis computation.[2] Our main result Theorem 1 is valid under the following assumption:

Assumption 1. *Let $f_1, \cdots, f_k$ be the polynomials having the desired root over $\mathbb{Z}$ for $k \geq 3$ computed using LLL reduction. Furthermore, let I be the ideal generated by these polynomials. Then, the algebraic variety of I is zero-dimensional. In particular, the common root can be extracted by computing a Gröbner basis on I.*

Since our result in Theorem 1 relies on this assumption, it is heuristic. However, our experiments show that this assumption holds in general (see Section 4). The computational complexity of a Gröbner basis computation can be bounded by a polynomial in $\log N$ assuming the number of variables and the maximal degree of input polynomials is fixed [10].

3   An Attack with Known LSBs

In this section, we prove our main Theorem 1.

---

[2] Resultant computation could be another option as well, but it was less efficient for our experiments.

Proof (Theorem 1). Multi-power RSA parameters satisfy the congruence $ed \equiv 1 \bmod \phi(N)$ with $\phi(N) = (p^r - p^{r-1})(q - 1)$. This implies the equation that $ed - 1 = k(p^r - p^{r-1})(q - 1)$ for some $k \in \mathbb{Z}$. Since we know an LSB part of d, we can write this as $eM\tilde{d} + ed_0 - 1 = k(p^r - p^{r-1})(q - 1)$ where $d = \tilde{d}M + d_0$ and M is a power of 2. Hence, we have the following polynomial

$$f_{eM}(x, y, z) = ed_0 - 1 - xN - xy^{r-1} + xy^{r-1}z + xy^r$$

carrying the root $(x_0, y_0, z_0) = (k, p, q)$ modulo eM. It is easy to see that $|x_0| < X := N^{\alpha+\beta-1}$, $|y_0| < Y := N^{\frac{1}{r+1}}$ and $|z_0| < Z := N^{\frac{1}{r+1}}$ neglecting small constants.

Let $m, t_1, t_2 \geq 0$ and define the following shift polynomials:

$$g_{i,j,k}(x, y, z) = x^j y^k z^{j+t_1} f_{eM}^i(x, y, z),$$
$$\text{where } i = 0, \cdots, m, \; j = 1, \cdots, m - i \text{ and } k = j, \cdots, j + 2r - 2,$$
$$g_{i,0,k}(x, y, z) = y^k z^{t_1} f_{eM}^i(x, y, z),$$
$$\text{where } i = 0, \cdots, m \text{ and } k = 0, \cdots, t_2.$$

Recall that $y_0^r z_0 = N$. Hence, we replace every occurrence of $y^r z$ with N in the shift polynomials. Denote new polynomials by $g'_{i,j,k}(x, y, z)$. Observe that choosing xy^r as the leading monomial of f_{eM}, the leading monomials in $g'_{i,j,k}$'s are of the form $x^{i+j} y^{k+ri-rl} z^{j+t_1-l}$, where $l = \min\{\lfloor \frac{k+ri}{r} \rfloor, j + t_1\}$.

Let a_ℓ denote the leading coefficient. Assuming $\gcd(a_\ell, eM) = 1$, we can multiply $g'_{i,j,k}$'s with the inverse a'_ℓ of their corresponding leading coefficient in $\mathbb{Z}/(eM)^m\mathbb{Z}$. Finally, the shift polynomials become

$$h_{i,j,k}(x, y, z) = a'_\ell \cdot g'_{i,j,k}(x, y, z) \cdot (eM)^{m-i}$$

which carry the root (x_0, y_0, z_0) modulo $(eM)^m$.

We let the coefficient vectors of $h_{i,j,k}(xX, yY, zZ)$ represent the basis vectors of a lattice L. Generation of L is summarized in Algorithm 1.

Note that each polynomial in H generated by Algorithm 1 introduces exactly one new monomial, which is appended to Ord that defines the monomial ordering. Hence, the matrix representing the lattice is lower triangular when each row is ordered with respect to Ord. As a result, the determinant of L is the product of the diagonal entries of the representation matrix.

$$\det(L) = \left(\prod_{i=0}^{m} \prod_{j=1}^{m-i} \prod_{k=j}^{j+2r-2} X^{i+j} Y^{k+ri-rl_1} Z^{j+t_1-l_1} (eM)^{m-i} \right)$$
$$\times \left(\prod_{i=0}^{m} \prod_{k=0}^{t_2} X^i Y^{k+ri-rl_2} Z^{t_1-l_2} (eM)^{m-i} \right),$$

where $l_1 = \min\{\lfloor \frac{k+ri}{r} \rfloor, j + a\}$ and $l_2 = \min\{\lfloor \frac{k+ri}{r} \rfloor, a\}$. Letting s_x, s_y, s_z and s_{eM} be the powers of X, Y, Z and eM in $\det(L)$, respectively, and denoting the dimension of the lattice by w, we obtain

---

Algorithm 1. Generating the Lattice L

Input: $r \geq 2$; $m, t_1, t_2 \geq 0$ and $f_{eM}(x, y, z)$

$G, H, Ord \leftarrow \emptyset$
for $i \in \{0, 1, \cdots, m\}$ **do**
   for $j \in \{1, 2, \cdots, m - i\}$ **do**
      for $k \in \{j, j + 1, \cdots, j + 2r - 2\}$ **do**
         Append $(x^j y^k z^{j+t_1} f_{eM}^i, i)$ to G
         $l \leftarrow \min \left\{ \lfloor \frac{k+ri}{r} \rfloor, j + t_1 \right\}$
         Append $x^{i+j} y^{k+ri-rl} z^{j+t_1-l}$ to Ord
      end for
   end for
end for
for $i \in \{0, 1, \cdots, m\}$ **do**
   for $k \in \{0, 1, \cdots, t_2\}$ **do**
      Append $(y^k z^{j+t_1} f_{eM}^i, i)$ to G
      $l \leftarrow \min \left\{ \lfloor \frac{k+ri}{r} \rfloor, j + t_1 \right\}$
      Append $x^{i+j} y^{k+ri-rl} z^{j+t_1-l}$ to Ord
   end for
end for
for each element (g, i) in G **do**
   Replace each occurrence of $y^r z$ with N in g
   $a_\ell' \leftarrow a_\ell^{-1} \bmod eM$, where a_ℓ is the leading coefficient of g
   Append $(a_\ell' \cdot g \cdot (eM)^{m-i})$ to H
end for
$i \leftarrow 1$
for each polynomial $h(x, y, z)$ in H **do**
   Set i-th row of L to the coefficient vector of $h(xX, yY, zZ)$ ordered w.r.t. Ord
   Increment i
end for

$$w = \sum_{i=0}^{m} \sum_{j=1}^{m-i} \sum_{k=j}^{j+2r-2} 1 + \sum_{i=0}^{m} \sum_{k=0}^{t_2} 1 = \frac{2r-1}{2} m^2 + t_2 m + o(m^2)$$

$$s_x = \sum_{i=0}^{m} \sum_{j=1}^{m-i} (2r-1)(i+j) + \sum_{i=0}^{m} \sum_{k=0}^{t_2} i = \frac{2r-1}{3} m^3 + \frac{t}{2} m^2 + o(m^3)$$

$$s_{eM} = \sum_{i=0}^{m} \sum_{j=1}^{m-i} (2r-1)(m-i) + \sum_{i=0}^{m} \sum_{k=0}^{t_2} (m-i) = \frac{2r-1}{3} m^3 + \frac{t}{2} m^2 + o(m^3)$$

Assuming $\frac{t_2}{r} \leq t_1 \leq m$, we get as an asymptotic result

$$s_y = \sum_{i=0}^{m} \sum_{j=1}^{m-i} \sum_{k=j}^{j+2r-2} (k + ri - rl_1) + \sum_{i=0}^{m} \sum_{k=0}^{t_2} (k + ri - rl_2)$$

$$\approx \frac{1}{2} \left(\frac{r^2 m^3}{3} - r^2 m^2 t_1 + r^2 m t_1^2 - \frac{r^2 t_1^3}{3} + r m^2 t_2 \right.$$

$$\left. -2rmt_1 t_2 + r t_1^2 t_2 + m t_2^2 - t_1 t_2^2 + \frac{t_2^3}{3r} \right) + o(m^3)$$

$$s_z = \sum_{i=0}^{m} \sum_{j=1}^{m-i} \sum_{k=j}^{j+2r-2} (j + t_1 - l_1) + \sum_{i=0}^{m} \sum_{k=0}^{t_2} (t_1 - l_2)$$

$$\approx \frac{1}{2} \left(\frac{(r-1)^2 m^3}{3r} + (r-1)^2 m^2 t_1 + rmt_1^2 \right.$$

$$\left. - \frac{rt_1^3}{3} + t_1^2 t_2 - \frac{t_1 t_2^2}{r} + \frac{t_2^3}{3r^2} \right) + o(m^3)$$

which are approximated as in [20].

Neglecting the low order terms as similarly done in related works, the conditions in Theorem 2 and Theorem 3 can be simplified to $det(L) < (eM)^{wm}$. In our case, we need

$$s_x(\alpha + \beta - 1) + (s_y + s_z) \left(\frac{1}{r+1} \right) + (s_{eM} - wm)(\alpha + \delta) < 0.$$

to be satisfied. Plugging in the values for s_x, s_y, s_z and s_{eM}, we obtain a polynomial $\rho'(r, \alpha, \beta, \delta)$ with parameters t_1, t_2 and m. Let $t_1 = \tau_1 m$ and $t_2 = \tau_2 m$, and terms of $o(m^3)$ contribute to an error term ϵ. Next, we take the partial derivative of ρ' with respect to τ_1 and τ_2, and find the values making the derivatives zero to obtain the maximum value of ρ'. Finally, for $\gamma := \beta - \delta$, when $\tau_1 = \frac{1 - r\gamma + r^2(1-\gamma)}{2r}$ and

$$\tau_2 = \frac{1 + r^3(1 - \gamma) - r^2(1 + 2\gamma) + r(1 - \gamma) + 2r\sqrt{r^2(1 - \gamma) + r(1 - 2\gamma) + 1 - \gamma}}{2r + 2}$$

both derivatives become zero. Plugging in these values in ρ', we get a function $\rho(r, \alpha, \beta, \delta)$. When the tuple $(r, \alpha, \beta, \delta)$ satisfy $\rho(r, \alpha, \beta, \delta) < 0$, Howgrave-Graham's theorem is satisfied. We can extract the root (k, p, q) under Assumption 1, and thus factor N in time polynomial in $\log N$.   □

Remark 1. We note that our definition of shift polynomials is similar to the one in [20]. The difference is that we work modulo eM instead of modulo e. Hence, the constant coefficient of f_{eM} changes. Equating $M = 1$ (i.e., $\delta = 0$), we obtain the result of Sarkar [20] as a corollary of Theorem 1.

Table 1. Numerical values satisfying $\rho < 0$ for different r and α values where $\beta = 1$

r	smallest δ value satisfying $\rho(r) < 0$ for $\alpha = 1$	smallest δ value satisfying $\rho(r) < 0$ for $\alpha = 0$
2	0.828	0.362
3	0.798	0.344
4	0.750	0.314
5	0.703	0.285
6	0.662	0.259
7	0.625	0.237

Unfortunately, the exact expression of ρ is too complicated to be stated here. Thus, in Table 1, we provide some numerical values for δ which yields $\rho < 0$ when β is fixed to 1. We remind that for $r = 2$ new attack regions are given in Table 1 when either d or e is full-sized.

4   Experimental Results

Table 2. Experimental results for $\alpha = \beta = 1$. $n = 2048$ bits for the last row and $n = 1024$ bits for the rest.

r	m	t_1	t_2	w	δ	LLL time (secs)	Gröbner Basis time (secs)
2	6	4	7	119	0.870	1930.21	3.00
2	7	4	8	156	0.860	6517.26	67.99
2	8	4	7	180	0.850	19619.96	1227.18
2	8	5	9	198	0.835	28684.34	358.80
2	9	5	9	235	0.830	63748.97	635.33
2	9	5	10	245	0.823	67480.18	149.56
3	7	4	9	220	0.952	26671.68	7358.66
2	8	5	9	198	0.840	90981.76	2246.77

In this section, we provide various experimental results. In all of our experiments, we fix d to be full-sized (i.e., $\beta = 1$) which is mostly the case in real-life applications. The values for p, q and d are chosen randomly (or d is the inverse of $2^{16} + 1$ modulo $\phi(N)$). The experiments are performed on Sage 6.5 running on Ubuntu 14.04 LTS with Intel Core i7-3770 CPU at 3.40GHz and 16GB RAM.

Our results are given in Tables 2 and 3. In all of our experiments, Gröbner basis computation yields to a polynomial of the form $y - p$ giving the factorization of N. For the case when $\alpha = \beta = 1$ (which is illustrated in Table 2), we would like to highlight that our result in a case is better than the theoretical bound $\delta \geq 0.828$. However, when e is chosen small (e.g., $e = 2^{16} + 1$), the modulus eM becomes very small when compared to the case $\alpha = \beta = 1$. Therefore, the low

Table 3. Experimental results for $e = 2^{16} + 1$, $\beta = 1$. $n = 2048$ bits for the last two rows and $n = 1024$ bits for the rest.

r	m	t_1	t_2	w	δ	LLL time (secs)	Gröbner Basis time (secs)
2	8	3	2	135	0.520	21234.57	4114.00
2	8	3	2	135	0.510	19082.57	4280.77
2	9	3	3	175	0.500	48950.79	9134.06
2	10	3	2	198	0.485	84090.70	15927.35
3	9	3	3	265	0.510	148030.34	56230.82
2	10	3	2	198	0.500	203293.58	45573.57
2	10	3	2	198	0.490	185964.22	40817.77

order terms ignored to simplify the condition to $det(L) < (eM)^{wm}$ have much higher effect in this case. Thus, the results are a little bit worse than the best possible bound of Theorem 1.

5   Conclusion and Discussion

In this paper, we show a new partial key exposure attack on multi-power RSA, where $N = p^r q$. The attack takes advantage of known LSBs. Our result in Theorem 1 generalizes the work of Sarkar [20]. Moreover, we provide experimental results justifying our claims. Our attack even works in the case when $e, d \approx N$. In fact, our experimental result is better than the theoretical bound for this case. This paves the way for a further study: investigating sublattices of the original lattice to improve the theoretical bound. However, this is a hard task because in this case the lattice will not be of full rank and calculating the determinant gets complicated.

One may wonder why our attack is not directly applicable to known MSBs case. Suppose that we know an MSB part d_0 of d. Then, we obtain the equation

$$ed_0 + e\tilde{d} - 1 = k(p^r - p^{r-1})(q - 1),$$

where $\tilde{d}$ represents the unknown part of d. Considering this equation as a polynomial, we get

$$F(w, x, y, z) = 1 - ed_0 - ew + x(N - y^r - y^{r-1}z + y^{r-1}).$$

Now e, N or ed_0 are possible choices of moduli. The case e is studied in [20] where one cannot benefit from partial knowledge of d as it vanishes. If N is chosen as the modulus, then the trick of replacing each term $y^r z$ with N and finding its inverse cannot be applied. That leaves us with the option to choose ed_0 as the modulus. This case actually corresponds to finding a small root of *integer* equations [4], not modular equations [5].

Observe that reducing F modulo ed_0 does not eliminate any variable. In particular, F_{ed_0} and F have the same monomials. Hence, the polynomials derived from LLL may just be those of the form $F \cdot g_i$ for nonzero polynomials g_i not carrying the desired root. More concretely, the attacker does not obtain any additional information at all although LLL-reduced polynomials carry the root since they have the factor F.

For a recent work, one may see Coron's works [7,8] about methods to ensure independence between the initial polynomial F and the polynomials derived after LLL reduction[3]. Unfortunately, the tricks used in this work cannot be directly applied with Coron's method. This issue raises questions about the validity of known MSBs attack shown in [12]. The authors do not specify any methodology guaranteeing the independence aforementioned. Their experiments for this case are very far away from the new attack region described by Theorem 1 in their

---

[3] This independence is also ensured in Coppersmith's method [4].

paper. Moreover, the authors also state that in some experiments, they just verified that the LLL-reduced polynomials contain the root. As we explained earlier, this does not have any implication for an attacker to be able to find the root.

Acknowledgments. Uzunkol's research is supported by the project (114C027) funded by EU FP7-The Marie Curie Action and TÜBİTAK (2236-CO-FUNDED Brain Circulation Scheme). His work is also partly supported by a joint research project funded by Bundesministerium für Bildung und Forschung (BMBF), Germany (01DL12038) and TÜBİTAK, Turkey (TBAG-112T011).

References

1. Blömer, J., May, A.: New partial key exposure attacks on RSA. In: Boneh, D. (ed.) CRYPTO 2003. LNCS, vol. 2729, pp. 27–43. Springer, Heidelberg (2003)
2. Boneh, D., Durfee, G.: Cryptanalysis of RSA with private key d less than $N^{0.292}$. IEEE Transactions on Information Theory **46**(4), 1339–1349 (2000)
3. Boneh, D., Durfee, G., Frankel, Y.: An attack on RSA given a small fraction of the private key bits. In: Ohta, K., Pei, D. (eds.) ASIACRYPT 1998. LNCS, vol. 1514, pp. 25–34. Springer, Heidelberg (1998)
4. Coppersmith, D.: Finding a small root of a bivariate integer equation; factoring with high bits known. In: Maurer, U.M. (ed.) EUROCRYPT 1996. LNCS, vol. 1070, pp. 178–189. Springer, Heidelberg (1996)
5. Coppersmith, D.: Finding a small root of a univariate modular equation. In: Maurer, U.M. (ed.) EUROCRYPT 1996. LNCS, vol. 1070, pp. 155–165. Springer, Heidelberg (1996)
6. Coppersmith, D.: Small solutions to polynomial equations, and low exponent RSA vulnerabilities. Journal of Cryptology **10**(4), 233–260 (1997)
7. Coron, J.-S.: Finding small roots of bivariate integer polynomial equations revisited. In: Cachin, C., Camenisch, J.L. (eds.) EUROCRYPT 2004. LNCS, vol. 3027, pp. 492–505. Springer, Heidelberg (2004)
8. Coron, J.-S.: Finding small roots of bivariate integer polynomial equations: a direct approach. In: Menezes, A. (ed.) CRYPTO 2007. LNCS, vol. 4622, pp. 379–394. Springer, Heidelberg (2007)
9. Ernst, M., Jochemsz, E., May, A., de Weger, B.: Partial key exposure attacks on RSA up to full size exponents. In: Cramer, R. (ed.) EUROCRYPT 2005. LNCS, vol. 3494, pp. 371–386. Springer, Heidelberg (2005)
10. Faugère, J.C.: A new efficient algorithm for computing Gröbner Bases without reduction to zero (F5). In: Proceedings of the 2002 International Symposium on Symbolic and Algebraic Computation, ISSAC 2002, New York, NY, USA, pp. 75–83. ACM (2002)
11. Howgrave-Graham, N.: Finding small roots of univariate modular equations revisited. In: Darnell, M. (ed.) Crytography and Coding. Lecture Notes in Computer Science, vol. 1355, pp. 131–142. Springer, Heidelberg (1997)
12. Huang, Z., Hu, L., Xu, J., Peng, L., Xie, Y.: Partial key exposure attacks on Takagi's variant of RSA. In: Boureanu, I., Owesarski, P., Vaudenay, S. (eds.) ACNS 2014. LNCS, vol. 8479, pp. 134–150. Springer, Heidelberg (2014)
13. Itoh, K., Kunihiro, N., Kurosawa, K.: Small secret key attack on a variant of RSA (due to Takagi). In: Malkin, T. (ed.) CT-RSA 2008. LNCS, vol. 4964, pp. 387–406. Springer, Heidelberg (2008)

14. Joye, M., Lepoint, T.: Partial key exposure on RSA with private exponents larger than N. In: Ryan, M.D., Smyth, B., Wang, G. (eds.) ISPEC 2012. LNCS, vol. 7232, pp. 369–380. Springer, Heidelberg (2012)
15. Kocher, P.C.: Timing attacks on implementations of Diffie-Hellman, RSA, DSS, and other systems. In: Koblitz, N. (ed.) CRYPTO 1996. LNCS, vol. 1109, pp. 104–113. Springer, Heidelberg (1996)
16. Lenstra Jr., A.K., Lenstra, H.W., Lovász, L.: Factoring polynomials with rational coefficients. Mathematische Annalen **261**(4), 515–534 (1982)
17. Lu, Y., Zhang, R., Lin, D.: New results on solving linear equations modulo unknown divisors and its applications. Cryptology ePrint Archive, Report 2014/343 (2014). http://eprint.iacr.org/
18. May, A.: Secret exponent attacks on RSA-type schemes with moduli $N = p^r q$. In: Bao, F., Deng, R., Zhou, J. (eds.) PKC 2004. LNCS, vol. 2947, pp. 218–230. Springer, Heidelberg (2004)
19. Nguyên, P.Q., Stehlé, D.: Floating-Point LLL revisited. In: Cramer, R. (ed.) EUROCRYPT 2005. LNCS, vol. 3494, pp. 215–233. Springer, Heidelberg (2005)
20. Sarkar, S.: Small secret exponent attack on RSA variant with modulus $N = p^r q$. Designs, Codes and Cryptography **73**(2), 383–392 (2014)
21. Takagi, T.: Fast RSA-type cryptosystem modulo $p^k q$. In: Krawczyk, H. (ed.) Advances in Cryptology - CRYPTO '98. Lecture Notes in Computer Science, vol. 1462, pp. 318–326. Springer, Heidelberg (1998)
22. Wiener, M.J.: Cryptanalysis of short RSA secret exponents. IEEE Transactions on Information Theory **36**, 553–558 (1990)

A Chomsky-Schützenberger Theorem
for Weighted Automata with Storage

Luisa Herrmann and Heiko Vogler[✉]

Department of Computer Science,
Technische Universität Dresden, D-01062 Dresden, Germany
{Luisa.Herrmann,Heiko.Vogler}@tu-dresden.de

Abstract. We enrich the concept of automata with storage by weights taken from any unital valuation monoid. We prove a Chomsky-Schützenberger theorem for the class of weighted languages recognizable by such weighted automata with storage.

1 Introduction

The classical Chomsky-Schützenberger theorem [3, Prop. 2] (for short: CS theorem) states that each context-free language is the homomorphic image of the intersection of a Dyck-language and a regular language. In [28] it was shown under which conditions the homomorphism can be non-erasing. In [23] the CS theorem was employed to specify a parser for context-free languages. The CS theorem has been extended to string languages generated by tree-adjoining grammars [32], multiple context-free languages [33], indexed languages [17][1], and yield images of simple context-free tree languages [25].

Already in [3] the CS theorem for context-free languages was proved in a special weight setting: each word in the language is associated with the number of its derivations. In [29] the CS theorem was shown for algebraic (formal) power series over commutative semirings. In [9] this result was generalized to algebraic power series over unital valuation monoids, called quantitative context-free languages; (unital) valuation monoids allow to describe, e.g., average consumption of energy. Also in [9] quantitative context-free languages were characterized by weighted pushdown automata over unital valuation monoids. Recently, the CS theorem has been proved for weighted multiple context-free languages over complete commutative strong bimonoids [6].

In the classical CS theorem, the set Y of letters occurring in the Dyck-language depends on the given context-free grammar or pushdown automaton. An alternative is to code Y by a homomorphism g over a two-letter alphabet and to obtain the following CS theorem [22, Thm. 10.4.3]: each context-free language L can be represented in the form $L = h(g^{-1}(D_2) \cap R)$ for some homomorphisms h and g and a regular language R; D_2 denotes the Dyck-language over a two letter alphabet. In the sequel we call this alternative the CS theorem.

---

[1] We are grateful to one of the reviewers for pointing out this reference to us.

© Springer International Publishing Switzerland 2015
A. Maletti (Ed.): CAI 2015, LNCS 9270, pp. 115–127, 2015.
DOI: 10.1007/978-3-319-23021-4_11

In this paper we prove a CS theorem for the class of weighted languages recognizable by weighted iterated pushdown automata over unital valuation monoids. A weighted language[2] is a mapping from Σ^* to some weight algebra. Intuitively, an iterated pushdown is a pushdown in which each square contains a pushdown in which each square contains a pushdown ... (and so on). The idea of iterated pushdowns goes back to [21,26,27]. It was proved in [11, Thm. 6] that the classes of languages accepted by iterated pushdown automata form a strict, infinite hierarchy with increasing nesting of pushdowns. In [5] it was proved that n-iterated pushdown automata characterize the n-th level of the OI-string language hierarchy [4,13,31] which starts at its first three levels with the regular, context-free, and indexed languages [1] (equivalently, OI-macro languages [16]).

We obtain the CS theorem for weighted iterated pushdown automata as application of the even more general, main result of our paper: the CS theorem for K-weighted automata with storage where K is an arbitrary unital valuation monoid. An automaton with storage S [30,19,12][3] is a one-way nondeterministic finite-state automaton with an additional storage of type S; a successful computation starts with the initial state and an initial configuration of S; in each transition the automaton can test the current storage configuration and apply an instruction to it. For instance, pushdown automata, n-iterated pushdown automata, stack automata [20], and nested stack automata [2] can be formulated as automata with storage. For a number of examples of storages we refer to [12] where these automata were called $\mathrm{REG}(S)$ r-acceptors. The concept of automata with storage is quite flexible: for instance, we can also express M-automata [24] where M is a (multiplicative) monoid, in a straightforward way as such automata with storage (cf. Ex. 4).

We extend the concept of automata with storage to that of K-weighted automata with storage where K is a unital valuation monoid; this extension is done in the same way as pushdown automata have been extended in [9] to weighted pushdown automata over unital valuation monoids. Then our main result states the following (cf. Thm. 11). Let $r : \Sigma^* \to K$ be recognizable by some K-weighted automaton over storage type S. Then there are a regular language R, a finite set Ω of pairs (each consisting of a predicate and an instruction), a configuration c of S, a letter-to-letter morphism g, and a (weighted) alphabetic morphism h such that $r = h(g^{-1}(\mathrm{B}(\Omega, c)) \cap R)$ where $\mathrm{B}(\Omega, c)$ is the set of all Ω-behaviours of c.

2   Preliminaries

Notations and Notions. The set of non-negative integers (including 0) is denoted by $\mathbb{N}$. Let $n \in \mathbb{N}$. Then $[n]$ denotes the set $\{i \in \mathbb{N} \mid 1 \leq i \leq n\}$. Thus $[0] = \emptyset$. Let A and B be sets. The set of all subsets (finite subsets) of A is denoted by $\mathcal{P}(A)$ ($\mathcal{P}_{\mathrm{fin}}(A)$, resp.). We denote the identity mapping on A by id_A. Let $f : A \to B$ be a mapping. We denote by $\mathrm{im}(f)$ the set $\{b \in B \mid \exists a \in A : f(a) = b\}$.

---

[2] or, equivalently, formal power series

[3] If we cite notions or definitions from [12], then we always refer to the version of 2014.

We fix a countably infinite set Λ and call its elements symbols. We call each finite subset Σ of Λ an *alphabet*. *In the rest of this paper, we let Σ and Δ denote alphabets unless specified otherwise.*

Unital Valuation Monoids. The concept of valuation monoid was introduced in [7,8] and extended in [9] to unital valuation monoid. A *unital valuation monoid* is a tuple $(K, +, \mathrm{val}, 0, 1)$ such that $(K, +, 0)$ is a commutative monoid and val: $K^* \to K$ is a mapping such that (i) $\mathrm{val}(a) = a$ for each $a \in K$, (ii) $\mathrm{val}(a_1, \ldots, a_n) = 0$ whenever $a_i = 0$ for some $i \in [n]$, (iii) $\mathrm{val}(a_1, \ldots, a_{i-1}, 1, a_{i+1}, \ldots, a_n) = \mathrm{val}(a_1, \ldots, a_{i-1}, a_{i+1}, \ldots, a_n)$ for any $i \in [n]$, and (iv) $\mathrm{val}(\varepsilon) = 1$.

A monoid $(K, +, 0)$ is *complete* if it has an infinitary sum operation $\sum_I \colon K^I \to K$ for each enumerable set I (for the axioms cf. [10]). We call a unital valuation monoid $(K, +, \mathrm{val}, 0, 1)$ complete if $(K, +, 0)$ has this property. We write $\sum_{i \in I} a_i$ instead of $\sum_I (a_i \mid i \in I)$.

We refer the reader to [9, Ex. 1 and 2] for a number of examples of unital valuation monoids. For instance, each complete semiring (in particular, the *Boolean semiring* $\mathbb{B} = (\{0, 1\}, \vee, \wedge, 0, 1)$) and each complete lattice is a complete unital valuation monoid. *In the rest of this paper, we let K denote an arbitrary unital valuation monoid $(K, +, \mathrm{val}, 0, 1)$ unless specified otherwise.*

Weighted Languages. A *K-weighted language over Σ* is a mapping of the form $r \colon \Sigma^* \to K$. We denote the set of all such mappings by $K\langle\langle \Sigma^* \rangle\rangle$. For every $r \in K\langle\langle \Sigma^* \rangle\rangle$, we denote the set $\{w \in \Sigma^* \mid r(w) \neq 0\}$ by $\mathrm{supp}(r)$.

A family $(r_i \mid i \in I)$ of K-weighted languages $r_i \in K\langle\langle \Sigma^* \rangle\rangle$ is *locally finite* if for each $w \in \Sigma^*$ the set $I_w = \{i \in I \mid r_i(w) \neq 0\}$ is finite. In this case or if K is complete, we define $\sum_{i \in I} s_i \in K\langle\langle \Sigma^* \rangle\rangle$ by $\left(\sum_{i \in I} s_i\right)(w) = \sum_{i \in I_w} s_i(w)$ for each $w \in \Sigma^*$.

Each $L \in \mathbb{B}\langle\langle \Sigma^* \rangle\rangle$ determines the set $\mathrm{supp}(L) \subseteq \Sigma^*$. Vice versa, each set $L \subseteq \Sigma^*$ determines the $\mathbb{B}$-weighted language $\chi_L \in \mathbb{B}\langle\langle \Sigma^* \rangle\rangle$ with $\chi_L(w) = 1$ if and only if $w \in L$. Thus, for every $L \subseteq \Sigma^*$, we have $\mathrm{supp}(\chi_L) = L$; and for every $L \in \mathbb{B}\langle\langle \Sigma^* \rangle\rangle$ we have $\chi_{\mathrm{supp}(L)} = L$. In the sequel we will not distinguish between these two points of view.

3   Weighted Automata with Storage

We take up the concept of automata with storage [30] and present it in the style of [12] (cf. [14,15] for further investigations). Moreover, we add weights to the transitions of the automaton where the weights are taken from some unital valuation monoid.

Storage Types: We recall the definition of storage type from [12,30] with a slight modification. A *storage type* S is a tuple (C, P, F, C_0) where C is a set (*configurations*), P is a set of total functions each having the type $p \colon C \to \{\text{true}, \text{false}\}$ (*predicates*), F is a set of partial functions each having the type $f \colon C \to C$ (*instructions*), and $C_0 \subseteq C$ (*initial configurations*).

Example 1. Let c be an arbitrary but fixed symbol. The *trivial storage type* is the storage type TRIV $= (\{c\}, \{p_{\text{true}}\}, \{f_{\text{id}}\}, \{c\})$ where $p_{\text{true}}(c) = \text{true}$ and $f_{\text{id}}(c) = c$. $\qquad\qquad\square$

Next we recall the pushdown operator P from [12, Def. 5.1] and [14, Def. 3.28]: if S is a storage type, then $P(S)$ is a storage type of which the configurations have the form of a pushdown; each cell contains a pushdown symbol and a configuration of S. Formally, let Γ be a fixed infinite set (*pushdown symbols*). Also, let $S = (C, P, F, C_0)$ be a storage type. The *pushdown of S* is the storage type $P(S) = (C', P', F', C_0')$ where

- $C' = (\Gamma \times C)^+$ and $C_0' = \{(\gamma_0, c_0) \mid \gamma_0 \in \Gamma, c_0 \in C_0\}$,
- $P' = \{\text{bottom}\} \cup \{(\text{top} = \gamma) \mid \gamma \in \Gamma\} \cup \{\text{test}(p) \mid p \in P\}$ such that for every $(\delta, c) \in \Gamma \times C$ and $\alpha \in (\Gamma \times C)^*$ we have

$$
\begin{aligned}
\text{bottom}\big((\delta, c)\alpha\big) &= \text{true} \ \text{ if and only if } \alpha = \varepsilon \\
(\text{top} = \gamma)\big((\delta, c)\alpha\big) &= \text{true} \ \text{ if and only if } \gamma = \delta \\
\text{test}(p)\big((\delta, c)\alpha\big) &= p(c)
\end{aligned}
$$

- $F' = \{\text{pop}\} \cup \{\text{stay}(\gamma) \mid \gamma \in \Gamma\} \cup \{\text{push}(\gamma, f) \mid \gamma \in \Gamma, f \in F\}$ such that for every $(\delta, c) \in \Gamma \times C$ and $\alpha \in (\Gamma \times C)^*$ we have

$$
\begin{aligned}
\text{pop}\big((\delta, c)\alpha\big) &= \alpha \ \text{ if } \alpha \neq \varepsilon \\
\text{stay}(\gamma)\big((\delta, c)\alpha\big) &= (\gamma, c)\alpha \\
\text{push}(\gamma, f)\big((\delta, c)\alpha\big) &= (\gamma, f(c))(\delta, c)\alpha \ \text{ if } f(c) \ \text{ is defined}
\end{aligned}
$$

and undefined in all other situations.

For each $n \geq 0$ we define $P^n(S)$ inductively as follows: $P^0(S) = S$ and $P^n(S) = P(P^{n-1}(S))$ for each $n \geq 1$.

Example 2. Intuitively, $P(\text{TRIV})$ corresponds to the usual pushdown storage except that there is no empty pushdown. For $n \geq 0$, we abbreviate $P^n(\text{TRIV})$ by P^n and call it the *n-iterated pushdown storage*. $\qquad\qquad\square$

Throughout this paper we let S denote an arbitrary storage type (C, P, F, C_0) unless specified otherwise.

Automata with Storage: An (S, Σ)-*automaton* is a tuple $\mathcal{A} = (Q, \Sigma, c_0, q_0, Q_f, T)$ where Q is a finite set (*states*), Σ is an alphabet (*terminal symbols*), $c_0 \in C_0$ (*initial configuration*), $q_0 \in Q$ (*initial state*), $Q_f \subseteq Q$ (*final states*), and $T \subseteq Q \times (\Sigma \cup \{\varepsilon\}) \times P \times Q \times F$ is a finite set (*transitions*). If $T \subseteq Q \times \Sigma \times P \times Q \times F$, then we call $\mathcal{A}$ *ε-free*.

The computation relation of $\mathcal{A}$ is the binary relation on the set $Q \times \Sigma^* \times C$ of $\mathcal{A}$-*configurations* defined as follows. For every transition $\tau = (q, x, p, q', f)$ in T we define the binary relation $\vdash^\tau$ on the set of $\mathcal{A}$-configurations: for every $w \in \Sigma^*$ and $c \in C$, we let $(q, xw, c) \vdash^\tau (q', w, f(c))$ if $p(c)$ is true and $f(c)$ is defined. The *computation relation of $\mathcal{A}$* is the binary relation $\vdash = \bigcup_{\tau \in T} \vdash^\tau$. The *language recognized by $\mathcal{A}$* is the set $L(\mathcal{A}) = \{w \in \Sigma^* \mid (q_0, w, c_0) \vdash^* (q_f, \varepsilon, c)$ for some $q_f \in Q_f, c \in C\}$.

A *computation* is a sequence $\theta = \tau_1 \ldots \tau_n$ of transitions τ_i ($i \in [n]$) such that there are $\mathcal{A}$-configurations $c_0, \ldots, c_n$ with $c_{i-1} \vdash^{\tau_i} c_i$. We abbreviate this computation by $c_0 \vdash^\theta c_n$. Let $q \in Q$, $w \in \Sigma^*$, and $c \in C$. A q-*computation on w and c* is a computation θ such that $(q, w, c) \vdash^\theta (q_f, \varepsilon, c')$ for some $q_f \in Q_f$, $c' \in C$. We denote the set of all q-computations on w and c by $\Theta_\mathcal{A}(q, w, c)$. Furthermore, we denote the set of all q_0-computations on w and c_0 by $\Theta_\mathcal{A}(w)$. Thus we have $L(\mathcal{A}) = \{w \in \Sigma^* \mid \Theta_\mathcal{A}(w) \neq \emptyset\}$.

We say that $\mathcal{A}$ is *ambiguous* if there is a $w \in \Sigma^*$ such that $|\Theta_\mathcal{A}(w)| \geq 2$. Otherwise $\mathcal{A}$ is *unambiguous*. A language $L \subseteq \Sigma^*$ is (S, Σ)-*recognizable* if there is an (S, Σ)-automaton $\mathcal{A}$ with $L(\mathcal{A}) = L$.

Example 3. (1) The TRIV-automata are (usual) finite-state automata, and P^1-automata are essentially pushdown automata. (2) For each $n \geq 1$, P^n-automata correspond to n-iterated pushdown automata of [26,27,11,5]. (3) Nested stack automata [2] correspond to NS(TRIV)-automata where NS is an operator on storage types (cf. [14, Def. 7.1]). In [14, Thm. 7.4] it was proved that, for every S, the storage types $P^2(S)$ and NS(S) are equivalent (cf. [14, Def. 4.6] for the definition of equivalence), which implies that the acceptance power of automata using these storage types is the same (cf. [14, Thm. 4.18] for this implication). □

Example 4. We indicate how to embed the concept of M-automata [24] where $(M, \cdot, 1)$ is a multiplicative monoid, into the setting of automata with storage. For this we define the storage type *monoid M*, denoted by MON(M), by (C, P, F, C_0) where $C = M$ and $C_0 = \{1\}$, $P = \{\text{true?}\} \cup \{1?\}$ with $\text{true?}(m) = \text{true}$, and $1?(m) = \text{true}$ if and only if $m = 1$, $F = \{[m] \mid m \in M\}$ and $[m] \colon M \to M$ is defined by $[m](m') = m' \cdot m$.

For a given M-automaton $\mathcal{A}$, we construct an equivalent MON(M)-automaton $\mathcal{B}$ as follows. If (q, x, q', m) is a transition of $\mathcal{A}$ (with states q, q', input symbol x, and $m \in M$), then $(q, x, \text{true?}, q', [m])$ is a transition of $\mathcal{B}$. Moreover, for each final state q of $\mathcal{A}$, the transition $(q, \varepsilon, 1?, q_f, [1])$ is in $\mathcal{B}$ where q_f is the only final state of $\mathcal{B}$. □

Weighted Automata with Storage: Next we define the weighted version of (S, Σ)-automata. The line of our definitions follows the definition of weighted pushdown automata in [9].

An (S, Σ)-*automaton with weights in K* is a tuple $\mathcal{A} = (Q, \Sigma, c_0, q_0, Q_f, T, \text{wt})$ where $(Q, \Sigma, c_0, q_0, Q_f, T)$ is an (S, Σ)-automaton (*underlying (S, Σ)-automaton*) and $\text{wt} \colon T \to K$ (*weight assignment*). If the underlying (S, Σ)-automaton is ε-free, then we call $\mathcal{A}$ ε-*free*. Let $\theta = \tau_1 \ldots \tau_n$ be a computation of $\mathcal{A}$. The *weight of θ* is the element in K defined by $\text{wt}(\theta) = \text{val}(\text{wt}(\tau_1), \ldots, \text{wt}(\tau_n))$.

An *(S,Σ,K)-automaton* is an (S, Σ)-automaton $\mathcal{A}$ with weights in K such that (i) $\Theta_\mathcal{A}(w)$ is finite for every $w \in \Sigma^*$ or (ii) K is complete. In this case the *weighted language recognized by $\mathcal{A}$* is the K-weighted language $\|\mathcal{A}\| \colon \Sigma^* \to K$ defined for every $w \in \Sigma^*$ by $\|\mathcal{A}\|(w) = \sum_{\theta \in \Theta_\mathcal{A}(w)} \text{wt}(\theta)$.

A weighted language $r \colon \Sigma^* \to K$ is (S, Σ, K)-*recognizable* if there is an (S, Σ, K)-automaton $\mathcal{A}$ such that $r = \|\mathcal{A}\|$.

Example 5. (1) Each $(S, \Sigma, \mathbb{B})$-automaton $\mathcal{A}$ can be considered as an (S, Σ)-automaton which recognizes $\mathrm{supp}(\|\mathcal{A}\|)$. (2) Apart from ε-moves, $(\mathrm{TRIV}, \Sigma, K)$-automata are the same as weighted finite automata over Σ and the valuation monoid K [9]. (3) The $(\mathrm{P}^1, \Sigma, K)$-automata are essentially the same as weighted pushdown automata over Σ and K [9] where acceptance with empty pushdown can be simulated in the usual way. Thus, for every $r \colon \Sigma^* \to K$ we have: r is the quantitative behaviour of a WPDA as defined in [9] if and only if r is $(\mathrm{P}^1, \Sigma, K)$-recognizable. $\qquad\qquad\square$

For $n \geq 0$, a *weighted n-iterated pushdown language over Σ and K* is a $(\mathrm{P}^n, \Sigma, K)$-recognizable weighted language.

4   Separating the Weights from an (S, Σ, K)-Automaton

In this section we will represent an (S, Σ, K)-recognizable weighted language as the homomorphic image of an (S, Δ)-recognizable language.

We recall from [9] the concept of (weighted) alphabetic morphism. First, we introduce monomes and alphabetic morphisms. A mapping $r \colon \Sigma^* \to K$ is called a *monome* if $\mathrm{supp}(r)$ is empty or a singleton. If $\mathrm{supp}(r) = \{w\}$, then we also write $r(w).w$ instead of r. We let $K[\Sigma \cup \{\varepsilon\}]$ denote the set of all monomes with support in $\Sigma \cup \{\varepsilon\}$.

Let Δ be an alphabet and $h \colon \Delta \to K[\Sigma \cup \{\varepsilon\}]$ be a mapping. The *alphabetic morphism induced by h* is the mapping $h' \colon \Delta^* \to K\langle\!\langle \Sigma^* \rangle\!\rangle$ such that for every $n \geq 0$, $\delta_1, \ldots, \delta_n \in \Delta$ with $h(\delta_i) = a_i.y_i$ we have $h'(\delta_1 \ldots \delta_n) = \mathrm{val}(a_1, \ldots, a_n).y_1 \ldots y_n$. Note that $h'(v)$ is a monome for every $v \in \Delta^*$, and $h'(\varepsilon) = 1.\varepsilon$. If $L \subseteq \Delta^*$ such that the family $(h'(v) \mid v \in L)$ is locally finite or if K is complete, we let $h'(L) = \sum_{v \in L} h'(v)$. In the sequel we will use the following convention. If we write "alphabetic morphism $h \colon \Delta \to K[\Sigma \cup \{\varepsilon\}]$", then we mean the alphabetic morphism induced by h.

We define a special case of alphabetic morphisms in which $K = \mathbb{B}$. If for every $\delta \in \Delta$ the support of $h(\delta)$ is $\{\sigma\}$ for some $\sigma \in \Sigma$, then we call h' a *letter-to-letter morphism*. Note that in this case the alphabetic morphism induced by h has the property that for every $v \in \Delta^*$, $\mathrm{supp}(h'(v))$ contains at most one element and if $\mathrm{supp}(h'(v)) = \{w\}$ for some $w \in \Sigma^*$, then the lengths of w and v are equal.

Theorem 6. *For every $r \in K\langle\!\langle \Sigma^* \rangle\!\rangle$ the following two statements are equivalent:*
(1) r is (S, Σ, K)-recognizable.
(2) There are an alphabet Δ, an unambiguous ε-free (S, Δ)-automaton $\mathcal{A}$, and an alphabetic morphism $h \colon \Delta \to K[\Sigma \cup \{\varepsilon\}]$ such that $r = h(L(\mathcal{A}))$.

Proof. (1) $\Rightarrow$ (2): This generalizes [9, Lm. 3] in a straightforward way. Let $\mathcal{B} = (Q, \Sigma, c_0, q_0, Q_f, T, \mathrm{wt})$ be an (S, Σ, K)-automaton. We construct the (S, T)-automaton $\mathcal{A} = (Q, T, c_0, q_0, Q_f, T')$ and the mapping $h \colon T \to K[\Sigma \cup \{\varepsilon\}]$ such that, if $\tau = (q, x, p, q', f)$ is in T, then (q, τ, p, q', f) is in T' and we define $h(\tau) = \mathrm{wt}(\tau).x$. Obviously, $\mathcal{A}$ is unambiguous and ε-free.

Let $w \in \Sigma^*$ and $\theta = \tau_1 \ldots \tau_n \in \Theta_{\mathcal{B}}(w)$. By definition of h, we have that $h(\theta) = \mathrm{val}(\mathrm{wt}(\tau_1), \ldots, \mathrm{wt}(\tau_n)).w$. Hence $\mathrm{wt}(\theta) = \big(h(\theta)\big)(w)$. Also, by definition of (S, Σ, K)-automata, the set $\Theta_{\mathcal{B}}(w)$ is finite if K is not complete. Thus the family $(h(\theta) \mid \theta \in L(\mathcal{A}))$ is locally finite if K is not complete. Then, for every $w \in \Sigma^*$, we have $\|\mathcal{B}\|(w) = \sum_{\theta \in \Theta_{\mathcal{B}}(w)} \mathrm{wt}(\theta) = \sum_{\theta \in \Theta_{\mathcal{B}}(w)} \big(h(\theta)\big)(w) \overset{(*)}{=} \sum_{\theta \in L(\mathcal{A})} \big(h(\theta)\big)(w) = \big(\sum_{\theta \in L(\mathcal{A})} h(\theta)\big)(w) = \big(h(L(\mathcal{A}))\big)(w)$ where $(*)$ holds because for every $\theta \in L(\mathcal{A})$ with $\theta \notin \Theta_{\mathcal{B}}(w)$, we have $\big(h(\theta)\big)(w) = 0$ and due to the fact that $\sum_{\theta \in L(\mathcal{A}),\, \theta \notin \Theta_{\mathcal{B}}(w)} 0 = 0$. Thus $\|\mathcal{B}\| = h(L(\mathcal{A}))$.

$(2) \Rightarrow (1)$: Let $\mathcal{A} = (Q, \Delta, c_0, q_0, Q_f, T)$ be an unambiguous ε-free (S, Δ)-automaton and $h \colon \Delta \to K[\Sigma \cup \{\varepsilon\}]$ an alphabetic morphism. Moreover, we assume that the family $(h(v) \mid v \in L(\mathcal{A}))$ is locally finite if K is not complete. We will construct an (S, Σ, K)-automaton $\mathcal{B}$ such that $\|\mathcal{B}\| = h(L(\mathcal{A}))$.

Our construction employs a similar technique of coding the preimage of h into the set of states as in [9, Lm. 4] in order to handle non-injectivity of h appropriately. However, we have to modify the construction slightly, because the straightforward generalization would require that S has an identity instruction (needed in the first step of the computation), which in general we do not assume. In our constructed automaton, the target state (and not, as in [9, Lm. 4], the source state) of each transition encodes a preimage of the symbol which is read by this transition.

Formally, we construct the (S, Σ, K)-automaton $\mathcal{B} = (Q', \Sigma, c_0, q_0', Q_f', T', \mathrm{wt})$ where $Q' = \{q_0'\} \cup \Delta \times Q$ with some element q_0' with $q_0' \notin \Delta \times Q$, $Q_f' = \Delta \times Q_f$, and T' and wt are defined as follows. Let $\delta \in \Delta$ and $h(\delta) = a.y$.

- If (q_0, δ, p, q, f) is in T, then $(q_0', y, p, (\delta, q), f)$ is in T', and its weight is a.
- If (q, δ, p, q', f) is in T, then $((\delta', q), y, p, (\delta, q'), f)$ is in T' for each $\delta' \in \Delta$, and its weight is a.

Let $w \in \Sigma^*$. First, let $v \in \Delta^*$ with $h(v) = z.w$ for some $z \in K$. We write $v = \delta_1 \ldots \delta_n \in \Delta^*$ with $n \geq 0$ and $\delta_i \in \Delta$. Let $h(\delta_i) = a_i.y_i$ for every $1 \leq i \leq n$. Thus $h(v) = \mathrm{val}(a_1, \ldots, a_n).y_1 \ldots y_n$ and $w = y_1 \ldots y_n$ and $z = \mathrm{val}(a_1, \ldots, a_n)$.

Let $\theta = \tau_1 \ldots \tau_n$ be a q_0-computation in $\Theta_{\mathcal{A}}(v)$. Clearly, for each $i \in [n]$, the second component of τ_i is δ_i. Then we construct the q_0'-computation $\theta' = \tau_1' \ldots \tau_n'$ in $\Theta_{\mathcal{B}}(y_1 \ldots y_n)$ inductively as follows:

- If $\tau_1 = (q_0, \delta_1, p_1, q_1, f_1)$, then we let $\tau_1' = (q_0', y_1, p_1, (\delta_1, q_1), f_1)$.
- If $1 < i \leq n$ and $\tau_i = (q_{i-1}, \delta_i, p_i, q_i, f_i)$, then we let $\tau_i' = ((\delta_{i-1}, q_{i-1}), y_i, p_i, (\delta_i, q_i), f_i)$.

Note that $\big(h(v)\big)(w) = \mathrm{val}(a_1, \ldots, a_n) = \mathrm{val}(\mathrm{wt}(\tau_1'), \ldots, \mathrm{wt}(\tau_n')) = \mathrm{wt}(\theta')$.

Conversely, for every q_0'-computation $\theta' = \tau_1' \ldots \tau_n'$ in $\Theta_{\mathcal{B}}(w)$ by definition of T' there are a uniquely determined $v \in \Delta^*$ and a uniquely determined q_0-computation $\theta = \tau_1 \ldots \tau_n$ in $\Theta_{\mathcal{A}}(v)$ such that θ' is the computation constructed above. Hence, for every $v \in \Delta^*$ and $w \in \Sigma^*$, if $h(v) = z.w$ for some $z \in K$, then $\Theta_{\mathcal{A}}(v)$ and $\Theta_{\mathcal{B}}(w)$ are in a one-to-one correspondence.

Thus, for every $w \in \Sigma^*$, we obtain $\big(h(L(\mathcal{A}))\big)(w) = \sum_{v \in L(\mathcal{A})} \big(h(v)\big)(w) = \sum_{\substack{v \in L(\mathcal{A}): \\ (h(v))(w) \neq 0}} \big(h(v)\big)(w)$. Since $\mathcal{A}$ is unambiguous this is equal to

$\sum_{v \in L(\mathcal{A}), \theta \in \Theta_{\mathcal{A}}(v):}$ wt(θ'). Since there is a one-to-one correspondence between
$(h(v))(w) \neq 0$
$\Theta_{\mathcal{A}}(v)$ and $\Theta_{\mathcal{B}}(w)$, this is equal to $\sum_{\theta' \in \Theta_{\mathcal{B}}(w)}$ wt$(\theta') = \|\mathcal{B}\|(w)$. Thus
$h(L(\mathcal{A})) = \|\mathcal{B}\|$. □

We could strengthen Theorem 6 by proving $(2') \Rightarrow (1)$ where $(2')$ is obtained
from (2) by dropping the ε-freeness of $\mathcal{A}$.

5   Separating the Storage from an (S, Δ)-Automaton

In this section we will characterize the language recognized by an ε-free (S, Δ)-
automaton $\mathcal{A}$ as the image of the set of behaviours of the initial configuration of
$\mathcal{A}$ under a simple transducer mapping. Note that $\mathcal{A}$ need not be unambiguous.
Our proof follows closely the technique in the proof of [14, Thm. 3.26].

Let c_0 be the initial configuration of $\mathcal{A}$ and θ a computation of $\mathcal{A}$, i.e., $\theta \in$
$\Theta_{\mathcal{A}}(q_0, w, c_0)$ for some w. By dropping from θ all references to states and to the
input, a sequence of pairs remains where each pair consists of a predicate and
an instruction. This sequence might be called a behaviour of c_0. Formally, let Ω
be a finite subset of $P \times F$,[4] $c \in C$, and $v = (p_1, f_1) \ldots (p_n, f_n) \in \Omega^*$. We say
that v is an Ω-behaviour of c if for every i with $i \in [n]$ we have (i) $p_i(c') = $ true
and (ii) $f_i(c')$ is defined where $c' = f_{i-1}(\ldots f_1(c) \ldots)$ (note that $c' = c$ for $i = 1$).
We denote the set of all Ω-behaviours of c by B(Ω, c). Note that each behaviour
of c is a path in the approximation of c according to [14, Def. 3.23].

An a-transducer [19] is a machine $\mathcal{M} = (Q, \Omega, \Delta, \delta, q_0, Q_f)$ where Q, Ω, and
Δ are alphabets (states, input/output symbols, resp.), $q_0 \in Q$ (initial state),
$Q_f \subseteq Q$ (final states), and δ is a finite subset of $Q \times \Omega^* \times Q \times \Delta^*$. We say that
$\mathcal{M}$ is a simple transducer (from Ω to Δ) if $\delta \subseteq Q \times \Omega \times Q \times \Delta$. The binary
relation $\vdash_{\mathcal{M}}$ on $Q \times \Omega^* \times \Delta^*$ is defined as follows: let $(q, ww', v) \vdash_{\mathcal{M}} (q', w', vv')$ if
$(q, w, q', v') \in \delta$. The mapping induced by $\mathcal{M}$, also denoted by $\mathcal{M}$, is the mapping
$\mathcal{M}: \Omega^* \to \mathcal{P}(\Delta^*)$ defined by $\mathcal{M}(w) = \{v \in \Delta^* \mid (q_0, w, \varepsilon) \vdash_{\mathcal{M}}^* (q, \varepsilon, v), q \in Q_f\}$.
If $\mathcal{M}$ is a simple transducer, then $\mathcal{M}(w)$ is finite for every w. For every $L \subseteq \Omega^*$
we define $\mathcal{M}(L) = \bigcup_{v \in L} \mathcal{M}(v)$.

Our goal is to prove the following theorem.

Theorem 7. *Let $S = (C, P, F, C_0)$ be a storage type. Moreover, let $L \subseteq \Delta^*$.
Then the following are equivalent:*
(1) L is recognizable by some ε-free (S, Δ)-automaton.
*(2) There are $c \in C$, a finite set $\Omega \subseteq P \times F$, and a simple transducer $\mathcal{M}$ from
Ω to Δ such that $L = \mathcal{M}(B(\Omega, c))$.*

We note that $(1) \Rightarrow (2)$ of Theorem 7 is similar to [19, Lm. 2.3] (after decomposing
the simple transducer $\mathcal{M}$ from Ω to Δ according to Theorem 9).

For the proof of this theorem, we define the concept of relatedness between
an ε-free (S, Δ)-automaton $\mathcal{A}$ and a simple transducer $\mathcal{M}$ with the following
intention:

---

[4] We recall that $S = (C, P, F, C_0)$ is an arbitrary storage type.

$\mathcal{A}$ allows a computation
$$(q_0, x_1, p_1, q_1, f_1)(q_1, x_2, p_2, q_2, f_2) \ldots (q_{n-1}, x_n, p_n, q_n, f_n) \ ,$$
for some states $q_1, \ldots, q_{n-1}$ if and only if
$$(q_0, (p_1, f_1)) \ldots (p_n, f_n), \varepsilon) \vdash^*_{\mathcal{M}} (q_n, \varepsilon, x_1 \ldots x_n) \ .$$
That is, while reading a behaviour of the initial configuration of $\mathcal{A}$, the simple transducer $\mathcal{M}$ produces a string which is recognized by $\mathcal{A}$. Formally, let $\mathcal{A} = (Q, \Delta, c, q_0, Q_f, T)$ be an ε-free (S, Δ)-automaton and $\mathcal{M} = (Q', \Omega, \Delta', \delta, q'_0, Q'_f)$ be a simple transducer. Then $\mathcal{A}$ *is related to* $\mathcal{M}$ if

- $Q = Q'$, $q_0 = q'_0$, $Q_f = Q'_f$,
- $\Delta = \Delta'$ and Ω is the set of all pairs (p, f) such that T contains a transition of the form (q, x, p, q', f) for some q, q', and x, and
- for every $q, q' \in Q$, $x \in \Delta$, $p \in P$, and $f \in F$ we have: $(q, x, p, q', f) \in T$ if and only if $(q, (p, f), q', x) \in \delta$.

Lemma 8. Let $\mathcal{A}$ be an ε-free (S, Δ)-automaton with initial configuration c and let $\mathcal{M}$ be a simple transducer from Ω to Δ. If $\mathcal{A}$ is related to $\mathcal{M}$, then $L(\mathcal{A}) = \mathcal{M}(\mathrm{B}(\Omega, c))$.

Proof. Let $\mathcal{A} = (Q, \Delta, c, q_0, Q_f, T)$ and $\mathcal{M} = (Q, \Omega, \Delta, \delta, q_0, Q_f)$. First we prove that $L(\mathcal{A}) \subseteq \mathcal{M}(\mathrm{B}(\Omega, c))$. Let $v \in L(\mathcal{A})$. Then $v = x_1 \ldots x_n$ for some $n \geq 0$ and $x_i \in \Delta$ for every $1 \leq i \leq n$. Moreover, there is a q_0-computation θ in $\Theta_{\mathcal{A}}(v)$ with $\theta = \tau_1 \ldots \tau_n$, such that $\tau_i \in T$ where $\tau_1 = (q_0, x_1, p_1, q_1, f_1)$, for every $2 \leq i \leq n$ we have $\tau_i = (q_{i-1}, x_i, p_i, q_i, f_i)$, and $q_n \in Q_f$. Since $\mathcal{A}$ is related to $\mathcal{M}$, we have $(q_{i-1}, (p_i, f_i), q_i, x_i) \in \delta$ for every $1 \leq i \leq n$. Hence $(q_0, w, \varepsilon) \vdash^*_{\mathcal{M}} (q_n, \varepsilon, x_1 \ldots x_n)$ with $w = (p_1, f_1) \ldots (p_n, f_n)$. Since $w \in \mathrm{B}(\Omega, c)$ is a behaviour of c, $v = x_1 \ldots x_n$, and $q_n \in Q_f$, we obtain that $v \in \mathcal{M}(\mathrm{B}(\Omega, c))$.

Next we prove that $\mathcal{M}(\mathrm{B}(\Omega, c)) \subseteq L(\mathcal{A})$. Let $v \in \mathcal{M}(\mathrm{B}(\Omega, c))$ with $v = x_1 \ldots x_n$ for some $n \geq 0$ and $x_i \in \Delta$ for every $1 \leq i \leq n$. Then there is a behaviour $w \in \mathrm{B}(\Omega, c)$ of c such that $v \in \mathcal{M}(w)$. Then there are $(p_i, f_i) \in \Omega$ with $1 \leq i \leq n$ such that $w = (p_1, f_1) \ldots (p_n, f_n)$. Moreover, there are $q_0, \ldots, q_n \in Q$ such that $(q_0, (p_1, f_1), q_1, x_1) \in \delta$, for every $2 \leq i \leq n$: $(q_{i-1}, (p_i, f_i), q_i, x_i) \in \delta$, and $q_n \in Q_f$. Since $\mathcal{A}$ is related to $\mathcal{M}$, we have $\tau_i = (q_{i-1}, x_i, p_i, q_i, f_i) \in T$. Since $w \in \mathrm{B}(\Omega, c)$, q_0 is the initial state of $\mathcal{A}$, and $q_n \in Q_f$, we have that $\tau_1 \ldots \tau_n \in \Theta_{\mathcal{A}}(v)$ and thus $v \in L(\mathcal{A})$. $\qquad\square$

Proof (of Theorem 7). (1) $\Rightarrow$ (2): Let L be recognizable by some ε-free (S, Δ)-automaton $\mathcal{A} = (Q, \Delta, c, q_0, Q_f, T)$. Let Ω be the set of all pairs (p, f) such that T contains a transition of the form (q, x, p, q', f) for some q, q', and x. We construct the simple transducer $\mathcal{M} = (Q, \Omega, \Delta, \delta, q_0, Q_f)$ by defining $(q, (p, f), q', x) \in \delta$ if and only if $(q, x, p, q', f) \in T$ for every $q, q' \in Q$, $x \in \Delta$, and $(p, f) \in \Omega$. Clearly, $\mathcal{A}$ is related to $\mathcal{M}$ and thus, by Lemma 8, we have that $L(\mathcal{A}) = \mathcal{M}(\mathrm{B}(\Omega, c))$.

(2) $\Rightarrow$ (1): Let $c \in C$, Ω a finite subset of $P \times F$, and $\mathcal{M} = (Q, \Omega, \Delta, \delta, q_0, Q_f)$ a simple transducer. First we reduce $\mathcal{M}$ to the simple transducer $\mathcal{M}' = (Q, \Omega', \Delta, \delta, q_0, Q_f)$ where Ω' is the set of all pairs (p, f) such that $(q, (p, f), q', x) \in \delta$ for some $q, q' \in Q$ and $x \in \Delta$. Obviously, $\delta \subseteq Q \times \Omega' \times Q \times \Delta$ and $\mathcal{M}(\mathrm{B}(\Omega, c)) = \mathcal{M}'(\mathrm{B}(\Omega', c))$.

Next we construct the ε-free (S, Δ)-automaton $\mathcal{A} = (Q, \Delta, c, q_0, Q_f, T)$ by defining $T = \{(q, x, p, q', f) \mid (q, (p, f), q', x) \in \delta\}$. Since $\mathcal{A}$ is related to $\mathcal{M}'$, we have that $L(\mathcal{A}) = \mathcal{M}'(\mathrm{B}(\Omega', c)) = \mathcal{M}(\mathrm{B}(\Omega, c))$ by Lemma 8.   □

6   The Main Result and Its Applications

For the proof of our CS theorem for weighted automata with storage, we first recall a result for simple transducers [18, proof of Thm. 2.1].

Theorem 9. *Let Ω be an alphabet and $L \subseteq \Omega^*$ and let $\mathcal{M} \colon \Omega^* \to \mathcal{P}_{fin}(\Delta^*)$ be induced by a simple transducer $\mathcal{M}$. Then there are an alphabet Φ, two letter-to-letter morphisms $h_1 \colon \Phi \to \mathbb{B}[\Omega]$ and $h_2 \colon \Phi \to \mathbb{B}[\Delta]$, and a regular language $R \subseteq \Phi^*$ such that $\mathcal{M}(L) = h_2(h_1^{-1}(L) \cap R)$.*

Next we show that a letter-to-letter morphism $h_2 \colon \Phi \to \mathbb{B}[\Delta]$ and an alphabetic morphism $h \colon \Delta \to K[\Sigma \cup \{\varepsilon\}]$ can be combined smoothly. We define the alphabetic morphism $(h \circ h_2) \colon \Phi \to K[\Sigma \cup \{\varepsilon\}]$ for every $x \in \Phi$ by $(h \circ h_2)(x) = h(\delta)$ if $h_2(x) = 1.\delta$ for some $\delta \in \Delta$ (recall that $|\operatorname{supp}(h_2(x))| = 1$).

Lemma 10. *Let $h_2 \colon \Phi \to \mathbb{B}[\Delta]$ be a letter-to-letter morphism and $h \colon \Delta \to K[\Sigma \cup \{\varepsilon\}]$ an alphabetic morphism. Moreover, let $H \subseteq \Phi^*$ be a language. If $(h(v) \mid v \in h_2(H))$ is locally finite, then $((h \circ h_2)(w) \mid w \in H)$ is locally finite.*

Proof. Let $u \in \Sigma^*$. By assumption, we have that $\{v \in h_2(H) \mid u \in \operatorname{supp}(h(v))\}$ is finite; let us denote this set by C_u. Since h_2 is letter-to-letter, we have that $\{y \in H \mid v \in h_2(y)\}$ is finite for each $v \in h_2(H)$. Then we have: $|\{w \in H \mid u \in \operatorname{supp}((h \circ h_2)(w))\}| = \sum_{v \in C_u} |\{y \in H \mid v \in h_2(y)\}|$. Hence, $\{w \in H \mid u \in \operatorname{supp}((h \circ h_2)(w)\}$ is finite.   □

Now we can prove the CS theorem for (S, Σ, K)-automata (cf. Fig.1).

Theorem 11. *Let $S = (C, P, F, C_0)$ be a storage type, Σ an alphabet, and K a unital valuation monoid. If $r \in K\langle\!\langle \Sigma^* \rangle\!\rangle$ is (S, Σ, K)-recognizable, then there are*
- *an alphabet Φ and a regular language $R \subseteq \Phi^*$,*
- *a finite set $\Omega \subseteq P \times F$ and a configuration $c \in C$,*
- *a letter-to-letter morphism $h_1 \colon \Phi \to \mathbb{B}[\Omega]$, and*
- *an alphabetic morphism $h' \colon \Phi \to K[\Sigma \cup \{\varepsilon\}]$*

such that $r = h'(h_1^{-1}(\mathrm{B}(\Omega, c)) \cap R)$.

Proof. By Theorem 6 there are an alphabet Δ, an ε-free (S, Δ)-automaton $\mathcal{A}$, and an alphabetic morphism $h \colon \Delta \to K[\Sigma \cup \{\varepsilon\}]$ such that $r = h(L(\mathcal{A}))$. Hence, if K is not complete, then $\Theta_{\mathcal{A}}(w)$ is finite for every $w \in \Sigma^*$, and $(h(v) \mid v \in L(\mathcal{A}))$ is locally finite. According to Theorem 7, there are $c \in C$, a finite set $\Omega \subseteq P \times F$, and a simple transducer $\mathcal{M}$ from Ω to Δ such that $L(\mathcal{A}) = \mathcal{M}(\mathrm{B}(\Omega, c))$. Due to Theorem 9, there are an alphabet Φ, two letter-to-letter morphisms $h_1 \colon \Phi \to \mathbb{B}[\Omega]$ and $h_2 \colon \Phi \to \mathbb{B}[\Delta]$, and a regular language $R \subseteq \Phi^*$ such that

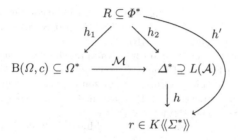

Fig. 1. An illustration of the proof of Theorem 11

$\mathcal{M}(\mathrm{B}(\Omega, c)) = h_2(h_1^{-1}(\mathrm{B}(\Omega, c)) \cap R)$. Let us denote the language $h_1^{-1}(\mathrm{B}(\Omega, c)) \cap R$ by H. Thus $L(\mathcal{A}) = h_2(H)$.

Since $(h(v) \mid v \in L(\mathcal{A}))$ is locally finite if K is not complete, we have by Lemma 10 that also $((h \circ h_2)(w) \mid w \in H)$ is locally finite if K is not complete. Thus $r = (h \circ h_2)(h_1^{-1}(\mathrm{B}(\Omega, c)) \cap R)$ and we can take $h' = (h \circ h_2)$. □

Finally we instantiate the storage type S in Theorem 11 in several ways and obtain the CS theorem for the corresponding class of (S, Σ, K)-recognizable weighted languages: (1) $S = \mathrm{P}^n$: K-weighted n-iterated pushdown languages. (2) $S = \mathrm{NS}(\mathrm{TRIV})$ where NS is the nested stack operator defined in [14, Def. 7.1]: K-weighted nested stack automata (cf. Ex. 3). (3) $S = \mathrm{SC}(\mathrm{TRIV})$ where SC is obtained from NS by forbidding instructions for creating and destructing nested stacks: K-weighted stack automata (weighted version of stack automata [20]). (4) $S = \mathrm{MON}(M)$ for some monoid M (cf. Ex. 4): K-weighted M-automata (weighted version of M-automata [24]).

In future investigations we will compare formally the CS theorem for quantitative context-free languages over Σ and K [9, Thm. 2(1) ⇔ (2)] with our Theorem 11 for (P^1, Σ, K)-recognizable weighted languages.

References

1. Aho, A.V.: Indexed grammars – an extension of context-free grammars. J. ACM **15**, 647–671 (1968)
2. Aho, A.V.: Nested stack automata. JACM **16**, 383–406 (1969)
3. Chomsky, N., Schützenberger, M.P.: The algebraic theory of context-free languages. In: Computer Programming and Formal Systems, pp. 118–161. North-Holland, Amsterdam (1963)
4. Damm, W.: The IO- and OI-hierarchies. Theoret. Comput. Sci. **20**, 95–207 (1982)
5. Damm, W., Goerdt, A.: An automata-theoretical characterization of the OI-hierarchy. Inform. Control **71**, 1–32 (1986)
6. Denkinger, T.: A Chomsky-Schützenberger representation for weighted multiple context-free languages. In: The 12th International Conference on Finite-State Methods and Natural Language Processing (FSMNLP 2015) (2015). (accepted for publication)

7. Droste, M., Meinecke, I.: Describing average- and longtime-behavior by weighted MSO logics. In: Hliněný, P., Kučera, A. (eds.) MFCS 2010. LNCS, vol. 6281, pp. 537–548. Springer, Heidelberg (2010)
8. Droste, M., Meinecke, I.: Weighted automata and regular expressions over valuation monoids. Intern. J. of Found. of Comp. Science **22**(8), 1829–1844 (2011)
9. Droste, M., Vogler, H.: The Chomsky-Schützenberger theorem for quantitative context-free languages. In: Béal, M.-P., Carton, O. (eds.) DLT 2013. LNCS, vol. 7907, pp. 203–214. Springer, Heidelberg (2013)
10. Eilenberg, S.: Automata, Languages, and Machines - Volume A. Pure and Applied Mathematics, vol. 59. Academic Press (1974)
11. Engelfriet, J.: Iterated pushdown automata and complexity classes. In: Proc. of STOCS 1983, pp. 365–373. ACM, New York (1983)
12. Engelfriet, J.: Context-free grammars with storage. Technical Report 86–11, University of Leiden (1986). see also: arXiv:1408.0683 [cs.FL] (2014)
13. Engelfriet, J., Schmidt, E.M.: IO and OI.I. J. Comput. System Sci. **15**(3), 328–353 (1977)
14. Engelfriet, J., Vogler, H.: Pushdown machines for the macro tree transducer. Theoret. Comput. Sci. **42**(3), 251–368 (1986)
15. Engelfriet, J., Vogler, H.: High level tree transducers and iterated pushdown tree transducers. Acta Inform. **26**, 131–192 (1988)
16. Fischer, M.J.: Grammars with macro-like productions. Ph.D. thesis, Harvard University, Massachusetts (1968)
17. Fratani, S., Voundy, E.M.: Dyck-based characterizations of indexed languages. published on arXiv http://arxiv.org/abs/1409.6112 (March 13, 2015)
18. Ginsburg, S., Greibach, S.A.: Abstract families of languages. Memoirs of the American Math. Soc. **87**, 1–32 (1969)
19. Ginsburg, S., Greibach, S.A.: Principal AFL. J. Comput. Syst. Sci. **4**, 308–338 (1970)
20. Greibach, S.A.: Checking automata and one-way stack languages. J. Comput. System Sci. **3**, 196–217 (1969)
21. Greibach, S.A.: Full AFLs and nested iterated substitution. Inform. Control **16**, 7–35 (1970)
22. Harrison, M.A.: Introduction to Formal Language Theory, 1st edn. Addison-Wesley Longman Publishing Co., Inc, Boston (1978)
23. Hulden, M.: Parsing CFGs and PCFGs with a Chomsky-Schützenberger representation. In: Vetulani, Z. (ed.) LTC 2009. LNCS, vol. 6562, pp. 151–160. Springer, Heidelberg (2011)
24. Kambites, M.: Formal languages and groups as memory. arXiv:math/0601061v2 [math.GR] (October 19, 2007)
25. Kanazawa, M.: Multidimensional trees and a Chomsky-Schützenberger-Weir representation theorem for simple context-free tree grammars. J. Logic Computation (2014)
26. Maslov, A.N.: The hierarchy of indexed languages of an arbitrary level. Soviet Math. Dokl. **15**, 1170–1174 (1974)
27. Maslov, A.N.: Multilevel stack automata. Probl. Inform. Transm. **12**, 38–42 (1976)
28. Okhotin, A.: Non-erasing variants of the Chomsky–Schützenberger theorem. In: Yen, H.-C., Ibarra, O.H. (eds.) DLT 2012. LNCS, vol. 7410, pp. 121–129. Springer, Heidelberg (2012)
29. Salomaa, A., Soittola, M.: Automata-Theoretic Aspects of Formal Power Series. Texts and Monographs in Computer Science. Springer-Verlag (1978)
30. Scott, D.: Some definitional suggestions for automata theory. J. Comput. System Sci. **1**, 187–212 (1967)

31. Wand, M.: An algebraic formulation of the Chomsky hierarchy. In: Manes, E.G. (ed.) Category Theory Applied to Computation and Control. LNCS, vol. 25, pp. 209–213. Springer, Heidelberg (1975)
32. Weir, D.J.: Characterizing Mildly Context-Sensitive Grammar Formalisms. Ph.D. thesis, University of Pennsylvania (1988)
33. Yoshinaka, R., Kaji, Y., Seki, H.: Chomsky-Schützenberger-type characterization of multiple context-free languages. In: Dediu, A.-H., Fernau, H., Martín-Vide, C. (eds.) LATA 2010. LNCS, vol. 6031, pp. 596–607. Springer, Heidelberg (2010)

EF+EX Forest Algebras

Andreas Krebs[1]($\boxtimes$) and Howard Straubing[2]

[1] Wilhelm-Schickard-Institut at Eberhard-Karls, Universität Tübingen,
Tübingen, Germany
krebs@informatik.uni-tuebingen.de
[2] Boston College, Chestnut Hill, USA

Abstract. We examine languages of unranked forests definable using
the temporal operators EF and EX. We characterize the languages definable in this logic, and various fragments thereof, using the syntactic forest
algebras introduced by Bojanczyk and Walukiewicz. Our algebraic characterizations yield efficient algorithms for deciding when a given language
of forests is definable in this logic. The proofs are based on understanding the wreath product closures of a few small algebras, for which we
introduce a general ideal theory for forest algebras. This combines ideas
from the work of Bojanczyk and Walukiewicz for the analogous logics on
binary trees and from early work of Stiffler on wreath product of finite
semigroups.

1 Overview

Understanding the expressive power of temporal and first-order logic on trees is
important in several areas of computer science, for example in formal verification.
Using algebraic methods, in particular, finite monoids, to understand the power
of subclasses of the regular languages of finite words has proven to be extremely
successful, especially in the characterization of regular languages definable in various fragments of first-order and temporal logics ([CPP93, TW96, Str94]). Here
we are interested in sets of of finite trees (or, more precisely, sets of finite forests),
where the analogous algebraic structures are forest algebras.

Bojanczyk *et. al.* [BW08, BSW12] introduced forest algebras, and under-
scored the importance of the wreath product decomposition theory of these
algebras in the study of the expressive power of temporal and first-order logics on finite unranked trees. For languages inside of CTL the associated forest
algebras can be built completely via the wreath product of copies of the forest
algebra $\mathcal{U}_2 = (\{0, \infty\}, \{1, 0, c_0\})$, where the vertical element 0 is the constant
map to ∞, and the vertical element c_0 is the constant map to 0 ([BSW12]). The
problem of effectively characterizing the wreath product closure of $\mathcal{U}_2$ is thus an
important open problem, equivalent to characterization of CTL. Note that if one
strips away the additive structure of $\mathcal{U}_2$, the wreath product closure is the family
of all finite aperiodic semigroups (the Krohn-Rhodes Theorem). Forest algebras
have been successfully applied to the obtain characterization of other logics on
trees; see, for example [BSS12, BS09].

© Springer International Publishing Switzerland 2015
A. Maletti (Ed.): CAI 2015, LNCS 9270, pp. 128–139, 2015.
DOI: 10.1007/978-3-319-23021-4_12

Here we study in detail the wreath product closures of proper subalgebras of $\mathcal{U}_2$. In one sense, this generalizes early work of Stiffler [Sti73], who carried out an analogous program for wreath products of semigroups. Along the way, we develop the outlines of a general ideal theory for forest algebras, which we believe will be useful in subsequent work. After developing the underlying algebraic theory, we give an application to logic, obtaining a characterization of the languages of unranked forests definable with the temporal operators EF and EX.

Bojanczyk and Walukiewicz [BW06] obtained similar results for binary trees, using methods quite different from ours. Esik [Ési05] considered the analogous logics for ranked trees, and proved similar decidability results with techniques very much in the same spirit as ours, relying on a version of the wreath product tree automata acting on ranked trees.

Much of our goal in presenting these results in the context of unranked forest algebras is to develop the outlines of a general ideal theory for these algebras, and to show its connection with wreath product decompositions. We believe this approach will prove useful in subsequent work.

2 Forest Algebras

2.1 Preliminaries

We refer the reader to [BW08, BSW12] for the definitions of abstract forest algebra, free forest algebra, and syntactic forest algebra. We denote the free forest algebra over a finite alphabet A by $A^{\Delta} = (H_A, V_A)$, where H_A denotes the monoid of forests over A, with concatenation as the operation, and V_A denotes the monoid of contexts over A, with composition as the operation. A subset L of H_A is called a forest language over A. We denote its syntactic forest algebra by (H_L, V_L), and its syntactic morphism by $\mu_L : A^{\Delta} \to (H_L, V_L)$.

For the most part, our principal objects of study are not the forest algebras themselves, but homomorphisms $\alpha : A^{\Delta} \to (H, V)$. It is important to bear in mind that each such homomorphism is actually a pair of monoid homomorphisms, one mapping H_A to H and the other mapping V_A to V. It should usually be clear from the context which of the two component homomorphisms we mean, and thus we denote them both by α. The 'freeness' of A^{Δ} is the fact that a homomorphism α into (H, V) is completely determined by giving its value, in V, at each $a \in A$.

A homomorphism α as above *recognizes* a language $L \subseteq H_A$ if there exists $X \subseteq H$ such that $\alpha^{-1}(X) = L$.

If $\alpha : A^{\Delta} \to (H, V)$ and $\beta : A^{\Delta} \to (H', V')$, are homomorphisms, we say that β *factors through* α if for all $s, s' \in H_A$, $\alpha(s) = \alpha(s')$ implies $\beta(s) = \beta(s')$. This is equivalent to the existence of a homomorphism ρ from the image of α into (H', V') such that $\beta = \rho\alpha$. A homomorphism α recognizes $L \subseteq H_A$ if and only if μ_L factors through α.([BW08]).

In the course of the paper we will see several *congruences* defined on free forest algebras. Such a congruence is determined by an equivalence relation $\sim$ on H_A

such that for any $p \in V_A$, $s \sim s'$ implies $ps \sim ps'$. This gives a well-defined action of V_A on the set of $\sim$-classes of H_A. We define an equivalence relation (also denoted $\sim$) on V_A by setting $p \sim p'$ if for all $s \in H_A$, $ps \sim p's$. The result is a quotient forest algebra $(H_A/\sim, V_A/\sim)$. In order to prove that an equivalence relation $\sim$ on H_A is a congruence, it is sufficient to verify that $s \sim s'$ and $t \sim t'$ implies $s + t \sim s' + t'$ and $as \sim as'$ for all $s, s', t, t' \in H_A$ and $a \in A$.

2.2   Horizontally Idempotent and Commutative Algebras

We now introduce an important restriction. Throughout the rest of the paper, we will assume that all of our finite forest algebras (H, V) have H idempotent and commutative; that is $h + h' = h' + h$ and $h + h = h$ for all $h, h' \in H$. This is a natural restriction when talking about classes of forest algebras arising in temporal logics, which is the principal application motivating this study.

When H is horizontally idempotent and commutative, the sum of all its elements is an absorbing element for the monoid. While an absorbing element in a monoid is ordinarily written 0, since we use additive notation for H, its identity is denoted 0, and accordingly we denote the absorbing element, which is necessarily unique, by ∞.

We say that two forests $s_1, s_2 \in H_A$ are *idempotent-and-commutative equivalent* if s can be transformed into t by a sequence of operations of the following three types: *(i)* interchange the order of two adjacent subtrees (that is, if $s = p(t_1 + t_2)$ for some context p and trees t_1, t_2, then we transform s to $p(t_2 + t_1)$); *(ii)* replace a subtree t by two adjacent copies (that is, transform pt to $p(t + t)$); *(iii)* replace two identical adjacent subtrees by a single copy (transform $p(t + t)$ to pt). Since operations *(ii)* and *(iii)* are inverses of one another, and operation *(i)* is its own inverse, this is indeed an equivalence relation.

We have the following obvious lemma:

Lemma 1. *Let $\alpha : A^\triangle \to (H, V)$ be a homomorphism, where H is horizontally idempotent and commutative. If $s, t \in H_A$ are idempotent-and-commutative equivalent, then $\alpha(s) = \alpha(t)$.*

There is a smallest nontrivial idempotent and commutative forest algebra, $\mathcal{U}_1 = (\{0, \infty\}, \{1, 0\})$. The horizontal and vertical monoids of $\mathcal{U}_1$ are isomorphic, but we use different names for the elements because of the additive notation for the operation in one of these monoids, and multiplicative notation in the other. We have not completely specified how the vertical monoid acts on the horizontal monoid—this is done by setting $0 \cdot x = \infty$ for $x \in \{0, \infty\}$.

2.3   1-Definiteness

In Section 5 we will discuss in detail the notion of definiteness in forest algebras; for this preliminary section, we will only need to consider a special case. A forest algebra homomorphism $\alpha : A^\triangle \to (H, V)$ is said to be *1-definite* if for $s \in H_A$, the value of $\alpha(s)$ depends only on the set of labels of the root nodes of s.

We define an equivalence relation $\sim_1$ on H_A by setting $s \sim_1 s'$ if and only if the sets of labels of root nodes of s and s' are equal. This defines a congruence on A^Δ. We denote the homomorphism from A^Δ onto the quotient under $\sim_1$ by $\alpha_{A,1}$. It is easy to show that a homomorphism $\alpha : A^\Delta \to (H, V)$ is 1-definite if and only if it factors through $\alpha_{A,1}$.

2.4 Wreath Products

We summarize the discussion of wreath products given in [BSW12]. The *wreath product* of two forest algebras $(H_1, V_1), (H_2, V_2)$ is $(H_1, V_1) \circ (H_2, V_2) = (H_1 \times H_2, V_1 \times V_2^{H_1})$, where the monoid structure of $H_1 \times H_2$ is the ordinary direct product, and the action is given by $(v_1, f)(h_1, h_2) = (v_1 h_1, f(h_1) h_2)$, for all $h_1 \in H_1$, $h_2 \in H_2$, $v_1 \in V_1$, and $f : H_1 \to V_2$. It is straightforward to verify that the resulting structure satisfies the axioms for a forest algebra. Note that if one forgets about the monoid structure on H_1 and H_2, this is just the ordinary wreath product of left transformation monoids. Because we use left actions rather than the right actions that are traditional in the study of monoid decompositions, we reverse the usual order of the factors. The projection maps $\pi : (h_1, h_2) \mapsto h_1, (v, f) \mapsto v$, define a homomorphism from the wreath product onto the left-hand factor.

2.5 Reachability

Let (H, V) be a finite forest algebra. For $h, h' \in H$ we write $h \le h'$ if $h = vh'$ for some $v \in V$, and say that h is *reachable* from h'. This gives a preorder on H. We set $h \cong h'$ if both $h \le h'$ and $h' \le h$. An equivalence class of $\cong$ is called a *reachability class*. The preorder consequently results in a partial order on the set of reachability classes of H. We always have $h + h' \le h$, because $h + h' = (1 + h')h$. If $h \in H$ and Γ is a reachability class of H then we write, for example, $h \ge \Gamma$ to mean that $\Gamma \le \Gamma'$, where Γ' is the class of h.

A *reachability ideal* in (H, V) is a subset I of H such that $h \in I$ and $h' \le h$ implies $h' \in I$. If we have a homomorphism $\alpha : A^\Delta \to (H, V)$ and a reachability ideal $I \subseteq H$, we define an equivalence relation $\sim_I$ on H_A by setting $s \sim_I s'$ if $\alpha(s) = \alpha(s') \notin I$, or if $\alpha(s), \alpha(s') \in I$. Easily $s \sim_I s'$ implies $ps \sim_I ps'$ for any $p \in V_A$. We thus obtain a homomorphism onto the quotient algebra $\alpha_I : A^\Delta \to (H/\sim_I, V/\sim_I)$ which factors through α. Note that I is, in particular, a two-sided ideal in the monoid H, and $H/\sim_I$ is identical to the usual quotient monoid $H/I = (H - I) \cup \{\infty\}$. We will thus use the notation $(H/I, V/I)$ for the quotient algebra, instead of $(H/\sim_I, V/\sim_I)$. If $\Gamma \subseteq H$ is a reachability class, then both $I_\Gamma = \{h \in H : h \not\ge \Gamma\}$ and $I_{\ge \Gamma} = \{h \in H : h \not\ge \Gamma\}$ are reachability ideals. We denote the associated quotients and projection homomorphisms by (H_Γ, V_Γ), α_Γ, $(H_{\ge \Gamma}, V_{\ge \Gamma})$, $\alpha_{\ge \Gamma}$.

Given the restriction that H is idempotent and commutative, the absorbing element ∞ is reachable from every element. The reachability class of ∞ is accordingly the unique minimal class, which we denote $\Gamma_{\min}$. A reachability class

Γ is *subminimal* if $\Gamma_{\min} < \Gamma$, but there is no class Λ with $\Gamma_{\min} < \Lambda < \Gamma$. The following lemma will be used several times.

Lemma 2. *Let* $\alpha : A^\Delta \to (H, V)$, *and let* $\Gamma_1, \ldots, \Gamma_r$ *be the subminimal reachability classes of* (H, V). *Then*

$$\alpha_{\Gamma_{\min}} : A^\Delta \to (H_{\Gamma_{\min}}, V_{\Gamma_{\min}})$$

factors through the direct product

$$\left(\prod_{j=1}^r \alpha_{\geq \Gamma_j} \right) : A^\Delta \to \prod_{j=1}^r (H_{\geq \Gamma_j}, V_{\geq \Gamma_j}).$$

Further each of the algebras $(H_{\geq \Gamma_j}, V_{\geq \Gamma_j})$ *has a unique subminimal reachability class.*

We will also need the following lemma, which concerns the behavior of reachability classes under homomorphisms.

Lemma 3. *Let* $\beta : (H_1, V_1) \to (H_2, V_2)$ *be a homomorphism of finite forest algebras. Let* $\Lambda \subseteq H_1$ *be a reachability class. There is a reachability class* Γ *of* (H_2, V_2) *such that* $\beta(\Lambda) \subseteq \Gamma$. *If* Λ *is a minimal class of* (H_1, V_1) *satisfying* $\beta(\Lambda) \subseteq \Gamma$, *and* β *is onto, then* $\beta(\Lambda) = \Gamma$. *If, further,* H_2 *is idempotent and commutative, then there is only one such minimal class* Λ.

3   Connections to Logic

For the definition of temporal logic and especially the temporal operators EF and EX we refer to [BSW12] as our approach closely follows the one given there.

Intuitively, when we interpret formulas in trees, $\mathsf{EF}\phi$ means 'at some time in the future ϕ' and $\mathsf{EX}\phi$ means 'at some next time ϕ'. When we interpret such formulas in forests, we are in a sense treating the forest as though it were a tree with a phantom root node. Observe that if $a \in A$, we do not interpret the formula a in forests at all. Thus a formula can have different interpretations depending on whether we view it as a tree or a forest formula. For example, as a forest formula $\mathsf{EX}a$ means 'there is a root node labeled a' while as a tree formula it means 'some child of the root is labeled a'. If ϕ is a forest formula, then we denote by L_ϕ the set of all $s \in H_A$ such that $s \models \phi$. L_ϕ is the *language defined by* ϕ.

Example 4. Consider the following property of forests over $\{a, b\}$: There is a tree component containing only as, and another tree component that contains at least one b. Now consider the set L of forests s that either have this property, or in which for some node x, the forest of strict descendants of x has the property. The property itself is defined by the forest formula

$$\psi : \mathsf{EX}(a \wedge \neg \mathsf{EF}b) \wedge \mathsf{EX}(b \vee \mathsf{EF}b)$$

and L is defined by $\psi \vee \mathsf{EF}\psi$. In Example 9, we discuss the syntactic forest algebra of L.

3.1   Correspondence of Operators with Wreath Products

The principal result of this paper is the algebraic characterization of the forest languages using the operators EF and EX, either separately or in combination. It will require some algebraic preparation, in Sections 4, 5 and 6 before we can give the precise statement of this theorem. The bridge between the logic and the algebra is provided by the next two propositions.

Let ϕ be a tree formula. Then ϕ can be written as a disjunction $\bigvee_{a \in A}(a \wedge \psi_a)$, where each ψ_a is a forest formula. Let $\Psi = \{\psi_a : a \in A\}$. We'll call Ψ the set of forest formulas of ϕ. We say that a homomorphism $\beta : A^\Delta \to (H, V)$ recognizes Ψ if the value of $\beta(s)$ determines exactly which formulas of Ψ are satisfied by s. To construct such a homomorphism, we can take the direct product of the syntactic algebras of L_ψ for $\psi \in \Psi$, and set β to be the product of the syntactic morphisms.

The following theorem, adapted from [BSW12], gives the connection between the EF operator and wreath products with $\mathcal{U}_1$:

Proposition 5. *(a) Suppose that ϕ is a tree formula, Ψ is the set of forest formulas of ϕ, and that Ψ is recognized by $\alpha : A^\Delta \to (H, V)$. Then $\mathsf{EF}\phi$ is recognized by a homomorphism $\beta : A^\Delta \to (H, V) \circ \mathcal{U}_1$, where $\pi\beta = \alpha$.*
(b) Suppose that $L \subseteq H_A$ is recognized by a homomorphism $\beta : A^\Delta \to (H, V) \circ \mathcal{U}_1$. Then L is a boolean combination of languages of the form $\mathsf{EF}(a \wedge \phi)$, where L_ϕ is recognized by $\pi\beta$.

Here we prove an analogous result for the temporal operator EX.

Proposition 6. *(a) Suppose that ϕ is a tree formula, Ψ is the set of forest formulas of ϕ, and that Ψ is recognized by $\alpha : A^\Delta \to (H, V)$. Then $\mathsf{EX}\phi$ is recognized by a homomorphism $\alpha \otimes \beta : A^\Delta \to (H, V) \circ (H', V')$, where $\beta : (A \times H)^\Delta \to (H', V')$ is 1-definite.*
(b) Suppose that $L \subseteq H_A$ is recognized by a homomorphism $\alpha \otimes \beta : A^\Delta \to (H, V) \circ (H', V')$, Suppose further that every language recognized by α is defined by a formula in some set Ψ of formulas. If $\beta : (A \times H)^\Delta \to (H', V')$ is 1-definite, then L is a boolean combination of languages of the form L_ψ and $\mathsf{EX}(a \wedge \psi)$, where $\psi \in \Psi$.

4   EF-algebras

Following [BSW12], we define:

Definition 7. *A finite forest algebra (H, V) is an EF-algebra if it satisfies the identities $h + h' = h' + h$, $vh + h = vh$ for all $h, h' \in H$ and $v \in V$. The second identity with $v = 1$ gives $h + h = h$. Thus every EF-algebra is horizontally idempotent and commutative.*

The following result is proved in [BSW12], and is the key element in the characterization of languages definable in one of the temporal logics we consider in Section 3. We will give a new proof, as it provides a good first illustration of how we use the reachability ideal theory introduced above in decomposition arguments.

Theorem 8. *Let* $\alpha : A^\Delta \rightarrow (H, V)$ *be a homomorphism onto a forest algebra.* (H, V) *is an* EF*-algebra if and only if* α *factors through a homomorphism* $\beta :$ $A^\Delta \rightarrow \mathcal{U}_1 \circ \cdots \circ \mathcal{U}_1.$

A classic result of Stiffler [Sti73] shows that a right transformation monoid (Q, M) divides an iterated wreath product of copies of the transformation monoid $U_1 = (\{0, 1\}, \{0, 1\})$ if and only if M is $\mathcal{R}$-trivial. In terms of transformation monoids this means there is no pair of distinct states $q \neq q' \in Q$ such that $qm = q', q'm' = q$ for some $m, m' \in M$. Since forest algebras are left transformation monoids, the analogous result would suggest that a forest algebra (H, V) divides an iterated wreath product of copies of $\mathcal{U}_1$ if and only if V is $\mathcal{L}$-trivial—that is, if and only if (H, V) has trivial reachability classes. We have already seen that this condition is necessary.

However, the following example shows that it is not sufficient.

Example 9. Figure 1 below defines the syntactic forest algebra of the language L of Example 4. The nodes in the diagram represent the elements of the horizontal monoid, and the arrows give the action of a generating set of letters $A = \{a, b\}$ on the horizontal monoid. The letter transitions, together with the conventions about idempotence and commutativity, and the meaning of 0 and ∞, completely determine the addition and the action. Since $\infty = a + b = a + ba \neq ba = b$, this is not an EF-algebra, but the reachability classes are singletons.

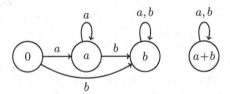

Fig. 1. An algebra with trivial reachability classes that is not an EF-algebra

5   Definiteness

5.1   Definite Homomorphisms

Let $k > 0$. A finite semigroup S is said to be *reverse k-definite* if it satisfies the identity $x_1 x_2 \cdots x_k y = x_1 \cdots x_k$. The reason for the word 'reverse' is that definiteness of semigroups was originally formulated in terms of right transformation monoids, so the natural analogue of definiteness in the setting of forest algebras corresponds to reverse definiteness in semigroups. Observe that the notions of definiteness and reverse definiteness in semigroups do not really make sense for monoids, since only the trivial monoid can satisfy the underlying identities. For much the same reason, we define definiteness for forest algebras not as a property of the algebras themselves, but of homomorphisms $\alpha : A^\Delta \to (H, V)$.

The *depth* of a context $p \in V_A$ is defined to be the depth of its hole; so for instance a context with its hole at a root node has depth 0. We say that the homomorphism α is k-definite, where $k > 0$, if for every $p \in V_A$ of depth at least k, and for all $s, s' \in H_A$, $\alpha(ps) = \alpha(ps')$. Easily, if α_1, α_2 are k-definite homomorphisms, then so are $\alpha_1 \times \alpha_2$ and $\psi\alpha_1$, where $\psi : (H, V) \to (H', V')$ is a homomorphism of forest algebras.

A context is *guarded* if it has depth at least 1, that is, if the hole is not at the root. We denote by V_A^{gu} the subsemigroup of V_A consisting of the guarded contexts.

Lemma 10. *Let $k > 0$. A homomorphism $\alpha : A^\Delta \to (H, V)$ is k-definite if and only if $\alpha(V_A^{gu})$ is a reverse k-definite semigroup.*

Definition 11. *An EX-homomorphism is a homomorphism that is k-definite for some $k \in \mathbb{N}$.*

5.2   Free k-definite Algebra

We construct what we will call *free k-definite algebra* over an alphabet A. This is a slight abuse of terminology, since as we noted above, it is the homomorphism into this algebra, and not the algebra itself, that is k-definite. We do this by recursively defining a sequence of congruences $\sim_k$ on A^Δ. If $k = 0$, then $\sim_0$ is just the trivial congruence that identifies all forests. If $k \geq 0$ and $\sim_k$ ha been defined then we associate to each forest $s = a_1 s_1 + \cdots a_r s_r$, where each $a_i \in A$, $s_i \in H_A$, the set

$$T_s^{k+1} = \{(a_i, [s_i]_{\sim_k}) : 1 \leq i \leq r\},$$

where $[]_{\sim_k}$ denotes the $\sim_k$-class of a forest. We then define $s \sim_{k+1} s'$ if and only if $T_s^{k+1} = T_{s'}^{k+1}$.

Proposition 12. *Let $k \geq 0$. Then $\sim_{k+1}$ refines $\sim_k$. $\sim_k$ is a congruence of finite index on A^Δ, with a horizontally idempotent and commutative quotient.*

Intuitively, $s \sim_k s'$ means that the forests s and s' are identical at the k levels closest to the root, up to idempotent and commutative equivalence. In fact, this intuition provides an equivalent characterization of $\sim_k$, which we give below. We omit the simple proof.

Lemma 13. *Let $s, s' \in H_A$ and $k > 0$. Let $\bar{s}, \bar{s}'$, denote, respectively, the forests obtained from s and s' by removing all the nodes at depth k or more. Then $s \sim_k s'$ if and only if $\bar{s}$ and $\bar{s}'$ are idempotent-and-commutative equivalent.*

Let us denote by $\alpha_{A,k}$ the homomorphism from A^Δ onto its quotient by $\sim_k$. In the case where $k = 1$, we will identify $H_A/\sim_1$ with the monoid $(\mathcal{P}(A), \cup)$, and the horizontal component of $\alpha_{A,1}$ with the map that sends each forest to the set of its root nodes.

The following theorem gives both the precise sense in which this is the 'free k-definite forest algebra', as well as the wreath product decomposition of k-definite homomorphisms into 1-definite homomorphisms into a forest algebra with horizontal monoid $\{0, \infty\}$.

Theorem 14. *Let $\alpha : A^\Delta \to (H, V)$ be a homomorphism onto a finite forest algebra. Let $k > 0$. The following are equivalent.*

(a) α is k-definite.
(b) α factors through $\alpha_{A,k}$.
(c) α factors through

$$\beta_1 \otimes \cdots \otimes \beta_k : A^\Delta \to (H_1, V_1) \circ \cdots \circ (H_k, V_k),$$

where each $\beta_i : (A \times H_1 \times \cdots \times H_{i-1})^\Delta \to (H_i, V_i)$ is 1-definite.
(d) α factors through an iterated wreath product of k 1-definite homomorphisms into $\mathcal{U}_2$.

6 (EF, EX)-algebras

6.1 The Principal Result

Definition 15. *An (EF, EX)-homomorphism $\alpha : A^\Delta \to (H, V)$ is one that factors through an iterated wreath product*

$$\beta_1 \otimes \cdots \otimes \beta_k,$$

where each β_i either maps into $\mathcal{U}_1$ or is 1-definite. By Theorem 14 we can suppose that each 1-definite β_i maps into $\mathcal{U}_2$.

The principal result of this paper is an effective necessary and sufficient condition for a homomorphism to be a (EF, EX)-homomorphism.

Definition 16. *Suppose $\alpha : A^\Delta \to (H, V)$. Let $s_1, s_2 \in H_A$, $k > 0$, and $\Gamma \subseteq H$ a reachability class for (H, V). We say that s_1, s_2 are (α, k, Γ)-confused, and write $s_1 \equiv_{\alpha,k,\Gamma} s_2$, if*

$$(s_1)^{\alpha_\Gamma} \sim_k (s_2)^{\alpha_\Gamma}, \quad \alpha(s_1), \alpha(s_2) \in \Gamma.$$

Observe that the equivalence relation $\sim_k$ in the first item is over the extended alphabet $A \times H_\Gamma$. It is worth emphasizing what $(s)^{\alpha_\Gamma}$ is when $\alpha(s) \in \Gamma$: We are tagging each node of x of s with the value $\alpha(t) \in H$ if the tree rooted at x is at and $\alpha(t) > \Gamma$, but we are tagging the node by ∞–effectively leaving the node untagged–if $\alpha(t) \in \Gamma$. Since $\alpha(s) \in \Gamma$, every node is of one of these two types.

Definition 17. *A homomorphism α is nonconfusing if and only if there exists $k > 0$ such that $\equiv_{\alpha,k,\Gamma}$ is equality for reachability classes Γ.*

In the full version of the paper [KS14] we show that it can be effectively determined if a forest algebra morphism is nonconfusing.

It follows from Proposition 12 that $\equiv_{\alpha,k+1,\Gamma}$ refines $\equiv_{\alpha,k,\Gamma}$, so that if α is nonconfusing with associated parameter k, then it is nonconfusing for all $m > k$.

Our main result is:

Theorem 18. *Let* $\alpha : A^\Delta \to (H, V)$ *be a homomorphism into a finite forest algebra. Then* α *is a* (EF, EX) *homomorphism if and only if it is nonconfusing.*

The proof of Theorem 18 will be given in the next two subsections.

Example 19. Consider once again the algebra of Examples 4 and 9 and the associated homomorphism α from $\{a, b\}^\Delta$. Since the algebra has trivial reachability classes, α is nonconfusing for all k, so Theorem 18 implies that α is an (EF, EX)-homomorphism. We will see in the course of the proof of the main theorem how the wreath product decomposition is obtained.

Example 20. Consider again the forest algebra $\mathcal{U}_2 = (\{0, \infty\}, \{1, c_\infty, c_0\})$, and the homomorphism α from $\{a, b, c\}^\Delta$ onto $\mathcal{U}_2$ that maps a to 1, b to c_0 and c to c_∞. There is a unique reachability class Γ, so for any forest s, $s^{\alpha\Gamma}$ is identical to s. Now observe that $a^k b \sim_k a^k c$, but that these are mapped to different elements under α. So by our main theorem, α is not an (EF, EX)-homomorphism.

6.2   Sufficiency of the Condition

We will use the ideal theory developed earlier to prove that every nonconfusing homomorphism factors through a wreath product decomposition of the required kind. The structure of our proof resembles the one given for Theorem 8. Once again, we proceed by induction on $|H|$. The base of the induction is the trivial case $|H| = 1$. Let us suppose that $\alpha : A^\Delta \to (H, V)$ is nonconfusing with parameter k, that $|H| > 1$, and that every nonconfusing homomorphism into a forest algebra with a smaller horizontal monoid factors through a wreath product of the required kind.

Let $\Gamma = \Gamma_{\min}$. Suppose first that $|\Gamma| > 1$. We claim that α factors through

$$\beta = \alpha_\Gamma \otimes \alpha_{B,k} : A^\Delta \to (H_\Gamma, V_\Gamma) \circ B^\Delta / \sim_k$$

where $B = A \times H_\Gamma$. Since $|H_\Gamma| < |H|$ and α_Γ is also nonconfusing, the induction hypothesis gives the desired decomposition of α. To establish the claim, let $s \in H_A$. Then

$$\beta(s) = (\alpha_\Gamma(s), [s^{\alpha\Gamma}]_{\sim_k}).$$

If $s \notin \Gamma$, then the value of the left-hand coordinate determines $\alpha(s)$. If $s \in \Gamma$, then by the nonconfusion condition, the value of the right-hand coordinate determines $\alpha(s)$. Thus α factors through β as required.

So let $|\Gamma| = 1$. Then $\Gamma = \{\infty\}$ and $(H_\Gamma, V_\Gamma) = (H, V)$. Lemma 2 implies that we can suppose (H, V) has a single subminimal reachability class, because each of the component homomorphisms in the direct product is nonconfusing, and the direct product factors through the wreath product.

Thus we have a unique minimal element ∞, and a unique subminimal ideal Γ'. We claim that α factors through

$$\beta = \alpha_1 \otimes \alpha_2 \otimes \alpha_3 : A^\Delta \to (H_{\Gamma'}, V_{\Gamma'}) \circ B^\Delta/\!\sim_k \circ\, \mathcal{U}_1,$$

where $\alpha_1 = \alpha_{\Gamma'}$ and $\alpha_2 = \alpha_{B,k}$, where $B = A \times H_{\Gamma'}$, and $\alpha_3 : (B \times 2^B)^\Delta \to \mathcal{U}_1$ will be defined below. To see how α_3 should be defined, let us consider what this homomorphism needs to tell us. If $\alpha(s) > \Gamma'$, then the first coordinate of $\beta(s)$ determines $\alpha(s)$. If $\alpha(s) \in \Gamma'$, then the first two components of $\beta(s)$ determine $\alpha(s)$, by nonconfusion. So we will use the third component to distinguish between $\alpha(s) \in \Gamma'$ and $\alpha(s) = \infty$. The value of the first component already determines whether or not $\alpha(s) \in \Gamma' \cup \{\infty\}$, so we really just need to be able to tell when $\alpha(s) = \infty$. There are several cases to consider, depending on whether or not s contains a tree t such that $\alpha(t) = \infty$. If not, then $s = t_1 + \cdots + t_r$, where $\alpha(t_i) \geq \Gamma'$ for all i. Observe that if this is the case, then the set of values $\{\alpha(t_1), \cdots, \alpha(t_r)\}$ is determined by the second component $\{[t_1^{\alpha_1}]_{\sim_k}, \ldots, [t_r^{\alpha_1}]_{\sim_k}\}$ of $\beta(s)$. If s contains a tree t such that $\alpha(t) = \infty$, pick such a tree at maximal depth. Then $t = a(t_1 + \cdots + t_r)$, where once again $\alpha(t_i) \geq \Gamma'$ for all i, and the set of values $\{\alpha(t_1), \cdots, \alpha(t_r)\}$ is determined by the second component of $\beta(s)$. We now specify the value of $\alpha_3(a, h, Q)$. As remarked above, Q determines a set of values all in Γ' or strictly higher. Let $h_Q \in H_A$ be the sum of these values. If either $h_Q = \infty$, or $ah_Q = \infty$, set $\alpha_3(a, h, Q) = 0$. Otherwise, $\alpha_3(a, h, q) = 1$.

The third component of $\beta(s)$ will be ∞ if and only if there is some subtree $a(t_1 + \cdots + t_r)$ such that

$$\alpha_3(a, \alpha_1(t_1 + \cdots + t_r), \{[t_1^{\alpha_1}]_{\sim_k}, \ldots, [t_r^{\alpha_1}]_{\sim_k}\}) = 0.$$

If we pick the subtree of maximal depth at which this occurs, then as argued above, $\alpha(s) = \infty$. The only other way we can have $\alpha(s) = \infty$ is if there is no such subtree, but $s = t_1 + \cdots + t_r$ where each $\alpha(t_i) \geq \Gamma'$ and the sum of these values is ∞. In this case, the fact that no such subtree exists is determined by the third coordinate of $\beta(s)$ being 1, and the set of $\alpha(t_i) \geq \Gamma'$ is determined by the second coordinate of $\beta(s)$. So in all cases $\beta(s)$ determines $\alpha(s)$.

6.3   Necessity of the Condition

To prove the converse, we have to show preservation of nonconfusion under quotients and wreath products with the allowable factors. This is carried out in the following three lemmas. Preservation under quotients (Lemma 21) is the most difficult of the three to show.

Lemma 21. *Let $\alpha : A^\Delta \to (H_1, V_1), \beta : A^\Delta \to (H_2, V_2)$, be homomorphisms onto finite forest algebras such that β factors through α. If α is nonconfusing then so is β.*

Lemma 22. *Suppose that $\alpha : A^\Delta \to (H, V) \circ \mathcal{U}_1$ is a homomorphism, and that $\beta = \pi\alpha$, where π is the projection morphism onto (H, V), is nonconfusing. Then α is nonconfusing.*

Lemma 23. *Suppose that $\alpha = \beta \otimes \gamma : A^\Delta \to (H, V) \circ (H', V')$ is a homomorphism, that β is nonconfusing, and that $\gamma : (A \times H)^\Delta \to (H', V')$ is 1-definite. Then α is nonconfusing.*

7 Results

Using the wreath product characterizations of EF-algebras, EX-homomorphisms, and (EF, EX)-homomorphisms of the previous three sections, we get:

Theorem 24. *Let A be a finite alphabet, and let $L \subseteq H_A$.*

(a) L is defined by an EF-formula if and only if (H_L, V_L) is an EF-algebra.

(b) L is defined by an EX-formula if and only if μ_L is an EX-homomorphism.

(c) L is defined by an EF + EX-formula if and only if
μ_L is an (EF, EX)-homomorphism.

(d) There are effective procedures for determining, given a finite tree automaton recognizing L, whether L is definable by an EF-, EX-, or EF + EX-formula, and for producing a defining formula in case one exists.

References

[BS09] Benedikt, M., Segoufin, L.: Regular tree languages definable in FO and in FO$_{mod}$. ACM Trans. Comput. Log. **11**(1) (2009)

[BSS12] Bojanczyk, M., Segoufin, L., Straubing, H.: Piecewise testable tree languages. Logical Methods in Computer Science **8**(3) (2012)

[BSW12] Bojanczyk, M., Straubing, H., Walukiewicz, I.: Wreath products of forest algebras with applications to tree logics. Logical Methods in Computer Science **8**(3) (2012)

[BW06] Bojanczyk, M., Walukiewicz, I.: Characterizing EF and EX tree logics. Theor. Comput. Sci. **358**(2–3), 255–272 (2006)

[BW08] Bojanczyk, M., Walukiewicz, I.: Forest algebras. In: Flum, J., Grädel, E., Wilke, T. (eds.) Logic and Automata. Texts in Logic and Games, vol. 2, pp. 107–132. Amsterdam University Press (2008)

[CPP93] Cohen, J., Perrin, D., Pin, J.-E.: On the expressive power of temporal logic. J. Comput. Syst. Sci. **46**(3), 271–294 (1993)

[Ési05] Ésik, Z.: An algebraic characterization of the expressive power of temporal logics on finite trees. In: 1st Int. Conf. Algebraic Informatics. Aristotle Univ. of Thessaloniki, pp. 53–110 (2005)

[KS14] Krebs, A., Straubing, H.: EF+EX forest algebras. CoRR, abs/1408.0809 (2014)

[Sti73] Stiffler, P.E.: Extension of the fundamental theorem of finite semigroups. Advances in Mathematics **11**(2), 159–209 (1973)

[Str94] Straubing, H.: Finite Automata, Formal Logic, and Circuit Complexity. Birkhäuser, Boston (1994)

[TW96] Thérien, D., Wilke, T.: Temporal logic and semidirect products: An effective characterization of the until hierarchy. In: FOCS, pp. 256–263. IEEE Computer Society (1996)

On Near Prime-Order Elliptic Curves
with Small Embedding Degrees

Duc-Phong Le[1]([✉]), Nadia El Mrabet[2], and Chik How Tan[1]

[1] Temasek Laboratories, National University of Singapore, Singapore, Singapore
{tslld,tsltch}@nus.edu.sg
[2] SAS team CMP, Ecole des Mines de St Etienne LIASD,
University Paris 8, Saint-Denis, France
nadia.el-mrabet@emse.fr

Abstract. In this paper, we extend the method of Scott and Barreto and present an *explicit* and *simple* algorithm to generate families of generalized MNT elliptic curves. Our algorithm allows us to obtain *all* families of generalized MNT curves with any given cofactor. Then, we analyze the complex multiplication equations of these families of curves and transform them into generalized Pell equations. As an example, we describe a way to generate Edwards curves with embedding degree 6, that is, elliptic curves having cofactor $h = 4$.

Keywords: Pairing friendly elliptic curve · MNT curves · Complex multiplication · Pell's equation

1 Introduction

Pairings used in cryptology are efficiently *computable* bilinear maps on torsion subgroups of points on an elliptic curve that map into the multiplicative group of a finite field. We call such a map a *cryptographic pairing*. The first notable application of pairings to cryptology was the work of Menezes, Okamato and Vanstone [15]. They showed that the discrete logarithm problem on a supersingular elliptic curve can be reduced to the discrete logarithm problem in a finite field through the Weil pairing. Then, Frey and Ruck [8] also consider this through the Tate pairing. Pairings were thus used as a means of attacking cryptosystems.

However, pairings on elliptic curves only become a great interest since their first application in constructing cryptographic protocols in [12]. Joux describes an one-round 3-party Diffie-Hellman key exchange protocol in 2000. Since then, the use of cryptographic protocols based on pairings has had a huge success with some notable breakthroughs such as practical Identity-based Encryption (IBE) schemes [5]. Unlike standard elliptic curve cryptosystems, pairing-based cryptosystems require elliptic curves with *special* properties, namely, the embedding

N. El Mrabet—This work was supported in part by the French ANR-12-INSE-0014 SIMPATIC Project.

A. Maletti (Ed.): CAI 2015, LNCS 9270, pp. 140–151, 2015.
DOI: 10.1007/978-3-319-23021-4_13

degree k is small enough[1]. Balasubramanian and Koblitz [2] showed that ordinary elliptic curves with such a property are *very rare*. An elliptic curve with such nice properties is called a *pairing-friendly* elliptic curve.

Miyaji, Nakabayashi and Takano introduced the concept of "family of pairing-friendly elliptic curves" in [16]. They provided families of *prime-order* elliptic curves with embedding degrees $k = 3, 4$ and 6, such that the number of points on these curves $E(\mathbb{F}_q)$ are prime. As analyzed in [17], these families of curves, so-called MNT curves, are more efficient than supersingular elliptic curves when implementing pairing-based cryptosystems. Later, Scott and Barreto [18], and Galbraith *et al.* [9] extended and introduced more MNT curves. These curves are of *near prime-order*. The number of points on these curves is $\#E(\mathbb{F}_q) = h \cdot r$, where r is a big prime number and the cofactor $h \geq 2$ is small. While Galbraith *et al.*'s method allows generating explicit families of curves, Scott-Barreto's method only generates particular elliptic curves.

In this paper we extend the method of Scott and Barreto in [18] and present an explicit, simple algorithm to generate families of ordinary elliptic curves of prime order (or near prime order with any cofactor) with small embedding degrees. Given an embedding degree k and a cofactor h, we demonstrate that our algorithm will output *all* possible families. We then point out a one-to-one correspondence between families of MNT curves having the same embedding degree and the same cofactor (Theorems 2, 3, and 4). We also analyze the complex multiplication equations of these families of curves and show how to transform these complex multiplication equations into generalized Pell equations that allow us to find particular curves. We illustrate our analysis for constructing Edwards curves with embedding degree 6.

The paper is organized as follows: Section 2 briefly recalls MNT curves, as well as methods to generate MNT curves with small cofactors. Section 3 presents our alternative method to generate such curves. We give our results in Section 4. We also discuss the Pell equation for some particular cases of MNT curves in this section. Finally, we conclude in Section 5.

2 Backgrounds

2.1 MNT Curves

An elliptic curve generated randomly would have a large embedding degree. As a consequence, a random elliptic curve would not be suitable for efficient computation of a pairing based protocol. Supersingular elliptic curves have small embedding degree. However, such curves are limited to embedding degree $k = 2$ for prime fields and $k \leq 6$ in general [15]. If we want to vary the embedding degree to achieve a high security level, we must construct *pairing-friendly ordinary elliptic curves*. However, a study by Balasubramanian and Koblitz in [2] showed

---

[1] Let q be a prime number or a power of a prime, let E be an elliptic curve defined over $\mathbb{F}_q$ with a subgroup of prime order r. Then the embedding degree is the smallest integer such that r divides $(q^k - 1)$.

that ordinary elliptic curves with such a small embedding degree are *very rare* and thus require specific constructions.

Using the Complex Multiplication method (CM for short) to construct elliptic curves, the ρ value satisfies that $1 \leq \rho \leq 2$, where the value ρ is defined as $\rho = \frac{\log(q)}{\log(r)}$. In order to save bandwidth during the calculation we are looking for ρ as small as possible. The most interesting construction of pairing-friendly elliptic curves is the one such that the result is a parameterization of a family of elliptic curves. Miyaji, Nakabayashi, and Takano [16] presented the first parameterized families that yield ordinary elliptic curves with embedding degree $k \in \{3, 4, 6\}$. These curves have a ρ-value equal to 1. The families are given by parameterization for q and t as polynomials in $\mathbb{Z}[x]$ with $\#E(\mathbb{F}_q) = n(x)$. Let $\Phi_k(x)$ be the k-th *cyclotomic polynomial*. Recall that $n(x) = q(x) + 1 - t(x)$, $n(x) \mid \Phi_k(q(x))$, and $n(x)$ represents primes in the MNT construction. Their results are summarized in Table 1.

Table 1. Parameters for MNT curves [16]

k	$q(x)$	$t(x)$
3	$12x^2 - 1$	$-1 \pm 6x$
4	$x^2 + x + 1$	$-x$ or $x + 1$
6	$4x^2 + 1$	$1 \pm 2x$

The construction of MNT curves is based on the Complex Multiplication method. That is, we have to find solutions (x_0, V_0) of the following CM equation:

$$DV^2 = 4q(x) - t^2(x)$$

for small values of D. The right-hand side of this equation is of quadratic form and can be transformed into a generalized Pell equation. Since the construction depends on solving a Pell-like equation, MNT curves of prime order are *sparse* [7]. It means that the equation admits only a few solutions.

2.2   MNT Curves with Small Cofactors

Let $E(\mathbb{F}_q)$ be a parameterized elliptic curve with cardinality $\#E(\mathbb{F}_q) = n(x)$. We call the cofactor of $E(\mathbb{F}_q)$, the integer h such that $n(x) = h \times r(x)$, where $r(x)$ is a polynomial representing primes. The original construction of MNT curves gives families of elliptic curves with cofactor $h = 1$. Scott-Barreto [18], and Galbraith-McKee-Valença [9] extended the MNT idea by allowing small values of the cofactor $h > 1$. This allows to find many more suitable curves with $\rho \approx 1$ than the original MNT construction. We recall the following proposition.

Proposition 1. *[7, Proposition2.4] Let k be a positive integer, $E(\mathbb{F}_q)$ be an elliptic curve defined over $\mathbb{F}_q$ with $\#E(\mathbb{F}_q) = q + 1 - t = hr$, where r is prime, and let t be the trace of $E(\mathbb{F}_q)$. Assume that $r \nmid kq$. Then $E(\mathbb{F}_q)$ has embedding degree k with respect to r if and only if $\Phi_k(q) \equiv 0 \pmod{r}$, or equivalently, if and only if $\Phi_k(t-1) \equiv 0 \pmod{r}$.*

Scott-Barreto's Method. Let $\Phi_k(x) = d \times r$ for some x. Scott-Barreto's method [18] first fixes small integers h and d and then substitutes $r = \Phi_k(t-1)/d$, where $t = x + 1$ to obtain the following CM equation:

$$DV^2 = 4h\frac{\Phi_k(x)}{d} - (x-1)^2. \tag{1}$$

Actually, Scott and Barreto used the fact that $\Phi_k(t-1) \equiv 0 \pmod{r}$. As above, the right-hand side of Equation (1) is quadratic, hence it can be transformed into a generalized Pell equation by a linear substitution (see [18, §2] for more details). Then, Scott-Barreto found integer solutions to this equation for small D and arbitrary V with the constraint $4h > d$. The Scott-Barreto method [18] presented generalized MNT elliptic curves with particular parameters. However it failed to give explicit families of generalized MNT elliptic curves.

Galbraith McKee and Valença's Method. Unlike Scott-Barreto's method, the mathematical analyses in [9] could lead to explicit families of generalized MNT curves. Galbraith et al. [9] extended the MNT method [16] and gave a complete characterization of MNT curves with small cofactors h. Actually, they used the fact that $\Phi_k(q) \equiv 0 \pmod{r}$. Similarly to the method in [16], Galbraith et al. defined λ by the equation $\Phi_k(q) = \lambda r$. For example, in the case $k = 6$, they required $\lambda r = \Phi_k(q) = q^2 - q + 1$. By using Hasse's bound, $|t| \leq 2\sqrt{q}$, they then analyzed and derived possible polynomials q, t from the equation $\Phi_k(q) = \lambda r$. Readers are referred to [9, Section 3] for a particular analysis in the case, in which the embedding degree is $k = 6$ and the cofactor is $h = 2$.

3   An Alternative Approach to Galbraith et al.'s Method

In this section, we present an alternative approach to generate explicit families of ordinary elliptic curves with embedding degree 3, 4, or 6 and small cofactors. Different from the analytic approach in [9], we obtain families of curves by presenting a very *simple* and *explicit* algorithm. Our analyses also show that this algorithm can find all families of generalized MNT elliptic curves with any given cofactor.

3.1   Preliminary Observations and Facts

Some well-known facts and observations that can be used to find families of curves are noted in this section. Similar to Scott-Barreto's method, we use the fact that $\Phi_k(t-1) \equiv 0 \bmod r$. Consider cyclotomic polynomials corresponding to embedding degrees $k = 3, 4, 6$:

$$\Phi_3(t(x) - 1) = t(x)^2 - t(x) + 1,$$
$$\Phi_4(t(x) - 1) = t(x)^2 - 2t(x) + 2,$$
$$\Phi_6(t(x) - 1) = t(x)^2 - 3t(x) + 3.$$

By setting $t(x) = ax + b$, we have the following equations:

$$\Phi_3(t(x) - 1) = a^2x^2 + a(2b - 1)x + \Phi_3(b - 1), \tag{2}$$
$$\Phi_4(t(x) - 1) = a^2x^2 + 2a(b - 1)x + \Phi_4(b - 1), \tag{3}$$
$$\Phi_6(t(x) - 1) = a^2x^2 + a(2b - 3)x + \Phi_6(b - 1). \tag{4}$$

Theorem 1. *The quadratic polynomials* $\Phi_3(t(x)-1)$, $\Phi_4(t(x)-1)$ *and* $\Phi_6(t(x)-1)$ *are irreducible over the rational field.*

Proof. We start with the following lemma.

Lemma 1. *Let* $f(x)$ *be a quadratic irreducible polynomial in* $\mathbb{Q}[x]$. *If we perform any* $\mathbb{Z}$*-linear change of variables* $x \mapsto ax + b$ *for any* $a \in \mathbb{Q} \setminus \{0\}$ *and* $b \in \mathbb{Q}$, $f(x)$ *will still be a quadratic irreducible polynomial in* $\mathbb{Q}[x]$.

Proof. If we assume that $f(ax + b)$ is not irreducible in $\mathbb{Q}[X]$, then as $f(x)$ is a quadratic polynomial it means that $f(ax + b)$ admits a decomposition of the form $f(ax + b) = c(x - c_1)(x - c_2)$, for $c, c_1, c_2 \in \mathbb{Q}$. The values c_1 and c_2 are rational roots of $f(ax + b) = 0$. It is easy to see that $ac_1 + b$ and $ac_2 + b$ would then be rational roots of $f(x) = 0$. □

We now prove Theorem 1. As the polynomial $\Phi_3(x) = x^2 - x + 1$ is irreducible in $\mathbb{Q}[x]$, according to Lemma 1 the polynomial $\Phi_3(t(x) - 1)$ is also irreducible in $\mathbb{Q}[x]$. The same argument ensures that $\Phi_4(t(x) - 1)$ and $\Phi_6(t(x) - 1)$ are irreducible in $\mathbb{Q}[x]$. □

Let a triple (t, r, q) parameterize a family of generalized MNT curves, and let h be a small cofactor. Let $n(x)$ be a polynomial representing the cardinality of elliptic curves in the family (t, r, q). That is, $n(x) = h \cdot r(x) = q(x) - t(x) + 1$. By [7, Definition 2.7], we have:

$$\Phi_k(t(x) - 1) = d \times r(x), \tag{5}$$

where $d \in \mathbb{Z}$, and $r(x)$ is a quadratic irreducible polynomial. By Hasse's bound, $4q(x) \geq t^2(x)$, we get the inequality:

$$4h \geq d \tag{6}$$

From equations (2)–(4), we can see that d is the greatest common divisor of the coefficients appearing in these equations. For instance, when $k = 3$, d is the GCD of $\Phi_3(b - 1)$, a^2, and $a(2b - 1)$. We recall the following well-known Lemma, which can be found in [10, Chapter V, §6]:

Lemma 2. *Let* d *be prime and* $k, n > 0$. *If* d *divides* $\Phi_k(n)$, *then* d *does not divide* n, *and either* d *divides* k *or* $d \equiv 1 \pmod{k}$.

The above lemma points out that if $\Phi_k(n)$ can be factorized by prime factors d_i, i.e. $\Phi_k(n) = \prod d_i$, then, either $d_i \mid k$ or $d_i \equiv 1 \pmod{k}$.

Lemma 3. *Given $t(x) = ax + b$, if d in Eq. (5) does not divide a, then d is square free.*

Proof. We know that $d \in \mathbb{Z}$, and d is the greatest common divisor of factors of $\Phi_k(t(x) - 1)$, i.e. d divides a^2, $2a(2b - 1)$ or $2a(b - 1)$ or $2a(2b - 3)$ and $\Phi_k(b - 1)$ (Equations (2)–(4)). Suppose that d is not square free, that is $d = p^2 \times d'$ with p a prime number greater or equal to 2. By Lemma 2, p does not divide $(b - 1)$ and either p divides k or $p \equiv 1 \pmod{k}$. We also assume that d divides a^2, but does not divide a, and hence $p^2 \nmid a$, and p is a prime factor of a.

- **k = 3**: As p divides $\Phi_3(b - 1) = b^2 - b + 1$ and p divides $2b - 1$ we have that p divides $(2b - 1) + \Phi_3(b - 1)$, i.e. p divides $b(b - 1)$. We know that p does not divide $(b - 1)$, thus p must divide b.
  We have $p \mid 2b - 1 = (b - 1) + b$, and $p \mid b$, hence p must divide $b - 1$. This is contradictory with Lemma 2. Thus, d is square free.
- **k = 4**: We have that p divides $2(b - 1)$, recall from Lemma 3 that p does not divide $(b - 1)$, then $p \mid 2$. However, we can show that $\Phi_4(b - 1) \equiv \{1, 2\} \pmod 4$. It is thus impossible to have $d = 2^2 \times d'$ and $d \mid \Phi_4(b - 1)$.
- **k = 6**: Likewise, as p divides $\Phi_6(b - 1) = b^2 - 3b + 3$ and $2b - 3$ we have that p divides $(2b - 3) + \Phi_3(b - 1) = b(b - 1)$. We know that p does not divide $(b - 1)$, then we have p divides b.
  We have p divides $2b - 3$, and p divides b. Then p must divides $2b - 3 + b = 3(b - 1)$, hence p divides 3. That is, $d = 3^2 \times d'$. But, by [11, Proposition 2.4], this cannot occur. Thus, d must be square free. $\qquad\square$

3.2 The Proposed Algorithm

We start this section by presenting the following definition:

Definition 1. *Let $r(x)$, $r'(x)$, $t(x)$ and $t'(x)$ be polynomials. We say that a pair $(t(x), r(x))$ is equivalent to $(t'(x), r'(x))$ if we can transform the first into the second by performing a $\mathbb{Z}$-linear change of variables $x \mapsto cx + d$.*

In principle, given an embedding degree k and a cofactor h, our method works as follows:

1. We first fix the Frobenius trace to be $t(x) = ax + b$, for $a \in \mathbb{Z} \setminus \{0\}$ and $b \in \mathbb{Z}$. The possible values of a, b for a given cofactor h are determined by Lemma 4.
2. Then, we determine d and $r(x)$ thanks to Equation (5).
3. For given d and $r(x)$, we determine $n(x)$ and $q(x)$.

Algorithm 1 explicitly describes our method. Given an embedding degree k and a cofactor h_{max}, we demonstrate that Algorithm 1 will output a list of *all* possible families of generalized MNT curves $(t(x), r(x), q(x))$ with the cofactors $h \le h_{max}$. Lemma 4 gives the boundary for the values a_{max}, b_{max} in order to find all the possible families of curves.

Algorithm 1. Generate families of generalized MNT curves

Input: An embedding degree k, a cofactor h_{max}.
Output: A list of polynomials $(t(x), r(x), q(x))$.

$L \leftarrow \{\}; \ T \leftarrow \{\}$;

for $a = -a_{max}$ **to** a_{max} **do**
  for $b = -b_{max}$ **to** b_{max} **do**
    $t(x) \leftarrow ax + b$;
    $f(x) \leftarrow \Phi_k(t(x) - 1)$;
    Let $f(x) = d \cdot r(x)$, where $d \in \mathbb{Z}$ and $r(x)$ is an irreducible quadratic
    polynomial;
    if *pair* $(t(x), r(x))$ *is not* equivalent *with any* $(t'(x), r'(x))$ *in* T **then**
      $T \leftarrow T + \{(d, t(x), r(x))\}$;
      for $h = \lceil d/4 \rceil$ **to** h_{max} **do**
        $q(x) \leftarrow h \cdot r(x) + t(x) - 1$;
        if $q(x)$ *is irreducible and* $gcd(q(x), r(x) : x \in \mathbb{Z}) = 1$ **then**
          $L \leftarrow L + \{(t(x), r(x), q(x), h)\}$;
        end
      end
    end
  end
end
return L

Lemma 4. *Given an embedding k, and a cofactor h_{max}, we have $a_{max} = 4h_{max}$, and $b_{max} < a_{max}$.*

Proof. We first demonstrate that $a_{max} = 4h_{max}$. Suppose that $d \mid a^2$, but $d \nmid a$, then by Lemma 3, d must be square free. This is a contradiction, thus we have $d \mid a$.

Suppose that the algorithm outputs a family of curves with $t(x) = ax + b$, and a is a multiple of d, that is, $a = m \times d$. By a $\mathbb{Z}$-linear transformation, we know that this family is equivalent to a family of curves with $t(x) = dx + b$. For the simplest form, the value of the coefficient a of polynomial $t(x)$ should be equal to d. Due to the inequality (6), the maximum value of a, $a_{max} = 4h_{max}$.

Likewise, if $b > a$, we can make a transformation $x \mapsto x + \lfloor b/a \rfloor$, and $b' = b \bmod a$. The value of b_{max} thus should be chosen less than a_{max}. $\qquad\square$

4 More Near Prime-Order Elliptic Curves

The families of elliptic curves obtained from Algorithm 1 for $k = 3, 4$ and 6 are presented in Tables 2, 3, and 4, respectively. Our algorithms execute an *exhaustive search* based on the given parameters, they can thus generate *all* families of elliptic curves of small embedding degrees 3, 4 and 6. In these tables, we present only families of curves with cofactors $1 \le h \le 6$, but it is worth to note that a family of curves with any cofactor can be easily found by adjusting the parameters of the algorithms.

4.1   k = 3

For the case of $k = 3$, our results are summarized curves in Table 2. We don't claim new explicit families in comparison to results in [9]. Our families of curves in the Table 2 can be obtained due to a linear transform of variables from Table 3 in [9] when $k = 3$. For example, for $h = 2$, our family $q(x) = 2x^2 + x + 1$, and $t(x) = -x$ is equivalent to the family $q(x) = 8x^2 + 2x + 1$, and $t(x) = -2x$ in [9, Table3]. Our algorithm just gives the polynomials $r(x)$ and $q(x)$ with the least value of coefficients.

Theorem 2. *Table 2 gives all families of elliptic curves of the embedding degree $k = 3$ with different cofactors $1 \leq h \leq 6$.*

Table 2. Valid q, r, t corresponding to $k = 3$

h	q	r	t	h	q	r	t
1	$3x^2 - 1$	$3x^2 + 3x + 1$	$-3x - 1$		$65x^2 + 22x + 1$	$13x^2 + 7x + 1$	$-13x - 3$
	$2x^2 + x + 1$	$x^2 + x + 1$	$-x$	5	$65x^2 + 48x + 8$	$13x^2 + 7x + 1$	$13x + 4$
2	$14x^2 + 3x - 1$	$7x^2 + 5x + 1$	$-7x - 2$		$95x^2 + 56x + 7$	$19x^2 + 15x + 3$	$-19x - 7$
	$14x^2 + 17x + 4$	$7x^2 + 5x + 1$	$7x + 3$		$95x^2 + 94x + 22$	$19x^2 + 15x + 3$	$19x + 8$
3	$3x^2 + 2x + 2$	$x^2 + x + 1$	$-x$		$6x^2 + 5x + 5$	$x^2 + x + 1$	$-x$
	$4x^2 + 3x + 3$	$x^2 + x + 1$	$-x$		$18x^2 + 15 + 4$	$3x^2 + 3x + 1$	$-3x - 1$
4	$12x^2 + 9x + 2$	$3x^2 + 3x + 1$	$-3x - 1$		$78x^2 + 29x + 2$	$13x^2 + 7x + 1$	$-13x - 3$
	$28x^2 + 13x + 1$	$7x^2 + 5x + 1$	$-7x - 2$	6	$78x^2 + 55x + 9$	$13x^2 + 7x + 1$	$13x + 4$
	$28x^2 + 27x + 6$	$7x^2 + 5x + 1$	$7x + 3$		$114x^2 + 71x + 10$	$19x^2 + 15x + 3$	$-19x - 7$
	$5x^2 + 4x + 4$	$x^2 + x + 1$	$-x$		$114x^2 + 109x + 25$	$19x^2 + 15x + 3$	$19x + 8$
5	$35x^2 + 18x + 2$	$7x^2 + 5x + 1$	$-7x - 2$		$126x^2 + 33x + 1$	$21x^2 + 9x + 1$	$-21x - 4$
	$35x^2 + 32x + 7$	$7x^2 + 5x + 1$	$7x + 3$		$126x^2 + 75x + 10$	$21x^2 + 9x + 1$	$21x + 5$

Proposition 2. *Let $q(x), r(x)$ and $t(x)$ be non-zero polynomials that parameterize a family of curves with embedding degree $k = 3$ and small cofactor $h \geq 1$. Then $q'(x) = q(x) - 2t(x) + 1$, $r(x)$, and $t'(x) = 1 - t(x)$ represent a family of curves with the same group order $r(x)$ and the same cofactor h.*

Proof. Let $q(x), r(x)$ and $t(x)$ parameterize a family of curves with embedding degree $k = 3$, the small cofactor $h \geq 1$, and let $n(x) = h \cdot r(x)$ represent the number of points on this family of curves. We have $\Phi_3(t(x) - 1) = t(x)^2 - t(x) + 1$ and $\Phi_3(t'(x) - 1) = \Phi_3(-t(x)) = t(x)^2 - t(x) + 1 = \Phi_3(t(x) - 1)$. Since $r(x) \mid \Phi_3(t(x) - 1)$, we have that $r(x) \mid \Phi_3(t'(x) - 1)$ and $q(x) = n(x) + t(x) - 1$. Thus, $q'(x) = q(x) - 2t(x) + 1 = n(x) - t(x) = n(x) + t'(x) - 1$. It is easy to verify that $q'(x)$ is the image of $q(x)$ by a $\mathbb{Z}$-linear transformation of $t(x) \mapsto 1 - t(x)$. According to Lemma 1, since $q(x)$ is irreducible then $q'(x)$ is irreducible. Let $n'(x) = n(x)$, then $q'(x)$ represent the characteristic of the family of curves.
   Now we need to prove that $q'(x)$ and $t'(x)$ satisfies Hasse's theorem, i.e. $t'(x)^2 \leq 4q'(x)$. Suppose that $t(x) = ax + b$, then $t'(x) = -ax - b + 1$. It is clear that the leading coefficient of $q'(x)$ is equal to that of $q(x)$. Since $h > m/4$, $4q(x)$ would be greater than $t^2(x)$ for some value of x. Thus, $q'(x)$ and $t'(x)$ satisfies Hasse's theorem whenever $q(x), t(x)$ do with some big enough values of x. $\square$

4.2   k = 4

For the case of $k = 4$, our results are summarized curves in Table 3. It seems that [9, Table3] gives more families than ours, but in fact several families of curves with a given cofactor in [9, Table3] are curves with a higher cofactor. Besides, some families of curves are equivalent by Definition 1, *e.g.*, two families $(t, q) = ((-10l-1), (60l^2+14l+1))$ and $((10l+4), (60l^2+46l+9))$ are equivalent. Thus, the number of their families obtained is not as much as they claimed.

Theorem 3. *Table 3 gives families of elliptic curves of the embedding degree $k = 4$ with small cofactors $1 \leq h \leq 6$.*

Table 3. Valid q, r, t corresponding to $k = 4$

h	q	r	t
1	$x^2 + x + 1$	$x^2 + 2x + 2$	$-x$
2	$4x^2 + 2x + 1$	$2x^2 + 2x + 1$	$-2x$
3	$3x^2 + 5x + 5$	$x^2 + 2x + 2$	$-x$
	$15x^2 + 7x + 1$	$5x^2 + 4x + 1$	$-5x - 1$
	$15x^2 + 13x + 3$	$5x^2 + 6x + 2$	$-5x - 2$
4	$8x^2 + 6x + 3$	$2x^2 + 2x + 1$	$-2x$
5	$5x^2 + 9x + 9$	$x^2 + 2x + 2$	$-x$
	$25x^2 + 15x + 3$	$5x^2 + 4x + 1$	$-5x - 1$
	$25x^2 + 25x + 7$	$5x^2 + 6x + 2$	$-5x - 2$

h	q	r	t
5	$65x^2 + 37x + 5$	$13x^2 + 10x + 2$	$-13x - 4$
	$65x^2 + 63x + 15$	$13x^2 + 10x + 2$	$13x + 6$
	$85x^2 + 23x + 1$	$17x^2 + 8x + 1$	$-17x - 3$
	$85x^2 + 57x + 9$	$17x^2 + 8x + 1$	$17x + 5$
6	$12x^2 + 10x + 5$	$2x^2 + 2x + 1$	$-2x$
	$60x^2 + 26x + 3$	$10x^2 + 6x + 1$	$-10x - 2$
	$60x^2 + 46x + 9$	$10x^2 + 6x + 1$	$10x + 4$
	$102x^2 + 31x + 2$	$17x^2 + 8x + 1$	$-17x - 3$
	$102x^2 + 65x + 10$	$17x^2 + 8x + 1$	$17x + 5$

Proposition 3. *Let non-zero polynomials $q(x), r(x)$ and $t(x)$ parameterize a family of curves with embedding degree $k = 4$ and the small cofactor h. Then $q'(x) = q(x) - 2t(x) + 2$, $r(x)$, and $t'(x) = 2 - t(x)$ represent a family of curves with the same embedding degree and the same cofactor.*

Proof. The proof of the Proposition 3 is similar to that of Proposition 2. Assume that $t(x) = ax + b$ and $t'(x) = 2 - t(x)$, we have $\Phi_4(t(x) - 1) = \Phi_4(t'(x) - 1) = t(x)^2 - 2t(x) + 2$. Likewise, we can get $q'(x) = q(x) - 2t(x) + 2 = n(x) + t'(x) - 1$. Polynomials $t'(x), q'(x)$ satisfy Hasse's theorem.   □

4.3   k = 6

Table 4 gives more explicit families than Table 3 of [9] for $k = 6$. For instance, when $h = 3$, we have one more family of pairing-friendly elliptic curves with $t(x) = -3x$, $q(x) = 9x^2 + 6x + 2$, and $r(x) = 3x^2 + 3x + 1$.

Theorem 4. *Table 4 gives families of elliptic curves of the embedding degree $k = 6$ with different cofactors $1 \leq k \leq 6$.*

Proposition 4. *Let non-zero polynomials $q(x), r(x)$ and $t(x)$ parameterize a family of curves with embedding degree $k = 6$ and the small cofactor $h \geq 2$. Then $q'(x) = q(x) - 2t(x) + 3$, $r(x)$, and $t'(x) = 3 - t(x)$ represent a family of curves with the same embedding degree and the same cofactor.*

Table 4. Valid q, r, t corresponding to $k = 6$

h	q	r	t		h	q	r	t
1	$x^2 + 1$	$x^2 + x + 1$	$-x + 1$			$15x^2 + 12x + 4$	$3x^2 + 3x + 1$	$-3x$
2	$2x^2 + x + 2$	$x^2 + x + 1$	$-x + 1$			$35x^2 + 18x + 3$	$7x^2 + 5x + 1$	$-7x - 1$
	$6x^2 + 3x + 1$	$3x^2 + 3x + 1$	$-3x$			$35x^2 + 32x + 8$	$7x^2 + 5x + 1$	$7x + 4$
3	$3x^2 + 2x + 3$	$x^2 + x + 1$	$-x + 1$		5	$65x^2 + 22x + 2$	$13x^2 + 7x + 1$	$-13x - 2$
	$9x^2 + 6x + 2$	$3x^2 + 3x + 1$	$-3x$			$65x^2 + 48x + 9$	$13x^2 + 7x + 1$	$13x + 5$
	$21x^2 + 8x + 1$	$7x^2 + 5x + 1$	$-7x - 1$			$95x^2 + 56x + 8$	$19x^2 + 5x + 3$	$-19x - 6$
	$21x^2 + 22x + 6$	$7x^2 + 5x + 1$	$7x + 4$			$95x^2 + 94x + 23$	$19x^2 + 5x + 3$	$19x + 9$
4	$4x^2 + 3x + 4$	$x^2 + x + 1$	$-x + 1$			$6x^2 + 5x + 6$	$x^2 + x + 1$	$-x + 1$
	$28x^2 + 13x + 2$	$7x^2 + 5x + 1$	$-7x - 1$			$18x^2 + 15x + 5$	$3x^2 + 3x + 1$	$-3x$
	$28x^2 + 27x + 7$	$7x^2 + 5x + 1$	$7x + 4$		6	$42x^2 + 23x + 4$	$7x^2 + 5x + 1$	$-7x - 1$
	$52x^2 + 15x + 1$	$13x^2 + 7x + 1$	$-13x - 2$			$42x^2 + 37x + 9$	$7x^2 + 5x + 1$	$7x + 4$
	$52x^2 + 41x + 8$	$13x^2 + 7x + 1$	$13x + 5$			$78x^2 + 29x + 3$	$13x^2 + 7x + 1$	$-13x - 2$
5	$5x^2 + 4x + 5$	$x^2 + x + 1$	$-x + 1$			$78x^2 + 55x + 10$	$13x^2 + 7x + 1$	$13x + 5$

Proof. The proof of the Proposition 4 is also similar to that of Proposition 2. Assume that $t(x) = ax + b$ and $t'(x) = 3 - t(x)$, we have $\Phi_6(t(x) - 1) = \Phi_6(t'(x) - 1) = t(x)^2 - 3t(x) + 3$. Similarly, we can get $q'(x) = q(x) - 2t(x) + 3 = n(x) + t'(x) - 1$. Polynomials $t'(x), q'(x)$ satisfy Hasse's theorem.   □

4.4 Solving the Pell Equations

For elliptic curves with embedding degrees $k = 3, 4, 6$ it is clear that the CM equation $DV^2 = 4q(x) - t^2(x)$ is quadratic. Such an equation can be transformed into a generalized Pell equation of the form $y^2 + DV^2 = f$. In [18], Scott and Barreto showed how to remove the linear term in the CM equation to get a generalized Pell equation. In this section, we generalize their idea to get Pell equations for families of elliptic curves presented in Tables 2, 3, and 4.

Let $t(x) = ax + b$, $\Phi_k(t(x) - 1) = d \cdot r(x)$, where $k = 3, 4, 6$ and $\#E(\mathbb{F}_q) = h \cdot r(x)$. Similarly to the analysis of Scott-Barreto in [18], we make a substitution $x = (y - a_k)/n$ to transform the CM equations to the generalized Pell equations, where $a_3 = 2h(2b - 1) - (b - 2)d$, $a_4 = 4h(b - 1) - (b - 2)d$, $a_6 = 2h(2b - 3) - (b - 2)d$ and $n = a(4h - d)$. We set $n' = n/a$, $g = dn'D$ and

$$f_3 = a_3^2 - (n'b)^2 + 4n'(b - 1)(h - d),$$
$$f_4 = a_4^2 - (n'b)^2 + 4n'(b - 1)(2h - d),$$
$$f_6 = a_6^2 - (n'b)^2 + 4n'(b - 1)(3h - d).$$

The CM equation is transformed to its Pell equation $y^2 - gV^2 = f_k$, where $k = 3, 4$, or 6^2. The works in [13],[6] investigated the problem on how solve Pell equations of MNT curves. We illustrate our method for $k = 6$ and $h = 4$.

---

[2] Note that we fix the typo in the value of f_k in [18, §2]. Indeed, f_k must be set to $a_k^2 - b^2$ instead of $a_k^2 + b^2$.

Case $k = 6$ and $h = 4$. Elliptic curves having cofactor $h = 4$ may be put in form $x^2 + y^2 = 1 + dx^2 y^2$ with d a non-square integer. Such curves called Edwards curves were introduced to cryptography by Bernstein and Lange [4]. They showed that the addition law on Edwards curves is faster than all previously known formulas. Edwards curves were later extended to the twisted Edwards curves in [3]. Readers also can see [1],[14] for efficient algorithms to compute pairings on Edwards curves. In this section, we give some facts to solve Pell equation for Edwards curves with embedding degree $k = 6$. We have:

$$y_1^2 - D_1' V^2 = -176, \tag{7}$$
$$y_2^2 - D_2' V^2 = -80, \tag{8}$$
$$y_3^2 - D_3' V^2 = -80, \tag{9}$$
$$y_4^2 - D_4' V^2 = 16, \tag{10}$$
$$y_5^2 - D_5' V^2 = 16, \tag{11}$$

where $y_i = (x - a_i)/b_i$, $D_i' = b_i D$, for $i \in [1,5]$, and $a_1 = -7$, $a_2 = -19$, $a_3 = -26$, $a_4 = -4$, $a_5 = -17$, $b_1 = 15$, $b_2 = 63$, $b_3 = 63$, $b_4 = 39$, $b_5 = 39$. Karabina and Teske [13, Lemma1] showed that if $4 \mid f_k$ then the set of solutions to $y^2 - D'V^2 = f_k$ does not contain any *ambiguous* class, i.e., there exists no primitive solution $\alpha = y + v\sqrt{D'}$ such that α and its *conjugate* $\alpha' = y - v\sqrt{D'}$ are in the same class. Equations (7)–(11) thus won't have any solution that contains an ambiguous class. If equations (7)–(11) have solutions with $y_i \equiv -a_i \bmod b_i$, and a fixed positive square-free integer D_i' relatively prime to b_i, for $1 \le i \le 5$, then t, r, q in Table 4 with $h = 4$ represent a family of pairing-friendly Edwards curves with embedding degree 6.

5   Conclusion

In this paper we extended Scott-Barreto's method and presented efficient and simple algorithms to obtain MNT curves with small cofactors. Our algorithm allows to find all possible families of generalized MNT curves. In the Propositions 2, 3 and 4 we point out a one-to-one correspondence between families of MNT curves having the same embedding degree and the same cofactor. If given a parameterization of a MNT curves, we can construct another MNT curve using a $\mathbb{Z}$-linear transformation. We also analyze the Complex Multiplication equations of MNT curves and point out how to transform these Complex Multiplication equations into generalized Pell equations. In addition, we give a method to generate Edwards curves with embedding degree 6.

Acknowledgments. The authors thank the anonymous referees for their detailed and valuable comments on the manuscript.

References

1. Arène, C., Lange, T., Naehrig, M., Ritzenthaler, C.: Faster computation of the Tate pairing. Journal of Number Theory $131(5)$, 842–857 (2011)
2. Balasubramanian, R., Koblitz, N.: The improbability that an elliptic curve has subexponential discrete log problem under the menezes - okamoto - vanstone algorithm. J. Cryptology, 141–145 (1998)
3. Bernstein, D.J., Birkner, P., Joye, M., Lange, T., Peters, C.: Twisted edwards curves. In: Vaudenay, S. (ed.) AFRICACRYPT 2008. LNCS, vol. 5023, pp. 389–405. Springer, Heidelberg (2008)
4. Bernstein, D.J., Lange, T.: Faster addition and doubling on elliptic curves. In: Kurosawa, K. (ed.) ASIACRYPT 2007. LNCS, vol. 4833, pp. 29–50. Springer, Heidelberg (2007)
5. Boneh, D., Franklin, M.: Identity-Based encryption from the weil pairing. In: Kilian, J. (ed.) CRYPTO 2001. LNCS, vol. 2139, pp. 213–229. Springer, Heidelberg (2001)
6. Fotiadis, G., Konstantinou, E.: On the efficient generation of generalized MNT elliptic curves Santa Barbara, California, USA. In: Muntean, T., Poulakis, D., Rolland, R. (eds.) CAI 2013. LNCS, vol. 8080, pp. 147–159. Springer, Heidelberg (2013)
7. Freeman, D., Scott, M., Teske, E.: A Taxonomy of Pairing-Friendly Elliptic Curves. J. Cryptol. 23, 224–280 (2010)
8. Frey, G., Rück, H.-G.: A remark concerning m-divisibility and the discrete logarithm in the divisor class group of curves. Math. Comput. $62(206)$, 865–874 (1994)
9. Galbraith, S.D., McKee, J.F., Valença, P.C.: Ordinary abelian varieties having small embedding degree. Finite Fields and their Applications $13(4)$, 800–814 (2007)
10. Grillet, P.A.: Abstract Algebra. Springer (July 2007)
11. Jameson, G.: The cyclotomic polynomials. http://www.maths.lancs.ac.uk/jameson/cyp.pdf
12. Joux, A.: A one round protocol for tripartite diffie–hellman. In: Bosma, W. (ed.) ANTS 2000. LNCS, vol. 1838, pp. 385–393. Springer, Heidelberg (2000)
13. Karabina, K., Teske, E.: On prime-order elliptic curves with embedding degrees k = 3, 4, and 6. In: van der Poorten, A.J., Stein, A. (eds.) ANTS-VIII 2008. LNCS, vol. 5011, pp. 102–117. Springer, Heidelberg (2008)
14. Le, D.-P., Tan, C.H.: Improved Miller's Algorithm for Computing Pairings on Edwards Curves. IEEE Transactions on Computers $63(10)$, 2626–2632 (2014)
15. Menezes, A., Vanstone, S., Okamoto, T.: Reducing elliptic curve logarithms to logarithms in a finite field. In: STOC 1991: Proceedings of the Twenty-third Annual ACM Symposium on Theory of Computing, pp. 80–89. ACM, New York (1991)
16. Miyaji, A., Nakabayashi, M., Takano, S.: New Explicit Conditions of Elliptic Curve Traces for FR-Reduction. IEICE Transactions on Fundamentals of Electronics, Communications and Computer Sciences $84(5)$, 1234–1243 (2001)
17. Page, D., Smart, N., Vercauteren, F.: A comparison of MNT curves and supersingular curves. Applicable Algebra in Engineering, Communication and Computing $17(5)$, 379–392 (2006)
18. Scott, M., Barreto, P.S.: Generating More MNT Elliptic Curves. Des. Codes Cryptography 38, 209–217 (2006)

Key-Policy Multi-authority Attribute-Based Encryption

Riccardo Longo[1](✉) , Chiara Marcolla[2], and Massimiliano Sala[1]

[1] Department of Mathematics, University of Trento, Via Sommarive, 14,
38123 Povo, Trento, Italy
{riccardolongomath,maxsalacodes}@gmail.com
[2] Department of Mathematics, University of Turin,
Via Carlo Alberto, 10, 10123 Turin, Italy
chiara.marcolla@gmail.com

Abstract. Bilinear groups are often used to create Attribute-Based Encryption (ABE) algorithms. In particular, they have been used to create an ABE system with multi authorities, but limited to the ciphertext-policy instance. Here, for the first time, we propose a multi-authority key-policy ABE system.

In our proposal, the authorities may be set up in any moment and without any coordination. A party can simply act as an ABE authority by creating its own public parameters and issuing private keys to the users. A user can thus encrypt data choosing both a set of attributes and a set of trusted authorities, maintaining full control unless all his chosen authorities collude against him.

We prove our system secure under the bilinear Diffie-Hellman assumption.

Keywords: ABE · Bilinear groups · Algebraic cryptography

1 Introduction

The key feature that makes the cloud so attracting nowadays is the great accessibility it provides: users can access their data through the Internet from anywhere. Unfortunately, at the moment the protection offered for sensitive information is questionable and access control is one of the greatest concerns. Illegal access may come from outsiders or even from insiders without proper clearance. One possible approach for this problem is to use Attribute-Based Encryption (ABE) that provides cryptographically enhanced access control functionality in encrypted data.

ABE developed from Identity Based Encryption, a scheme proposed by Shamir [18] in 1985 with the first constructions obtained in 2001 by Boneh and Franklin [4]. The use of bilinear groups, in particular the Tate and Weil pairings on elliptic curves [4], was the winning strategy that finally allowed to build schemes following the seminal idea of Shamir. Bilinear groups came in nicely when a preliminary version of ABE was invented by Sahai and Waters

© Springer International Publishing Switzerland 2015
A. Maletti (Ed.): CAI 2015, LNCS 9270, pp. 152–164, 2015.
DOI: 10.1007/978-3-319-23021-4_14

[17] in 2005. Immediately afterwards, Goyal, Pandey, Sahai, and Waters [7] formulated the two complimentary forms of ABE which are nowadays standard: ciphertext-policy ABE and key-policy ABE. In a ciphertext-policy ABE system, keys are associated with sets of attributes and ciphertexts are associated with access policies. In a KP-ABE system, the situation is reversed: keys are associated with access policies and ciphertexts are associated with sets of attributes. Several developments in efficiency and generalizations have been obtained for key-policy ABE, e.g. [1], [8], [16]. A first implementation of ciphertext-policy ABE has been achieved by Bethencourt et al. [3] in 2007 but the proofs of security of the ciphertext-policy ABE remained unsatisfactory since they were based on an assumption independent of the algebraic structure of the group (the generic group model). It is only with the work of Waters [20] that the first non-restricted ciphertext-policy ABE scheme was built with a security dependent on variations of the DH assumption on bilinear groups. Related to the work we propose in this paper is the construction for multiple authorities (ciphertext-policy ABE) that have been proposed in [5], [6] and [11].

However, before the present paper no multi-authority KP-ABE scheme has appeared in the literature with a proof of security.

Our Construction. In this paper we present the first multi-authority KP-ABE scheme. In our system, after the creation of an initial set of common parameters, the authorities may be set up in any moment and without any coordination. A party can simply act as an ABE authority by creating a public parameters and issuing private keys to different users (assigning access policies while doing so). A user can encrypt data under any set of attributes specifying also a set of *trusted* authorities, so the encryptor maintains high control. Also, the system does not require any central authority. Our scheme has both very short single-authority keys, that compensate the need of multiple keys (one for authority), and also very short ciphertexts. Moreover, the pairing computations in the bilinear group are involved only during the decryption phase, obtaining this way significant advantages in terms of encryption times.

Even if the authorities are collaborating, the existence of just one non-cheating authority guarantees that no illegitimate party (including authorities) has access to the encrypted data.

We prove our scheme secure using the classical bilinear Diffie-Hellman assumption.

Organization. This paper is organized as follows. In Section 2 we present the main mathematical tools used in the construction of multi authority KP-ABE scheme. In Section 3 we explain in detail our multi authority KP-ABE scheme and its security is proven under standard, non-interactive assumptions in the selective set model. Finally conclusions are drawn in Section 4.

2   Preliminaries

We do not prove original results here, we only provide what we need for our construction. See the cited references for more details on these arguments.

Let G_1, G_2 be groups of the same prime order p.

Definition 1 (Pairing). *A symmetric pairing is a bilinear map e such that $e : G_1 \times G_1 \to G_2$ has the following properties:*

- *Bilinearity: $\forall g, h \in G_1, \forall a, b \in \mathbb{Z}_p,\ e(g^a, h^b) = e(g, h)^{ab}$.*
- *Non-degeneracy: for g generator of G_1, $e(g, g) \neq 1$.*

Definition 2 (Bilinear Group). G_1 *is a Bilinear group if the conditions above hold and both the group operations in G_1 and G_2 as well as the bilinear map e are efficiently computable.*

Let $a, b, s, z \in \mathbb{Z}_p$ be chosen at random and g be a generator of the bilinear group G_1. The decisional bilinear Diffie-Hellman (BDH) problem consists in constructing an algorithm $\mathcal{B}(A = g^a, B = g^b, S = g^s, T) \to \{0, 1\}$ to efficiently distinguish between the tuples $(A, B, S, e(g, g)^{abs})$ and $(A, B, S, e(g, g)^z)$ outputting respectively 1 and 0. The advantage of $\mathcal{B}$ is:

$$Adv_{\mathcal{B}} = \left| \Pr\left[\mathcal{B}(A, B, S, e(g, g)^{abs}) = 1 \right] - \Pr\left[\mathcal{B}(A, B, S, e(g, g)^z) = 1 \right] \right|$$

where the probability is taken over the random choice of the generator g, of a, b, s, z in $\mathbb{Z}_p$, and the random bits possibly consumed by $\mathcal{B}$ to compute the response.

Definition 3 (BDH Assumption). *The decisional BDH assumption holds if no probabilistic polynomial-time algorithm $\mathcal{B}$ has a non-negligible advantage in solving the decisional BDH problem.*

Access structures define who may and who may not access the data, giving the sets of attributes that have clearance.

Definition 4 (Access Structure). *An access structure $\mathbb{A}$ on a universe of attributes U is the set of the subsets $S \subseteq U$ that are authorized. That is, a set of attributes S satisfies the policy described by the access structure $\mathbb{A}$ if and only if $S \in \mathbb{A}$.*

They are used to describe a policy of access, that is the rules that prescribe who may access to the information. If these rules are constructed using only AND, OR and THRESHOLD operators on the attributes, then the access structure is *monotonic*.

Definition 5 (Monotonic Access Structure). *An access structure $\mathbb{A}$ is said to be monotonic if given $S_0 \subseteq S_1 \subseteq U$ it holds*

$$S_0 \in \mathbb{A} \implies S_1 \in \mathbb{A}$$

An interesting property is that monotonic access structures (i.e. access structures $\mathbb{A}$ such that if S is an authorized set and $S \subseteq S'$ then also S' is an authorized set) may be associated to linear secret sharing schemes (LSSS). In this setting the parties of the LSSS are the attributes of the access structure.

A LSSS may be defined as follows (adapted from [2]).

Definition 6 (Linear Secret-Sharing Schemes (LSSS)). *A secret-sharing scheme Π over a set of parties P is called linear (over $\mathbb{Z}_p$) if*

1. *The shares for each party form a vector over $\mathbb{Z}_p$.*
2. *There exists a matrix M with l rows and n columns called the share-generating matrix for Π. For all $i \in \{1, \ldots, l\}$ the i-th row of M is labeled via a function ρ, that associates M_i to the party $\rho(i)$. Considering the vector $v = (s, r_2, \ldots, r_n) \in \mathbb{Z}_p^n$, where $s \in \mathbb{Z}_p$ is the secret to be shared, and $r_i \subset \mathbb{Z}_p$, with $i \in \{2, \ldots, n\}$ are randomly chosen, then Mv is the vector of l shares of the secret s according to Π. The share $(Mv)_i = M_i v$ belongs to party $\rho(i)$.*

It is shown in [2] that every linear secret sharing-scheme according to the above definition also enjoys the linear reconstruction property, defined as follows: suppose that Π is an LSSS for the access structure $\mathbb{A}$. Let $S \in \mathbb{A}$ be any authorized set, and let $I \subseteq \{1, \ldots, l\}$ be defined as $I = \{i : \rho(i) \in S\}$. Then, there exist constants $w_i \in \mathbb{Z}_p$, with $i \in I$ such that, if λ_i are valid shares of any secret s according to Π, then

$$\sum_{i \in I} w_i \lambda_i = s \tag{1}$$

Furthermore, it is shown in [2] that these constants w_i can be found in time polynomial in the size of the share-generating matrix M.

Note that the vector $(1, 0, \ldots, 0)$ is the target vector for the linear secret sharing scheme. Then, for any set of rows I in M, the target vector is in the span of I if and only if I is an authorized set. This means that if I is not authorized, then for any choice of $c \in \mathbb{Z}_p$ there will exist a vector u such that $u_1 = c$ and

$$M_i \cdot w = 0 \quad \forall i \in I$$

In the first ABE schemes the access formulas are typically described in terms of access trees. The appendix of [11] is suggested for a discussion of how to perform a conversion from access trees to LSSS.

See [7], [2] and [13] for more details about LSSS and access structures.

3 Our Construction

This section is divided in three parts. We start with definitions of Multi-Authority Key-Policy ABE and of CPA selective security. In the second part we present in detail our first scheme and, finally, we prove the security of this scheme under the classical BDH assumption in the selective set model.

A security parameter will be used to determine the size of the bilinear group used in the construction, this parameter represents the order of complexity of the assumption that provides the security of the scheme. Namely, first the complexity is chosen thus fixing the security parameter, then this value is used to compute the order that the bilinear group must have in order to guarantee the desired complexity, and finally a suitable group is picked and used.

3.1   Multi Authority KP-ABE Structure and Security

In this scheme, after the common universe of attributes and bilinear group are agreed, the authorities set up independently their master key and public parameters. The master key is subsequently used to generate the private keys requested by users. Users ask an authority for keys that embed a specific access structure, and the authority issues the key only if it judges that the access structure suits the user that requested it. Equivalently an authority evaluates a user that requests a key, assigns an access structure, and gives to the user a key that embeds it. When someone wants to encrypt, it chooses a set of attributes that describes the message (and thus determines which access structures may read it) and a set of trusted authorities. The ciphertext is computed using the public parameters of the chosen authorities, and may be decrypted only using a valid key for each of these authorities. A key with embedded access structure $\mathbb{A}$ may be used to decrypt a ciphertext that specifies a set of attributes S if and only if $S \in \mathbb{A}$, that is the structure considers the set authorized.

This scheme is secure under the classical BDH assumption in the selective set model, in terms of chosen-ciphertext indistinguishability.

The security game is formally defined as follows.

Let $\mathcal{E} = (\mathsf{Setup}, \mathsf{Encrypt}, \mathsf{KeyGen}, \mathsf{Decrypt})$ be a MA-KP-ABE scheme for a message space $\mathcal{M}$, a universe of authorities X and an access structure space $\mathcal{G}$ and consider the following MA-KP-ABE experiment $\mathsf{MA\text{-}KP\text{-}ABE\text{-}Exp}_{\mathcal{A}, \mathcal{E}}(\lambda, U)$ for an adversary $\mathcal{A}$, parameter λ and attribute universe U:

Init. The adversary declares the set of attributes S and the set of authorities $A \subseteq X$ that it wishes to be challenged upon. Moreover it selects the *honest authority* $k_0 \in A$.

Setup. The challenger runs the Setup algorithm, initializes the authorities and gives to the adversary the public parameters.

Phase I. The adversary issues queries for private keys of any authority, but k_0 answers only to queries for keys for access structures $\mathbb{A}$ such that $S \notin \mathbb{A}$. On the contrary the other authorities respond to every query.

Challenge. The adversary submits two equal length messages m_0 and m_1. The challenger flips a random coin $b \in \{0, 1\}$, and encrypts m_b with S for the set of authorities A. The ciphertext is passed to the adversary.

Phase II. Phase I is repeated.

Guess. The adversary outputs a guess b' of b.

Definition 7 (MA-KP-ABE Selective Security). *The MA-KP-ABE scheme $\mathcal{E}$ is CPA selective secure (or secure against chosen-plaintext attacks)* *for attribute universe* U *if for all probabilistic polynomial-time adversaries* $\mathcal{A}$, *there exists a negligible function* negl *such that:*

$$\Pr[\mathsf{MA\text{-}KP\text{-}ABE\text{-}Exp}_{\mathcal{A},\mathcal{E}}(\lambda, U) = 1] \leq \frac{1}{2} + negl(\lambda).$$

3.2   The Scheme

The scheme plans a set X of independent authorities, each with their own para-meters, and it sets up an encryption algorithm that lets the encryptor choose a set $A \subseteq X$ of authorities, and combines the public parameters of these in such a way that an authorized key for each authority in A is required to successfully decrypt.

Our scheme consists of three randomized algorithms (Setup, KeyGen, Encrypt) plus the decryption Decrypt. The techniques used are inspired from the scheme of Goyal et al. in [7]. The scheme works in a bilinear group $\mathbb{G}_1$ of prime order p, and uses LSSS matrices to share secrets according to the various access structures. Attributes are seen as elements of $\mathbb{Z}_p$.

The description of the algorithms follows.

Setup($U, g, \mathbb{G}_1$) $\rightarrow$ ($\mathsf{PK}_k, \mathsf{MK}_k$). Given the universe of attributes U and a genera-tor g of $\mathbb{G}_1$ each authority sets up independently its parameters. For $k \in X$ the Authority k chooses uniformly at random $\alpha_k \in \mathbb{Z}_p$, and $z_{k,i} \in \mathbb{Z}_p$ for each $i \in U$. Then the public parameters PK_k and the master key MK_k are:

$$\mathsf{PK}_k = (e(g, g)^{\alpha_k}, \{g^{z_{k,i}}\}_{i \in U}) \qquad \mathsf{MK}_k = (\alpha_k, \{z_{k,i}\}_{i \in U})$$

KeyGen$_k$($\mathsf{MK}_k, (M_k, \rho_k)$) $\rightarrow$ SK_k. The key generation algorithm for the authority k takes as input the master secret key MK_k and an LSSS access structure (M_k, ρ_k), where M_k is an $l \times n$ matrix on $\mathbb{Z}_p$ and ρ_k is a function which associates rows of M_k to attributes. It chooses uniformly at random a vector $v_k \in \mathbb{Z}_p^n$ such that $v_{k,1} = \alpha_k$. Then it computes the shares $\lambda_{k,i} = M_{k,i}v_k$ for $1 \leq i \leq l$ where $M_{k,i}$ is the i-th row of M_k. Then the private key SK_k is:

$$\mathsf{SK}_k = \left\{ K_{k,i} = g^{\frac{\lambda_{k,i}}{z_{k,\rho_k(i)}}} \right\}_{1 \leq i \leq l}$$

Encrypt($m, S, \{\mathsf{PK}_k\}_{k \in A}$) $\rightarrow$ CT. The encryption algorithm takes as input the pub-lic parameters, a set S of attributes and a message m to encrypt. It chooses $s \in \mathbb{Z}_p$ uniformly at random and then computes the ciphertext as:

$$\mathsf{CT} = \left(S, C' = m \cdot \left(\prod_{k \in A} e(g, g)^{\alpha_k} \right)^s, \{C_{k,i} = (g^{z_{k,i}})^s\}_{k \in A, \, i \in S} \right)$$

Decrypt(CT, $\{SK_k\}_{k \in A}$) → m'. The input is a ciphertext for a set of attributes S and a set of authorities A and an authorized key for every authority cited by the ciphertext. Let (M_k, ρ_k) be the LSSS associated to the key k, and suppose that S is authorized for each $k \in A$. The algorithm for each $k \in A$ finds $w_{k,i} \in \mathbb{Z}_p, i \in I_k$ such that

$$\sum_{i \in I_k} \lambda_{k,i} w_{k,i} = \alpha_k \tag{2}$$

for appropriate subsets $I_k \subseteq S$ and then proceeds to reconstruct the original message computing:

$$
\begin{aligned}
m' &= \frac{C'}{\prod_{k \in A} \prod_{i \in I_k} e(K_{k,i}, C_{k,\rho_k(i)})^{w_{k,i}}} \\
&= \frac{m \cdot (\prod_{k \in A} e(g,g)^{\alpha_k})^s}{\prod_{k \in A} \prod_{i \in I_k} e\left(g^{\frac{\lambda_{k,i}}{z_{k,\rho_k(i)}}}, (g^{z_{k,\rho_k(i)}})^s\right)^{w_{k,i}}} \\
&= \frac{m \cdot e(g,g)^{s(\sum_{k \in A} \alpha_k)}}{\prod_{k \in A} e(g,g)^{s \sum_{i \in I_k} w_{k,i}\lambda_{k,i}}} \\
&\overset{*}{=} \frac{m \cdot e(g,g)^{s(\sum_{k \in A} \alpha_k)}}{e(g,g)^{s(\sum_{k \in A} \alpha_k)}} = m
\end{aligned}
$$

Where $\overset{*}{=}$ follows from property (2).

3.3   Security

The scheme is proved secure under the BDH assumption (Definition 3) in a selective set security game in which every authority but one is supposed curious (or corrupted or breached) and then it will issue even keys that have enough clearance for the target set of attributes, while the honest one issues only unauthorized keys. Thus if at least one authority remains trustworthy the scheme is secure.

The security is provided by the following theorem.

Theorem 1. *If an adversary can break the scheme, then a simulator can be constructed to play the Decisional BDH game with a non-negligible advantage.*

Proof. Suppose there exists a polynomial-time adversary $\mathcal{A}$, that can attack the scheme in the Selective-Set model with advantage ϵ. Then a simulator $\mathcal{B}$ can be built that can play the Decisional BDH game with advantage $\epsilon/2$. The simulation proceeds as follows.

Init. The simulator takes in a BDH challenge g, g^a, g^b, g^s, T. The adversary gives the algorithm the challenge access structure S.

Setup. The simulator chooses random $r_k \in \mathbb{Z}_p$ for $k \in A \setminus \{k_0\}$ and implicitly sets $\alpha_k = -r_k b$ for $k \in A \setminus \{k_0\}$ and $\alpha_{k_0} = ab + b\sum_{k \in A \setminus \{k_0\}} r_k$ by computing:

$$e(g,g)^{\alpha_{k_0}} = e(g^a, g^b) \prod_{k \in A \setminus \{k_0\}} (g^b, g^{r_k})$$

$$e(g,g)^{\alpha_k} = e(g^b, g^{-r_k}) \qquad \forall k \in A \setminus \{k_0\}$$

Then it chooses $z'_{k,i} \in \mathbb{Z}_p$ uniformly at random for each $i \in U, k \in A$ and implicitly sets

$$z_{k,i} = \begin{cases} z'_{k,i} & \text{if } i \in S \\ bz'_{k,i} & \text{if } i \notin S \end{cases}$$

Then it can publish the public parameters computing the remaining values as:

$$g^{z_{k,i}} = \begin{cases} g^{z'_{k,i}} & \text{if } i \in S \\ (g^b)^{z'_{k,i}} & \text{if } i \notin S \end{cases}$$

Phase I. In this phase the simulator answers private key queries. For the queries made to the authority k_0 the simulator has to compute the $K_{k_0,i}$ values of a key for an access structure (M, ρ) with dimension $l \times n$ that is not satisfied by S. Therefore for the properties of an LSSS it can find a vector $y \in \mathbb{Z}_p^n$ with $y_1 = 1$ fixed such that

$$M_i y = 0 \qquad \forall i \text{ such that } \rho(i) \in S \tag{3}$$

Then it chooses uniformly at random a vector $v \in \mathbb{Z}_p^n$ and implicitly sets the shares of $\alpha_{k_0} = b(a + \sum_{k \in A \setminus \{k_0\}} r_k)$ as

$$\lambda_{k_0,i} = b\sum_{j=1}^n M_{i,j}(v_j + (a + \sum_{k \in A \setminus \{k_0\}} r_k - v_1)y_j)$$

Note that $\lambda_{k_0,i} = \sum_{j=1}^n M_{i,j} u_j$ where $u_j = b(v_j + (a + \sum_{k \in A \setminus \{k_0\}} r_k - v_1)y_j)$ thus $u_1 = b(v_1 + (a + \sum_{k \in A \setminus \{k_0\}} r_k - v_1)1) = ab + b\sum_{k \in A \setminus \{k_0\}} r_k = \alpha_{k_0}$ so the shares are valid. Note also that from (3) it follows that

$$\lambda_{k_0,i} = b\sum_{j=1}^n M_{i,j} v_j \qquad \forall i \text{ such that } \rho(i) \in S$$

Thus if i is such that $\rho(i) \in S$ the simulator can compute

$$K_{k_0,i} = (g^b)^{\frac{\sum_{j=1}^n M_{i,j} v_j}{z'_{k_0,\rho(i)}}} = g^{\frac{\lambda_{k_0,i}}{z_{k_0,\rho(i)}}}$$

Otherwise, if i is such that $\rho(i) \notin S$ the simulator computes

$$K_{k_0,i} = g^{\frac{\sum_{j=1}^n M_{i,j}(v_j + (r-v_1)y_j)}{z'_{k_0,\rho(i)}}} (g^a)^{\frac{\sum_{j=1}^n M_{i,j} y_j}{z'_{k_0,\rho(i)}}} = g^{\frac{\lambda_{1,i}}{z_{k_0,\rho(i)}}}$$

Remembering that in this case $z_{k_0,\rho(i)} := bz'_{k_0,\rho(i)}$. Finally for the queries to the other authorities $k \in A \setminus \{k_0\}$, the simulator chooses uniformly at random a vector $t_k \in \mathbb{Z}_p^n$ such that $t_{k,1} = -r_k$ and implicitly sets the shares $\lambda_{k,i} = b \sum_{j=1}^n M_{i,j} t_{k,j}$ by computing

$$
K_{k,i} = \begin{cases} (g^b)^{\frac{\sum_{j=1}^n M_{i,j} t_{k,j}}{z'_{k,\rho(i)}}} = g^{\frac{b \sum_{j=1}^n M_{i,j} t_{k,j}}{z'_{k,\rho(i)}}} = g^{\frac{\lambda_{k,i}}{z_{k,\rho(i)}}} & \text{if } i \in S \\ g^{\frac{\sum_{j=1}^n M_{i,j} t_{k,j}}{z'_{k,\rho(i)}}} = g^{\frac{b \sum_{j=1}^n M_{i,j} t_{k,j}}{bz'_{k,\rho(i)}}} = g^{\frac{\lambda_{k,i}}{z_{k,\rho(i)}}} & \text{if } i \notin S \end{cases}
$$

Challenge. The adversary gives two messages m_0, m_1 to the simulator. It flips a coin μ. It creates:

$$
C' = m_\mu \cdot T \overset{*}{=} m_\mu \cdot e(g,g)^{abs}
$$

$$
= m_\mu \cdot \left(e(g,g)^{(ab+b(\sum_{k \in A \setminus \{k_0\}} r_k))} \prod_{k \in A \setminus \{k_0\}} e(g,g)^{br_k} \right)^s
$$

$$
C_{k,i} = (g^s)^{z'_{k,\rho(i)}} = g^{s z_{k,\rho(i)}} \qquad k \in A, \quad i \in S
$$

Where the equality $\overset{*}{=}$ holds if and only if the BDH challenge was a valid tuple (i.e. T is non-random).

Phase II. During this phase the simulator acts exactly as in *Phase I*.

Guess. The adversary will eventually output a guess μ' of μ. The simulator then outputs 0 to guess that $T = e(g,g)^{abs}$ if $\mu' = \mu$; otherwise, it outputs 1 to indicate that it believes T is a random group element in $\mathbb{G}_2$. In fact when T is not random the simulator $\mathcal{B}$ gives a perfect simulation so it holds:

$$
Pr\left[\mathcal{B}\left(y, T = e(g,g)^{abs} \right) = 0 \right] = \frac{1}{2} + \epsilon
$$

On the contrary when T is a random element $R \in \mathbb{G}_2$ the message m_μ is completely hidden from the adversary point of view, so:

$$
Pr\left[\mathcal{B}\left(y, T = R \right) = 0 \right] = \frac{1}{2}
$$

Therefore, $\mathcal{B}$ can play the decisional BDH game with non-negligible advantage $\frac{\epsilon}{2}$.

4   Related Works and Final Comments

Our scheme gives a solution addressing the problem of faith in the authority, specifically the concerns arisen by key escrow and clearance check. Key escrow is a setting in which a party (in this case the authority) may obtain access to private keys and thus it can decrypt any ciphertext. Normally the users have faith in the authority and assume that it will not abuse its powers. The problem arises when the application does not plan a predominant role and there are trust issues selecting any third party that should manage the keys. In this situation the authority is seen as *honest but curious*, in the sense that it will provide correct keys to users (then it is not malicious) but will also try to access to data beyond its competence. It is clear that as long as a single authority is the unique responsible to issue the keys, there is no way to prevent key escrow. Thus the need for multi-authority schemes arises.

The second problem is more specific for KP-ABE. In fact, the authority has to assign to each user an appropriate access structure that represents what the user can and cannot decrypt. Therefore, the authority has to be trusted not only to give correct keys and to not violate the privacy, but also to perform correct checks of the users' clearances and to assign correct access structures accordingly. Therefore, in addition to satisfying the requirements of not being *malicious* and not being *curious*, the authority must also not have been *breached*, in the sense that a user's keys must embed access structures that faithfully represent that user's level of clearance, and that no one has access to keys with a higher level of clearance than the one they are due. In this case, to add multiple authorities to the scheme gives to the encryptor the opportunity to request more guarantees about the legitimacy of the decryptor's clearance. In fact, each authority checks the users independently, so the idea is to request that the decryption proceeds successfully only when a key for each authority of a given set A is used. This means that the identity of the user has been checked by every selected authority, and the choice of these by the encryptor models the trust that he has in them. Note that if these authorities set up their parameters independently and during encryption these parameters are bound together irrevocably, then no authority can single-handedly decrypt any ciphertext and thus key escrow is removed. So our KP-ABE schemes guarantee a protection against both breaches and curiosity.

The scheme proposed has very short single-keys (just one element per row of the access matrix) that compensates for the need of multiple single-keys (one for cited authority) in the decryption. Ciphertexts are also very short (the number of elements is linear in the number of authorities times the number of attributes under which it has been encrypted) thus the scheme is efficient under this aspect. Moreover, there are *no* pairing computations involved during encryption and this means significant advantages in terms of encryption times. Decryption time is not constant in the number of pairings (e.g. as in the scheme presented in [8] or the one in [20]) but requires $\sum_{k \in A} l_k$ pairings where A is the set of authorities involved in encryption and l_k is the number of rows of the access matrix of the key given by authority k, so to maintain the efficiency of the scheme only a few authorities should be requested by the encryptor.

Taking a more historical perspective, the problem of multi-authority ABE is not novel and a few solution have been proposed. The problem of building ABE systems with multiple authorities was proposed by Sahai and Waters. This problem with the presence of a central authority was firstly considered by Chase [5] and then improved by Chase and Chow [6], constructing simple-threshold schemes in the case where attributes are divided in disjoint sets, each controlled by a different authority. These schemes are also shown to be extensible from simple threshold to KP-ABE, but retaining the partition of attributes and requiring the involvement of every authority in the decryption. In those works the main goal is to relieve the central authority of the burden of generating key material for every user and add resiliency to the system. Multiple authorities manage the attributes, so that each has less work and the whole system does not get stuck if one is down. Another approach has been made by Lin et al. [12] where a central authority is not needed but a parameter directly sets the efficiency and number of users of the scheme.

More interesting results have been achieved for CP schemes, in which the partition of the attributes makes more sense, for example [15]. The most recent and interesting result may be found in [11], where Lewko and Waters propose a scheme where is not needed a central authority or coordination between the authorities, each controlling disjoint sets of attributes. They used composite bilinear groups and via Dual System Encryption (introduced by Waters [19] with techniques developed with Lewko [10]) proved their scheme fully secure following the example of Lewko et al. [9]. They allow the adversary to statically corrupt authorities choosing also their master key. Note however that they did not specifically address key escrow but distributed workload.

Our results of this article retain relevance since they address a different setting. In fact, with this extensions the differences in the situations of ciphertext-policy ABE and KP-ABE model become more distinct. For example a situation that suits the scheme proposed here, but not the one of Lewko and Waters is the following. Consider company branches dislocated on various parts of the world, each checking its personnel and giving to each an access policy (thus acting as authorities). This scheme allows encryptions that may be decrypted by the manager of the branch (simply use only one authority as in classic ABE) but also more secure encryptions that require the identity of the decryptor to be guaranteed by more centers, basing the requirements on which branches are still secure and/or where a user may actually authenticate itself.

Moreover, we observe that although the scheme of [11] is proven fully secure (against selective security), the construction is made in composite bilinear groups. It is in fact compulsory when using Dual System encryption, but this has drawbacks in terms of group size (integer factorization has to be avoided) and the computations of pairings and group operations are less efficient. This fact leads to an alternative construction in prime order groups in the same paper, that however is proven secure only in the generic group and random oracle model. These considerations demonstrate that our construction in prime groups under basic assumptions retains validity and interest.

Acknowledgments. Most results in this paper are contained in the first's author Msc. thesis [14] who wants to thank his supervisors: the other two authors.

References

1. Attrapadung, N., Herranz, J., Laguillaumie, F., Libert, B., De Panafieu, E., Ràfols, C., et al.: Attribute-based encryption schemes with constant-size ciphertexts. Theoretical Computer Science **422**, 15–38 (2012)
2. Beimel, A.: Secure schemes for secret sharing and key distribution. Ph.D. thesis, Technion-Israel Institute of technology, Faculty of computer science (1996)
3. Bethencourt, J., Sahai, A., Waters, B.: Ciphertext-policy attribute-based encryption. In: Proc. of SP 2007, pp. 321–334 (2007)
4. Boneh, D., Franklin, M.: Identity-Based encryption from the weil pairing. In: Kilian, J. (ed.) CRYPTO 2001. LNCS, vol. 2139, pp. 213–229. Springer, Heidelberg (2001)
5. Chase, M.: Multi-authority attribute based encryption. In: Vadhan, S.P. (ed.) TCC 2007. LNCS, vol. 4392, pp. 515 534. Springer, Heidelberg (2007)
6. Chase, M., Chow, S.S.: Improving privacy and security in multi-authority attribute-based encryption. In: Proceedings of the 16th ACM Conference on Computer and Communications Security, pp. 121–130. ACM (2009)
7. Goyal, V., Pandey, O., Sahai, A., Waters, B.: Attribute-based encryption for fine-grained access control of encrypted data. In: Proc. of CCS 2006, pp. 89–98 (2006)
8. Hohenberger, S., Waters, B.: Attribute-Based encryption with fast decryption. In: Kurosawa, K., Hanaoka, G. (eds.) PKC 2013. LNCS, vol. 7778, pp. 162–179. Springer, Heidelberg (2013)
9. Lewko, A., Okamoto, T., Sahai, A., Takashima, K., Waters, B.: Fully secure functional encryption: attribute-based encryption and (hierarchical) inner product encryption. In: Gilbert, H. (ed.) EUROCRYPT 2010. LNCS, vol. 6110, pp. 62–91. Springer, Heidelberg (2010)
10. Lewko, A., Waters, B.: New techniques for dual system encryption and fully secure HIBE with short ciphertexts. In: Micciancio, D. (ed.) TCC 2010. LNCS, vol. 5978, pp. 455–479. Springer, Heidelberg (2010)
11. Lewko, A., Waters, B.: Decentralizing attribute-based encryption. In: Paterson, K.G. (ed.) EUROCRYPT 2011. LNCS, vol. 6632, pp. 568–588. Springer, Heidelberg (2011)
12. Lin, H., Cao, Z., Liang, X., Shao, J.: Secure threshold multi authority attribute based encryption without a central authority. Information Sciences **180**(13), 2618–2632 (2010)
13. Liu, Z., Cao, Z.: On efficiently transferring the linear secret-sharing scheme matrix in ciphertext-policy attribute-based encryption. IACR Cryptology ePrint Archive (2010)
14. Longo, R.: Attribute Based Encryption with Algebraic Methods. Master's thesis (laurea magistrale), University of Trento, Department of Mathematics (2012)
15. Müller, S., Katzenbeisser, S., Eckert, C.: Distributed attribute-based encryption. In: Lee, P.J., Cheon, J.H. (eds.) ICISC 2008. LNCS, vol. 5461, pp. 20–36. Springer, Heidelberg (2009)
16. Ostrovsky, R., Sahai, A., Waters, B.: Attribute-based encryption with non-monotonic access structures. In: Proc. of CCS 2007, pp. 195–203 (2007)
17. Sahai, A., Waters, B.: Fuzzy identity-based encryption. In: Cramer, R. (ed.) EUROCRYPT 2005. LNCS, vol. 3494, pp. 457–473. Springer, Heidelberg (2005)

18. Shamir, A.: Identity-Based cryptosystems and signature schemes. In: Blakely, G.R., Chaum, D. (eds.) CRYPTO 1984. LNCS, vol. 196, pp. 47–53. Springer, Heidelberg (1985)
19. Waters, B.: Dual system encryption: realizing fully secure IBE and HIBE under simple assumptions. In: Halevi, S. (ed.) CRYPTO 2009. LNCS, vol. 5677, pp. 619–636. Springer, Heidelberg (2009)
20. Waters, B.: Ciphertext-Policy attribute-based encryption: an expressive, efficient, and provably secure realization. In: Catalano, D., Fazio, N., Gennaro, R., Nicolosi, A. (eds.) PKC 2011. LNCS, vol. 6571, pp. 53–70. Springer, Heidelberg (2011)

Extended Explicit Relations Between Trace, Definition Field, and Embedding Degree

Atsuko Miyaji[1,2], Xiaonan Shi[1], and Satoru Tanaka[1(✉)]

[1] Japan Advanced Institute of Science and Technology,
Asahidai 1–1, Nomi-shi, Ishikawa 923–1292, Japan
stanaka@jaist.ac.jp
[2] CREST, JST, Kawaguchi Center Building 4-1-8, Honcho,
Kawaguchi-shi, Saitama 332–0012, Japan

Abstract. An elliptic curve cryptosystem (ECC) is one of public key cryptosystem, whose security is based on elliptic curve discrete logarithm problem (ECDLP). An elliptic curve is uniquely determined by mathematical parameters such as j-invariant of an elliptic curve. By giving trace of elliptic curve, t, a definition field $\mathbb{F}_p$, and discriminant D, an elliptic curve with order $\sharp E(\mathbb{F}_p) = n$ is determined. Therefore it is an open problem to determine explicit relations between the mathematical parameters and the embedding degrees k. Hirasawa and Miyaji presented concrete relations between the mathematical parameters and the embedding degrees. In this research, a new explicit relation between elliptic-curve parameters and embedding degrees is investigated by generalizing their research.

Keywords: Elliptic curve · Embedding degree · Trace

1 Introduction

An elliptic curve cryptosystems (ECC) is one of public key cryptosystems, whose security is based on elliptic curve discrete logarithm problem (ECDLP) and which can achieve high security with a small key size. Another advantage of ECC is to use the Weil and Tate pairings to construct cryptographic protocols such as one-round key exchange [12], identity-based encryption [4], and short digital signatures [5]. These cryptosystems using pairings are called pairing-based cryptosystems. One of important issues on ECC is to evaluate the security level by using mathematical parameters such as j-invariant, definition field or trace. Remark that after MOV or FR-reduction [8,14] were proposed, one of security level of ECC is an embedding degree k, which means the security of ECDLP on $\mathbb{F}_p$ is equal to that of DLP on $\mathbb{F}_{p^k}$.

Only elliptic curves known to admit subgroups with appropriate k were supersingular curves, which are susceptible to discrete logarithm algorithms [16] until Miyaji, Nakabayashi and Takano [15] proposed ordinary elliptic curves of prime order and embedding degree $k \in \{3, 4, 6\}$ in 2001. Later, Brezing and Weng [6]

© Springer International Publishing Switzerland 2015
A. Maletti (Ed.): CAI 2015, LNCS 9270, pp. 165–175, 2015.
DOI: 10.1007/978-3-319-23021-4_15

presented an algebraic method, which generates ordinary elliptic curves with the ratio $\rho = \log p^m / \log \ell$ significantly less than 2, and the best known their results achieved $\rho \sim 5/4$ when $k = 8$ and 24. Barreto, Naehrig [3] and Freeman [7] then showed the constructions of elliptic curves of prime order with $k = 12$ and 10 using factorization of cyclotomic polynomials presented in [9], respectively.

Besides the above methods, Hirasawa and Miyaji [10] investigated concrete relations between definition field $\mathbb{F}_{p^m}$, order $\sharp E(\mathbb{F}_{p^m})$, trace t, and embedding degree k of $E(\mathbb{F}_{p^m})$ based on the elementary number theory, they gave two algorithms for searching mathematical elliptic curve parameters with a predetermined embedding degree k. In their results, k is determined as $k = 2^{r+1}L$ by r, L, where r can be any non-negative number but L can only be an odd prime. As a result, there are restrictions on the values of k. Their results focus on the most optimal case in cryptography, that is, prime order $\sharp E(\mathbb{F}_{p^m}) = n$ satisfies $n \mid \phi_k(p^m)$ where $\phi_k(x)$ is the cyclotomic polynomial. It is interesting to investigate another condition with an embedding degree k being even and $\rho > 1$ being as small as possible (say, $\rho \leq 5/4$) [3].

In this paper, we extend the methods of Hirasawa and Miyaji in two directions. First, we consider L to be positive integers ($L \geq 2$) from odd prime numbers such that k can be taken as all even values, which are advantageous from the point of view of efficient implementation of the pairing algorithm since the use of prime or odd k discourages many optimizations that are only possible for even k [2]. The second direction is to obtain more elliptic curve parameters by incorporating cofactors into the analysis. We present two searching algorithms dealing with searching elliptic curve parameters according to our extensional results. We also examine the CM discriminants D of our results and compare with the previous Hirasawa-Miyaji results as D is significant for constructions of elliptic curves.

This paper is organized as follows. Section 2 describes preliminary knowledge on elliptic curves and reviews Hirasawa and Miyaji's results. Section 3 shows our main contribution. We discuss the extensions of Hirasawa and Miyaji's results and present two searching algorithms according to our new relation. Section 5 compares our results with Hirasawa and Miyaji's results by discussing experimental results. We conclude and summarize our results in Section 6.

2 Preliminaries

Let $\mathbb{F}_{p^m}$ be a finite field for a prime p and a positive integer m. Let E be an elliptic curve defined over $\mathbb{F}_{p^m}$. The embedding degree is defined as follows:

Definition 1. *Let $E(\mathbb{F}_{p^m})$ be an elliptic curve defined over $\mathbb{F}_{p^m}$ whose group order is $\sharp E(\mathbb{F}_{p^m})$ is divisible by a prime ℓ. Then $E(\mathbb{F}_{p^m})$ has an **embedding degree** k with respect to ℓ if k is the smallest integer such that $\ell \mid p^{mk} - 1$.*

If $E(\mathbb{F}_{p^m})$ has a subgroup $\langle G \rangle$ with $\sharp \langle G \rangle = \ell$ and $\gcd(\ell, p) = 1$, then the Weil pairing e_ℓ is defined from the ℓ-torsion points on E, to an ℓth root of unity in $\mathbb{F}_{p^{mk}}^*$. An extension degree k is called an embedding degree. MOV-reduction and FR-reduction [8] embed the Elliptic Curve Discrete Logarithm Problem (ECDLP) in $E(\mathbb{F}_{p^m})$ to Discrete Logarithm Problem (DLP) in $\mathbb{F}_{p^{mk}}$, where there are attacks with the sub-exponential time. Therefore, as for pairing-based cryptosystems, we would like to build elliptic curves with embedding degrees k that are large enough for the DLP in the embedding field to be difficult, but small enough for the pairing to be efficiently computable. Unfortunately, however, for most non-supersingular curves, the embedding degree is enormous [1]. This is an open problem to give an explicit condition between embedding degrees k and the mathematical parameters such as the definition field $\mathbb{F}_{p^m}$ and $\sharp E(\mathbb{F}_{p^m})$.

The k-th cyclotomic polynomial $\phi_k(x)$ is defined as the minimum polynomial of the primitive k-th root of unity [13]. The embedding degree k of a subgroup $\langle G \rangle$ of $E(\mathbb{F}_{p^m})$ with $ord(G) = \ell$ is equal to the minimal number such that $\ell \mid \phi_k(p^m)$, i.e., $\phi_k(p^m) \equiv 0 \pmod{\ell}$. Let $t = p^m + 1 - \sharp E(\mathbb{F}_{p^m})$ be the trace of $E(\mathbb{F}_{p^m})$. Then, $(t-1) \equiv p^m \pmod{\ell}$, which implies $(t-1)^k \equiv p^{mk} \equiv 1 \pmod{\ell}$. Thus, $(t-1)$ must be a k-th root of unity modulo ℓ. That is, $E(\mathbb{F}_{p^m})$ has embedding degree k with respect to ℓ is equivalent to $\phi_k(t-1) \equiv 0 \pmod{\ell}$.

Hirasawa and Miyaji [10] presented an explicit relation between the definition field $\mathbb{F}_{p^m}$, $\sharp E(\mathbb{F}_{p^m})$ and embedding degree k by extending Hitt's results on hyperelliptic curve [11]. They also gave algorithms for searching elliptic-curve parameters with pre-determined embedding degree. We present their results, which will later be extended in this paper.

The following lemma determines order of an element a in $\mathbb{Z}_n$, which is denoted by $ord_n(a)$.

Lemma 1 ([10]). *Let $r, a, \lambda \in \mathbb{Z}$ ($r, \lambda \geq 0$), L be an odd prime, and $n = \frac{a^{2^r L}+1}{\lambda(a^{2^r}+1)}$. If $a^{2^r} \not\equiv -1 \pmod{n}$, then $ord_n(a) = 2^{r+1}L$.*

They also presented the following theorem that determines embedding degree k by describing a relation between embedding degree k and order n in $\mathbb{Z}_n$ $ord_n(a)$.

Theorem 1 ([10]). *Let $r, m, \lambda \in \mathbb{Z}(r, m, \lambda \geq 0)$. Let L and p be odd primes, $n = \frac{a^{2^r L}+1}{\lambda(a^{2^r}+1)}$, $D = gcd(ord_n(p), m)$. Then, the following two relations hold:*

1. *if $\sharp E(\mathbb{F}_{p^m}) = \frac{p^{2^r L}}{\lambda(p^{2^r}+1)} = n$ is a prime and $p^{2^r} \not\equiv -1 \pmod{n}$, then the embedding degree k of $E(\mathbb{F}_{p^m})$ is given by*
   - *$k = 2^{r+1-i}L$ when $D = 2^i (0 \leq i \leq r+1)$; and*
   - *$k = 2^{r+1-i}L$ when $D = 2^i L (0 \leq i \leq r+1)$;*
2. *if $\sharp E(\mathbb{F}_{p^m}) = \frac{(t-1)^{2^r L}}{\lambda((t-1)^{2^r}+1)} = n$ is a prime and $(t-1)^{2^r} \not\equiv -1 \pmod{n}$, then the embedding degree of $E(\mathbb{F}_{p^m})$ k is given by $k = 2^{r+1}L$.*

3 The New Relations

Lemma 1 and Theorem 1 cover only the cases where L is an odd prime. Extension degrees that Theorem 1 can construct are limited to $k = 2^{\alpha+1}q$ for $\mathbb{Z} \ni \alpha \geq 0$ and an odd prime q. This implies that the following extension degrees k can not be constructed by Theorem 1:

1. $k = 2^{\alpha+1} q\eta$ for $\mathbb{Z} \ni \alpha \geq 0$, an odd prime q, and an odd number η.
2. $k = 2^{\alpha+1}$ for $\mathbb{Z} \ni \alpha \geq 0$.

In this paper, we will extend conditions of L from odd primes to odd numbers and even numbers:

$$L = \{q\} \longrightarrow L = \{q\eta\} \cup \{2\beta \, (\mathbb{Z} \ni \beta \geq 2)\}.$$

Thus, we can construct elliptic curves with the above two types of embedding degrees k. Each embedding degree k has a different feature. For example. it is well known that even values of k are advantageous from the point of view of efficient implementation of pairing algorithms [3]. Thus, it is meaningful to extend conditions of embedding degrees k.

We discuss the case where L is odd composite numbers or powers of 2 in Lemma 2 or 3, respectively. In addition, the results in [10] investigates cases of prime-order elliptic curves, that is $n = \sharp E(\mathbb{F}_{p^m})$ is prime, i.e., $n \mid p^{mk} - 1$. This is the most relevant case in pairing-friendly elliptic curve cryptosystems. This, however, reduces the existence possibility of elliptic curves with given embedding degrees k. We extend prime-order elliptic curves to 'nearly-prime-order' elliptic curves. More specifically, we deal with elliptic curves E with $\sharp E(\mathbb{F}_{p^m}) = n = h \cdot \ell$ for the largest prime divisor ℓ of n and $\ell \mid p^{mk} - 1$ and a reasonably small cofactor h.

Lemma 2. *Let $r, a, \lambda \in \mathbb{Z}$ $(r \geq 0, \lambda > 0, a \neq 0)$, $L = q\eta$ be an odd positive integer, where q and η are a prime divisor and an odd divisor of L, respectively. If both $\ell = \frac{a^{2^r L}+1}{\lambda(a^{2^r \eta}+1)}$ and $a^{2^r \eta} \not\equiv -1 \pmod{\ell}$ hold, then $\mathrm{ord}_\ell(a) = 2^{r+1}L$.*

proof: *From $\ell = \frac{a^{2^r L}+1}{\lambda(a^{2^r \eta}+1)}$, we have $\lambda(a^{2^r \eta} + 1)\ell = a^{2^r L} + 1$. So, the following holds:*

$$a^{2^r L} \equiv -1 \pmod{\ell}. \tag{1}$$

This implies $a^{2^{r+1}L} \equiv 1 \pmod{\ell}$. Therefore, $\mathrm{ord}_\ell(a) \mid 2^{r+1}L$. On the other hand, from the above fact of $a^{2^r L} \equiv -1 \pmod{\ell}$, $\mathrm{ord}_\ell(a) \nmid 2^r L$, that is, both $\mathrm{ord}_\ell(a) \neq 2^j$ and $\mathrm{ord}_\ell(a) \neq 2^j L$ hold for $0 \leq \forall j \leq r$.

Suppose that $\mathrm{ord}_\ell(a) \mid 2^{r+1}\eta$, that is, $a^{2^{r+1}\eta} \equiv 1 \pmod{\ell}$. From the fact of $L = q\eta$, we can get the following sequences: $a^{2^r L} = a^{2^{r+1}\eta} a^{2^r(L-2\eta)} = a^{2^r(L-2\eta)}$. Continuing the same procedures for $L, L - 2\eta, L - 2 \cdot 2\eta, L - 2 \cdot 3\eta \cdots$, it reaches to $L - 2\frac{q-1}{2}\eta = q\eta - (q-1)\eta = \eta$ since q is an odd prime. Then, combining the equation (1), we get $a^{2^r L} \equiv \cdots \equiv a^{2^r \eta} \equiv -1 \pmod{\ell}$. This, however, contradicts the condition of Lemma 2 of $a^{2^r \eta} \not\equiv -1 \pmod{\ell}$. Hence, $\mathrm{ord}_\ell(a) \nmid 2^{r+1}\eta$. This implies that $\mathrm{ord}_\ell(a) \neq 2^j \eta$ $(0 \leq \forall j \leq r + 1)$, and $\mathrm{ord}_\ell(a) \neq 2^{r+1}$. Therefore, we have proved $\mathrm{ord}_\ell(a) = 2^{r+1}q\eta = 2^{r+1}L$ since q is an odd prime. $\square$

Lemma 3. *Let $r, a, \lambda \in \mathbb{Z}$. Let $L = 2\beta$ be an even positive integer for $\mathbb{Z} \ni \beta \geq 2$, where the next condition of $[(a > 0) \wedge (\lambda > 0)] \vee [(a < 0) \wedge (\text{even } \beta) \wedge (\lambda > 0)] \vee [(a < 0) \wedge (r > 0) \wedge (\lambda > 0)] \vee [(a < 0) \wedge (r = 0) \wedge (\lambda < 0)]$ holds. If $\ell = \frac{a^{2^r \beta}+1}{\lambda}$, then $\mathrm{ord}_\ell(a) = 2^r L$.*

proof: *From $\ell = \frac{a^{2^r\beta}+1}{\lambda}$, we have $\lambda\ell = a^{2^r\beta}+1$. So, $a^{2^r\beta} \equiv -1 \pmod{\ell}$, and, thus $a^{2^r2\beta} \equiv 1 \pmod{\ell}$ holds. Therefore, $\mathrm{ord}_\ell(a) \mid 2^rL$, $\mathrm{ord}_\ell(a) \neq 2^j$ $(0 \leq \forall j \leq r)$, and $\mathrm{ord}_\ell(a) \neq 2^j\beta$ $(0 \leq \forall j \leq r)$ hold. Thus, the only possible order of $\mathrm{ord}_\ell(a)$ is 2^rL, then the lemma follows.* □

Remark 1. Conditions of a are determined by those to lead a positive integer ℓ in Lemmas 2 and 3, respectively. In Lemma 2, when $a > 0$, obviously, $\ell = \frac{a^{2^rL}+1}{\lambda(a^{2^r\eta}+1)}$ is positive for any $r \geq 0$ and any $L > 0$. When $a < 0$ and $r = 0$, both $a^{2^rL}+1$ and $a^{2^r\eta}+1$ are negative since both L and η are odd numbers. Therefore, ℓ is positive. When $a < 0$ and $r > 0$, both $a^{2^rL}+1$ and $a^{2^r\eta}+1$ are positive. Thus, ℓ is positive. This discussion leads the conditions of a in Lemma 2, that is $a \neq 0$.

In the case of Lemma 3, when $a < 0$, $r = 0$, $\lambda < 0$ and $L = 2\beta$ with an odd number β, $\ell = \frac{a^{2^r\beta}+1}{\lambda} > 0$. When $a < 0$, $r = 0$, $\lambda > 0$ and $L = 2\beta$ with an even number β, $\ell = \frac{a^{2^r\beta}+1}{\lambda} > 0$. When $r > 0$ and $\lambda > 0$, $\ell = \frac{a^{2^r\beta}+1}{\lambda} > 0$, no matter the sign of a. Therefore, the conditions of a in Lemma 3, $[(a > 0) \wedge (\lambda > 0)] \vee [(a < 0) \wedge (\text{even } \beta) \wedge (\lambda > 0)] \vee [(a < 0) \wedge (r > 0) \wedge (\lambda > 0)] \vee [(a < 0) \wedge (r = 0) \wedge (\lambda < 0)]$ are leaded.

These discussion becomes important in Algorithms 1 and 3

We are interested in a relation between trace t, group order ℓ, and embedding degree k. Thus if we apply $a = t - 1$ to Lemmas 2 and 3, we have the following theorem that describes such a relation.

Theorem 2. *Let $r, \lambda \in \mathbb{Z}$ $(r \geq 0, \lambda > 0)$. Let L be a positive integer, t be the trace and k be the embedding degree of $E(\mathbb{F}_p)$ with order $\sharp E(\mathbb{F}_p) = n = h\ell$, then the following relations holds:*

1. *If $L = q\eta$ is an odd number for a prime q, $\ell = \frac{(t-1)^{2^rL}+1}{\lambda((t-1)^{2^r\eta}+1)}(t - 1 \neq 0)$, and $(t - 1)^{2^r\eta} \not\equiv -1 \pmod{\ell}$, then $k = 2^{r+1}L$.*

2. *If $L = 2\beta$ is an even number for $\beta \geq 2$ and $\ell = \frac{(t-1)^{2^r\beta}+1}{\lambda}(t - 1 \neq 0)$ with either condition of $[(t > 1) \wedge (\lambda > 0)] \vee [(t < 1) \wedge (\text{even } \beta) \wedge (\lambda > 0)] \vee [(t < 1) \wedge (r > 0) \wedge (\lambda > 0)] \vee [(t < 1) \wedge (r = 0) \wedge (\lambda < 0)]$, then $k = 2^rL$.*

proof: *As for 1, by substituting $a = t - 1$ into Lemma 2, we obtain $\ell = \frac{(t-1)^{2^rL}+1}{\lambda((t-1)^{2^r x}+1)}$. Since $t = p + 1 - n$, we have $t - 1 \equiv p \pmod{\ell}$. Since k is the smallest positive integer such that $p^k = 1 \pmod{\ell}$, $(t - 1)^k \equiv p^k \equiv 1 \pmod{\ell}$ holds. Therefore, we get embedding degree $k = \mathrm{ord}_\ell(p) = \mathrm{ord}_\ell(t - 1) = 2^{r+1}L$.*

As for 2, the proof follows in the same way as 1. □

4 Searching Algorithm

Based on the parametrization given in Theorem 2, we present two searching algorithms Algorithms 1 and 3, which for input of embedding degree k, the maximum cofactor $h_{\max}$, the initial maximum trace $t_{\max}$, and the security level Lev, outputs elliptic curve parameters such as definition field $\mathbb{F}_p$, order of elliptic curve $\sharp E(\mathbb{F}_p) = n = h\ell$, and the trace $t = p + 1 - n$. Algorithm 1 corresponds to the first case of Theorem 2; and Algorithm 3 corresponds to the second case of Theorem 2.

We also examine the ratio $\rho = \frac{\log p}{\log \ell}$, which is set to be the ideal range of $1 \leq \rho < 2$ as for secure and efficient implementation. In addition, CM discriminant D is also computed.

Algorithm 1. Searching $\sharp E(\mathbb{F}_p) = n = h \cdot \ell$ with $k = 2^{r+1}L$ (L is odd).

Require: $k = 2^{r+1}L$, $t_{\min}$, $t$max, $h$max, and Lev,
Ensure: $\mathbb{F}_p$, $\#E(\mathbb{F}_p) = n = h\ell$, ratio ρ
1: Set S to a set of prime divisors of L.
2: **for** all $q \in S$ **do**
3: $\quad \eta \leftarrow L/q$.
4: $\quad$ **while** $|t| \in [t_{\min}, t$max$]$ **do**
5: $\quad\quad \Gamma \leftarrow \frac{(t-1)^{2^r L}+1}{(t-1)^{2^r \eta}+1}$.
6: $\quad\quad$ **if** $(t-1)^{2^r \eta} \equiv -1 \pmod{\ell}$ **then**
7: $\quad\quad\quad$ go to step 4 for the next t.
8: $\quad\quad$ **end if**
9: $\quad\quad$ **if** The largest prime divisor ℓ of Γ satisfies $\ell <$ Lev **then**
10: $\quad\quad\quad$ go to step 4 for the next t.
11: $\quad\quad$ **end if**
12: $\quad\quad$ **for** $h = 1$ to $h$max **do**
13: $\quad\quad\quad$ Set $n \leftarrow h \cdot \ell$, $p \leftarrow n + t - 1$.
14: $\quad\quad\quad$ **if** p is a prime. **then**
15: $\quad\quad\quad\quad$ output $\{t, p_1, n, \ell\}$.
16: $\quad\quad\quad$ **end if**
17: $\quad\quad$ **end for**
18: $\quad$ **end while**
19: **end for**

Remark 2. In Algorithm 1, traces t can be taken positive and negative if $t - 1 \neq 0$ as we have discussed in Remark 1. $\Gamma = \frac{(t-1)^{2^r L}+1}{(t-1)^{2^r x}+1}$ has same value for t and $-t+2$ if $r > 0$, so after finding desired largest prime divisor ℓ of Γ for a certain t, we are able to apply this ℓ to $-t + 2$. When trace is equal to $-t + 2$, we have $p = n + (-t + 2) - 1 = n - t + 1$. Thus, $|t| \in [t_{\min}, t_0]$ needs to be checked but some negative t can be skipped. This short-cut algorithm will be shown in Algorithm 2.

The Algorithms 1 and 3 output all possible elliptic curve parameters for one taken trace t. We discuss some details of the two algorithms as follows:

Remark 3. In Algorithm 3, traces $t - 1$ can be taken positive and negative ($t - 1 \neq 0$) only if ($r > 0$) or ($r = 0$ and x is even) as we have discussed in Remark 1.

Algorithm 2. Searching $\sharp E(\mathbb{F}_p) = n = h \cdot \ell$ with $k = 2^{r+1}L$ (L is odd) and $r > 0$.

Require: $k = 2^{r+1}L$, $t_{\min}$, $t_{\max}$, $h_{\max}$, and Lev,

Ensure: $\mathbb{F}_p$, $\#E(\mathbb{F}_p) = n = h\ell$, ratio ρ

1: Set S to a set of prime divisors of L.

2: **for** all $q \in S$ **do**

3:     $\eta \leftarrow L/q$.

4:     **for** $t = t_{\max}$ to $t_{\min}$ **do**

5:         $\Gamma \leftarrow \frac{(t-1)^{2^r L} + 1}{(t-1)^{2^r \eta} + 1}$.

6:         **if** $(t-1)^{2^r \eta} \equiv -1 \pmod{\ell}$ **then**

7:             Go to step 4 for the next t.

8:         **end if**

9:         **if** The largest prime divisor ℓ of Γ satisfies $\ell <$ Lev **then**

10:             Go to step 4 for the next t.

11:         **end if**

12:         **for** $h = 1$ to $h_{\max}$ **do**

13:             Set $n \leftarrow h \cdot \ell$, $p_1 \leftarrow n + t - 1$.

14:             **if** p_1 is a prime. **then**

15:                 Output $\{t, p_1, n, \ell \}$.

16:             **end if**

17:             **if** $t \geq t_{\min} + 2$ **then**

18:                 Set $p_2 \leftarrow n + (2 - t) - 1 = n - t + 1$.

19:                 **if** p_2 is a prime. **then**

20:                     Output $\{2 - t, p_2, n, \ell \}$.

21:                 **end if**

22:             **end if**

23:         **end for**

24:     **end for**

25:     Set $t \leftarrow -t_{\max}$ or $-t_{\max} + 1$ and do the same procedure of steps 4 to 16.

26: **end for**

In this condition, we are able to search curve parameters with the corresponding negative trace $(-t + 2)$ of a positive t in same way as we have discussed in Remark 2. However, if ($r = 0$ and x is odd), negative traces lead to negative $\Gamma = \frac{(t-1)^{2^r L} - 1}{(t-1)^{2^r x} - 1}$, and thus we can not check p for negative traces.

5   Comparison

We compare our results with Hirasawa, Miyaji's results through implementing Algorithms 1 and 3 for $t \in [2^{\lceil \log p^m / (2^r(L-x)) \rceil}, 2^{\lceil \log p^m / (2^r(L-x)) \rceil} + 10000]$. Our implementation assumption is that $6 \leq k \leq 30$, $3 \leq L \leq 15$, and $0 \leq B \leq \frac{\log p^m}{2}$ when n is near prime. And we do comparison in the following two directions:

1. the extension of the embedding degree k to be arbitrary even values.
2. the extension that taking order n to be near prime.

Algorithm 3. Searching $\sharp E(\mathbb{F}_p) = n = h \cdot \ell \; k = 2^r L (L = 2\beta$ is even).

Require: $k = 2^r L = 2^{r+1} \beta$, $t_{\max}$, $h_{\max}$, and Lev,
Ensure: $\mathbb{F}_p$, $\#E(\mathbb{F}_p) = n = h\ell$, ratio ρ
1: Set $\beta \leftarrow L/2$.
2: **while** $|t| \in [t_{\min}, t_{\max}]$ **do**
3:     Set $\Gamma \leftarrow (t-1)^{2^r \beta} + 1$.
4:     **if** The largest prime divisor ℓ of Γ satisfies $\ell <$ Lev **then**
5:         Go to step 4 for the next t.
6:     **end if**
7:     **for** $h = 1$ to $h_{\max}$ **do**
8:         Set $n \leftarrow h \cdot \ell$, $p_1 \leftarrow n + t - 1$.
9:         **if** p_1 is a prime. **then**
10:           Output $\{t, p_1, n, \ell\}$.
11:         **end if**
12:     **end for**
13: **end while**

Hirasawa and Miyaji presented Table 1 for showing the total number of elliptic-curve parameters $\sharp\{n, p^m, t\}$ searched by their algorithm with fixed k (r, L) under running 10,000 times of different t. These parameters correspond to 160-bit ordinary elliptic curves of prime order n where the size of n should be equal or larger than 150 bits. Since the discriminant D is an essential parameter related to construction of elliptic curve, that is, when D is not large, say $\sim 10^9$, the elliptic curve can be constructed by using CM method. Therefore we investigate the size of D of their searched parameters and add to Table 1.

Table 1. $\sharp$parameters with different $k = 2^{r+1}L$ over $t \in [2^{\lceil 160/(2^r(L-1))\rceil}$, $2^{\lceil 160/(2^r(L-1))\rceil} + 10^5]$ (Experimental results in [10])

160-bit prime p				
k	r	L	$\sharp\{n, p^m, t\}$	size of D
10	0	5	225	141 ∼ 160
12	1	3	136	137 ∼ 160
14	0	7	135	138 ∼ 160
20	1	5	180	137 ∼ 160
24	2	3	84	139 ∼ 160

First, we examine all even values of k in the range of $k \in [8, 30]$ without changing the condition of n, i.e., n is still prime. The number of parameters corresponding to 160-bit elliptic curves under running 10^5 kinds of t are shown in Table 2.

For the value of k, 8, 16, 18 and 30 are only available by using our results. Furthermore, it becomes possible for factorizing fixed k to different ways. For example, $k = 12$ can be factored as $12 = 4 \cdot 3(\{r, L\} = \{1, 3\})$, and $12 = 2 \cdot 6(\{r, L\} = \{1, 6\})$. This implies the number of parameters can be increased by different factorizations of embedding degree k. However, we conjecture that factoring k with L to be prime probably can achieve more results. The reason might be that when $n = \frac{(t-1)^{2^r L}+1}{(t-1)^{2^r}+1}$ is 160 bits, the probability that the prime divisor $l = n/\lambda$ larger than 150 bits is much higher than when 160-bit $n = \frac{(t-1)^{2^r L}-1}{(t-1)^{2^r x}-1}$.

Table 2. ♯parameters over $t \in [2^{\lceil 160/(2^r(L-x))\rceil}, 2^{\lceil 160/(2^r(L-x))\rceil} + 10^5]$ ($k = 2^{r+1}L$ for odd L or $k = 2^r L$ for even L, n is prime)

160-bit prime p				
k	r	L	$\sharp\{n, p^m, t\}$	size of D
6	0	6	89	$134 \sim 160$
8	1	4	149	$143 \sim 160$
10	0	10	97	$137 \sim 160$
12	1	6	29	$135 \sim 160$
14	0	14	9	$138 \sim 160$
16	1	8	111	$148 \sim 160$
18	0	9	174	$141 \sim 160$
20	1	10	9	$136 \sim 160$
22	0	22	10	$137 \sim 160$
24	1	12	90	$142 \sim 160$
26	0	26	31	$137 \sim 160$
28	1	14	57	$136 \sim 160$
30	0	15	109	$137 \sim 160$

Then we choose same values of k in Table 2 and investigate the case where n is near prime and $n = h \cdot l$ with l is prime, for all integer $h \leq 10^3$. Since comparison should be taken in same security level, l is chosen as the same size

Table 3. ♯parameters, size of D, and ρ over $t \in [2^{\lceil 160/(2^r(L-x))\rceil}, 2^{\lceil 160/(2^r(L-x))\rceil} + 10^5]$. ($k = 2^{r+1}L$ for odd L or $k = 2^r L$ for even L, n is prime)

160-bit prime p					
k	r	L	$\sharp\{l, n, p^m, D\}$	size of D	$\rho(average)$
6	0	3	17138	$138 \sim 172$	1.05316
8	1	4	41640	$132 \sim 172$	1.05318
10	0	5	22916	$142 \sim 172$	1.05324
12	1	3	72974	$141 \sim 172$	1.05324
14	0	7	34185	$145 \sim 174$	1.05258
16	1	8	33261	$139 \sim 172$	1.05310
18	0	9	32490	$141 \sim 174$	1.05276
20	1	5	54842	$142 \sim 174$	1.05304
22	0	11	56300	$138 \sim 186$	1.05153
24	2	3	121913	$137 \sim 174$	1.05309
26	0	13	170618	$131 \sim 212$	1.04739
28	1	7	169781	$142 \sim 212$	1.04739
30	0	15	N/A	N/A	N/A

of n in Table 1. The results of running 10^5 kinds of t based on all proposed algorithms are shown in Table 3. By the Table 3, the number of elliptic-curve parameters significantly increased by transforming n is prime to n is near prime ($\rho \sim 1$), however it is rare enough to find such curve has order of nearly prime.

6   Conclusion

We have generalized the Hirasawa-Miyaji method to search mathematical parameters with pre-determined arbitrary even k of near prime order elliptic curves. This allows us to more flexibly choose k and find more elliptic-curve parameters with ρ closer to one.

References

1. Balasubramanian, R., Koblitz, N.: The improbability that an elliptic curve has subexponential discrete log problem under the Menezes-Okamoto-Vanstone algorithm. Journal of Cryptology **11**(2), 141–145 (1998)
2. Barreto, P.S.L.M., Lynn, B., Scott, M.: On the selection of pairing-friendly groups. In: Cohen, H. (ed.) 10th International Workshop on Revised Selected Papers of Selected Areas in Cryptography, SAC 2003. LNCS, vol. 3006, pp. 17–25. Springer, Heidelberg (2004)
3. Barreto, P.S.L.M., Naehrig, M.: Pairing-Friendly Elliptic Curves of Prime Order. In: Preneel, B., Tavares, S. (eds.) SAC 2005. LNCS, vol. 3897, pp. 319–331. Springer, Heidelberg (2006)
4. Boneh, D., Boyen, X.: Efficient selective identity-based encryption without random oracles. Journal of Cryptology **24**(4), 659–693 (2011)
5. Boneh, D., Lynn, B., Shacham, H.: Short Signatures from the Weil Pairing. In: Boyd, C. (ed.) ASIACRYPT 2001. LNCS, vol. 2248, pp. 514–532. Springer, Heidelberg (2001)
6. Brezing, F., Weng, A.: Elliptic curves suitable for pairing based cryptography. Designs, Codes and Cryptography **37**(1), 133–141 (2005)
7. Freeman, D.: Constructing Pairing-Friendly Elliptic Curves with Embedding Degree 10. In: Hess, F., Pauli, S., Pohst, M. (eds.) ANTS 2006. LNCS, vol. 4076, pp. 452–465. Springer, Heidelberg (2006)
8. Frey, G., Rück, H.G.: A remark concerning m-divisibility and the discrete logarithm in the divisor class group of curves. Mathematics of Computation **62**, 865–874 (1994)
9. Galbraith, S.D., McKee, J.F., Valença, P.C.: Ordinary abelian varieties having small embedding degree. Proceedings of Workshop on Mathematical Problems and Techniques in Cryptology **13**(4), 29–45 (2004)
10. Hirasawa, S., Miyaji, A.: New concrete relation between trace, definition field, and embedding degree. IEICE Transactions on Funamentals of Electronics, Communications and Computer Science 94-A(6), 1368–1374 (2011)
11. Hitt, L.: On the Minimal Embedding Field. In: Takagi, T., Okamoto, E., Okamoto, T., Okamoto, T. (eds.) Pairing 2007. LNCS, vol. 4575, pp. 294–301. Springer, Heidelberg (2007)

12. Joux, A.: A one round protocol for tripartite Diffie-Hellman. In: Bosma, W. (ed.) 4th International Symposium on Proceedings of Algorithmic Number Theory, ANTS-IV, pp. 385–394. Springer, Heidelberg (2000)
13. Lang, S.: Algebra, 3rd edn. Addison-Wesley (1993)
14. Menezes, A., Okamoto, T., Vanstone, S.: Reducing elliptic curve logarithms to logarithms in a finite field. IEEE Transactions on Infomation Theory **39**, 1639–1646 (1993)
15. Miyaji, A., Nakabayashi, M., Takano, S.: New explicit conditions of elliptic curve traces for fr-reduction. IEICE Transactions on Fundamentals of Electronics, Communications and Computer Sciences E84-A(5), 1234–1243 (2001)
16. Schirokauer, O., Weber, D., Denny, T.F.: Discrete logarithms: The effectiveness of the index calculus method. In: Cohen, H. (ed.) Proceedings of Algorithmic Number Theory - ANTS-II, Second International Symposium. LNCS, vol. 1122, pp. 337–361. Springer, Heidelberg (1996)

Complexity of Uniform Membership of Context-Free Tree Grammars

Johannes Osterholzer[✉]

Faculty of Computer Science,
Technische Universität Dresden, 01062 Dresden, Germany
johannes.osterholzer@tu-dresden.de

Abstract. We show that the uniform membership problem of context-free tree grammars is PSPACE-complete. The proof of the upper bound is by construction of an equivalent pushdown tree automaton representable in polynomial space. With this technique, we also give an alternative proof that the respective non-uniform membership problem is in NP. A corollary for uniform membership of ε-free indexed grammars is obtained.

1 Introduction

Context-free tree grammars (cftg) [3,10] generalize the concept of context-free rewriting to the realm of tree languages. They have been studied, among others, for their close connection to indexed grammars: their yield languages are precisely the indexed languages [1,10]. Recently, there has been renewed interest in cftg within the area of natural language processing, as they – and related formalisms such as tree adjoining grammars – allow modelling particular linguistic phenomena.

In this paper, we investigate the computational complexity of the uniform membership problem of cftg. In Section 5, this problem is shown to be PSPACE-complete. In order to prove containment in PSPACE, an equivalent pushdown tree automaton (pta) $M^\dagger$ [4] is constructed from G in a succession of intermediate steps (Section 4). We demonstrate that $M^\dagger$ can be implemented in polynomial space. The idea behind $M^\dagger$ is taken from Aho's proof that the indexed languages are context-sensitive [1, Sec. 5]. Note that in [1], the construction is given directly by means of a rather complex Turing machine, and without proof of correctness. In contrast, by employing pta, we can provide a formal proof; moreover, we think that this presentation is easier to understand. As a corollary, we establish the PSPACE-completeness of uniform membership of ε-free indexed grammars.

To show that the constructed pta $M^\dagger$ is also of potential interest besides the paper's main theorem, we use $M^\dagger$ in Section 6 for an alternative proof of the fact that the non-uniform membership problem of cftg is in NP. Note that this result already follows from the containment of the indexed languages in NP, whose proof in [11] rests, however, upon the correctness of the Turing machine mentioned above. In [7], containment in NP was proven for the class of output languages of compositions of macro tree transducers, which contains the context-free tree languages properly.

© Springer International Publishing Switzerland 2015
A. Maletti (Ed.): CAI 2015, LNCS 9270, pp. 176–188, 2015.
DOI: 10.1007/978-3-319-23021-4_16

Recall that there are two restricted modes of derivation for cftg, the OI and the IO mode. In fact, the OI mode is equivalent to the unrestricted mode used in this paper [3]. For complexity results on cftg under the IO mode cf. [2,14].

2 Preliminaries

The set of natural numbers with zero is denoted by $\mathbb{N}$, and the set $\{1, \ldots, n\}$ by $[n]$, for every $n \in \mathbb{N}$. Note that $[0] = \emptyset$, the empty set. Let A be a set. Given relations $R, S \subseteq A \times A$, their product $R \circ S$ is the relation $\{(a, c) \in A \times A \mid \exists b \in A : (a, b) \in R, (b, c) \in S\}$. An alphabet is a finite nonempty set. The set of words over A is A^*, the empty word is ε, and $A^+ = A^* \setminus \{\varepsilon\}$. Let $w = a_1 \ldots a_n$ with $a_1, \ldots, a_n \in A$ for some $n \in \mathbb{N}$. Then $|w| = n$, and $\widetilde{w} = a_n \ldots a_1$, the reversal of w.

An alphabet Σ equipped with a function $\mathrm{rk}_\Sigma \colon \Sigma \to \mathbb{N}$ is a *ranked alphabet*. Let Σ be a ranked alphabet. When Σ is obvious, we write rk instead of rk_Σ. Let $k \in \mathbb{N}$. Then $\Sigma^{(k)} = \mathrm{rk}^{-1}(k)$. We often write $\sigma^{(k)}$ and mean that $\mathrm{rk}(\sigma) = k$. We assume tacitly that there are some $\alpha^{(0)}$ and $\sigma^{(n)} \in \Sigma$ such that $n \geq 2$. Let U be a set and Λ denote $\Sigma \cup U \cup C$, where C is made up of the three symbols '(', ')', and ','. The set $T_\Sigma(U)$ of *trees (over Σ indexed by U)* is the smallest set $T \subseteq \Lambda^*$ such that $U \subseteq T$, and for every $k \in \mathbb{N}$, $\sigma \in \Sigma^{(k)}$, and $\xi_1, \ldots, \xi_k \in T$, we also have that $\sigma(\xi_1, \ldots, \xi_k) \in T$. A tree $\alpha()$ is abbreviated by α, a tree $\gamma(\xi)$ by $\gamma\xi$, and $T_\Sigma(\emptyset)$ by T_Σ. Let $\xi, \zeta \in T_\Sigma(U)$. The set of *positions (Gorn addresses) of* ξ is denoted by $\mathrm{pos}(\xi) \subseteq \mathbb{N}^*$. The *size* $|\xi|$ of ξ is $|\mathrm{pos}(\xi)|$. Denote the *label* of ξ at its position w by $\xi(w)$, and the *subtree* of ξ at w by $\xi|_w$. The result of *replacing* the subtree $\xi|_w$ in ξ by ζ is $\xi[\zeta]_w$. Given $k \in \mathbb{N}$, $\xi, \xi_1, \ldots, \xi_k \in \Lambda^*$, and pairwise different $u_1, \ldots, u_k \in U$, denote by $\xi[u_1/\xi_1, \ldots, u_k/\xi_k]$ the result of *substituting* every occurrence of u_i in ξ with ξ_i, where $i \in [k]$. If $\xi, \xi_1, \ldots, \xi_k$ are trees in $T_\Sigma(U)$, then so is $\xi[u_1/\xi_1, \ldots, u_k/\xi_k]$. We will use the sets of *variables* $X = \{x_1, x_2, \ldots\}$ and $Y = \{y\}$. For each $k \in \mathbb{N}$, let $X_k = \{x_i \mid i \in [k]\}$. Unless specified, Σ and N denote arbitrary ranked alphabets, and Γ an arbitrary alphabet.

We presuppose the basic definitions and results from computational complexity theory, cf. e.g. [8]. In particular, we will use the same concept of reduction as in [8]; i.e., many-one reductions that are computable by a deterministic multi-tape Turing machine with work tape space in $\mathcal{O}(\log n)$. Functions that are computable in this manner are *logspace-computable*. Assuming a reasonable encoding, operations on trees such as determining the j-th subtree of a node, or substitution at a given position, are logspace-computable, cf. [6, Lem. 2].

3 Context-Free Tree Grammars and Pushdown Automata

A *context-free tree grammar (cftg)* over Σ is a tuple $G = (N, \Sigma, S, P)$ such that Σ and N are disjoint ranked alphabets (of *terminal* resp. *nonterminal symbols*), $S \in N^{(0)}$, and P is a finite set of *productions* of the form $A(x_1, \ldots, x_k) \to \xi$ for some $k \in \mathbb{N}$, $A \in N^{(k)}$, and $\xi \in T_{N \cup \Sigma}(X_k)$. Let $G = (N, \Sigma, S, P)$ be a cftg. Given $\zeta_1, \zeta_2 \in T_{N \cup \Sigma}$, we write $\zeta_1 \Rightarrow_G \zeta_2$ if there are $(A(x_1, \ldots, x_k) \to \xi) \in P$ and $w \in \mathrm{pos}(\zeta_1)$ such that $\zeta_1(w) = A$ and $\zeta_2 = \zeta_1[\xi[x_1/\zeta_1|_{w1}, \ldots, x_k/\zeta_1|_{wk}]]_w$.

Table 1. Membership problems of cftg (over Σ)

	Uniform Membership	Membership
Input	cftg G over Σ, $\xi \in T_\Sigma$	$\xi \in T_\Sigma$
Question	Is $\xi \in L(G)$?	Is $\xi \in L(G)$ for a fixed cftg G over Σ?

The *tree language of* G, denoted by $L(G)$, is the set $\{\xi \in T_\Sigma \mid S \Rightarrow^*_G \xi\}$. In this situation, we call $L(G)$ a *context-free tree language*. The *size* of G, denoted by $|G|$, is $|N| + \sum_{(l \to r) \in P}(|l| + |r|)$.

In this work, we will investigate the *uniform membership problem*, as well as the *(non-uniform) membership problem* of cftg (over Σ), as defined in Tab. 1.

A *pushdown tree system (pts)* is a tuple $M = (Q, \Sigma, \Gamma, q_0, R)$ such that Q is an alphabet (of *states*), Σ is a ranked alphabet, Γ is a nonempty set, $q_0 \in Q$, and R is a set of *rules* of the following three forms:

$$(i)\; q(y) \to \sigma(p_1(y), \ldots, p_k(y)), \qquad (ii)\; q(y) \to p(\gamma y), \qquad (iii)\; q(\gamma y) \to p(y),$$

where $y \in Y$, $\sigma \in \Sigma^{(k)}$ for some $k \in \mathbb{N}$, $q, p, p_1, \ldots, p_k \in Q$, and $\gamma \in \Gamma$. We call M a *pushdown tree automaton (pta)* when Γ (and thus R) is finite.[1] The set of rules from R of form *(i)* (resp. *(ii)*, *(iii)*) is denoted by R_Σ (resp. by $R_\uparrow$, $R_\downarrow$), and their elements are called *stay*, *push*, and *pop rules*.

Given a pts $M = (Q, \Sigma, \Gamma, q_0, R)$, let $\mathcal{C}_M = \{q(\eta) \mid q \in Q, \eta \in \Gamma^*\}$ and $\mathcal{DF}_M = T_\Sigma(\mathcal{C}_M)$, the sets of *configurations* and *derivation forms* of M. Given a rule $\rho \in R$ of the form $l \to r$, and $\zeta_1, \zeta_2 \in \mathcal{DF}_M$, we write $\zeta_1 \Rightarrow^\rho_M \zeta_2$ if there are some $w \in \mathrm{pos}(\zeta_1)$ and $\eta \in \Gamma^*$ such that $\zeta_1|_w = l[y/\eta]$, $\zeta_2 = \zeta_1\big[r[y/\eta]\big]_w$, and w is the leftmost position in ζ_1 which is labeled by an element of Q. We let $\Rightarrow_M = \bigcup_{\rho \in R} \Rightarrow^\rho_M$. Let $\zeta, \zeta' \in \mathcal{DF}_M$. A *derivation* of ζ' from ζ in M is a sequence $\rho_1 \ldots \rho_m \in R^*$ such that there are $\zeta_0, \ldots, \zeta_m \in \mathcal{DF}_M$ where $\zeta = \zeta_0$, $\zeta_{i-1} \Rightarrow^{\rho_i}_M \zeta_i$ for every $i \in [m]$, and $\zeta' = \zeta_m$. If d is of this form, we write $\zeta_0 \Rightarrow^d_M \zeta_m$. The set of all derivations of ζ' from ζ in M is denoted by $\mathcal{D}_M(\zeta, \zeta')$. We let $\mathcal{D}_M = \bigcup_{\zeta \in \mathcal{C}_M, \xi \in T_\Sigma} \mathcal{D}_M(\zeta, \xi)$. The *tree language of* M, denoted by $L(M)$, is the set $\{\xi \in T_\Sigma \mid q_0(\varepsilon) \Rightarrow^*_M \xi\}$, and the *size of* M, which is denoted by $|M|$, is $|Q| + |\Gamma| + \sum_{(l \to r) \in R}(|l| + |r|)$.

Let Z_1 be a cftg or a pts, and Z_2 be a cftg or a pts. Then Z_1 and Z_2 are *equivalent* if $L(Z_1) = L(Z_2)$. If not specified otherwise, G will denote an arbitrary cftg (N, Σ, S, P) in the sequel, and M an arbitrary pta $(Q, \Sigma, \Gamma, q_0, R)$.

As proven in [4, Thm. 1], pta accept exactly the context-free tree languages. A close inspection of the proof shows that all constructions are logspace-computable.

---

[1] In fact, pta as given here are in two ways a special case of the *restricted pushdown tree automata (rpta)* of [4]. First, the pushdowns of rpta are monadic trees from $T_\Gamma(\{Z\})$ for some distinct nullary symbol Z, while pta use words over Γ. Both approaches are clearly equivalent. Second, the rules of pta are more restricted than those of rpta. However, in the construction of an equivalent rpta from a cftg in [4, Thm. 3], only rules of type *(i)*–*(iii)* are created, so the restriction has no impact.

Lemma 1 ([4]). *Let $L \subseteq T_\Sigma$. There is a cftg G such that $L = L(G)$ iff there is a pta M such that $L = L(M)$. Also, M is logspace-computable from G, and vice versa.*

4   Compact pts and Finite Representations

Both in our analysis of uniform and non-uniform membership of cftg, the proof of containment in the respective complexity class rests on the successive application of certain transformations to the pta M that is obtained from an input cftg G by using Lem. 1.

We introduce these transformations in the following subsections, drawing upon ideas Aho used in the construction of the linear bounded automaton that demonstrates the context-sensitivity of the indexed languages [1, Sec. 5]. We apply these ideas directly at the level of pta instead of building a complex Turing machine, and provide formal proofs of correctness.

4.1   Augmented pta

Later on, we will require for our proofs that the amount of steps in a derivation of a tree $\xi \in T_\Sigma$ in the pta M is bounded. However, for an arbitrary pta, it seems difficult to find the respective bound. This is due to the presence of *unnecessary turns* of M. Take, e.g., the following derivation in some pta M:

$$q(\gamma) \Rightarrow_M q_1(\delta\gamma) \Rightarrow_M q_2(\tau\delta\gamma) \Rightarrow_M q_3(\delta\gamma) \Rightarrow_M q_4(\gamma) \Rightarrow_M p(\varepsilon).$$

Clearly, the turn $q(\gamma) \Rightarrow_M^* q_4(\gamma)$ was in some sense unnecessary, and we could have avoided it if there was already some rule $q(\gamma y) \to p(y)$ in R, because then $q(\gamma) \Rightarrow_M p(\varepsilon)$. The existence of such rules to avoid unnecessary turns is exactly what constitutes an *augmented* pta. Formally, a pta M is augmented if for every $q_1, q_2, q_3 \in Q$ and $\gamma \in \Gamma$ such that $q_1(\varepsilon) \Rightarrow_M q_2(\gamma) \Rightarrow_M q_3(\varepsilon)$, and for every rule $q_3(l) \to r$ in R, the rule $q_1(l) \to r$ is also in R.

Lemma 2. *For every pta M, an equivalent augmented pta M' is constructible in polynomial time.*

Proof. Given a pta M, define the pta $M' = (Q, \Sigma, \Gamma, q_0, R')$, where R' results from the following fixed-point iteration. Initially, let $R' = R$. Then, while there are $q_1, q_2, q_3 \in Q$, $\gamma \in \Gamma$, and $(q_3(l) \to r) \in R'$ such that $q_1(\varepsilon) \Rightarrow_{M'} q_2(\gamma) \Rightarrow_{M'} q_3(\varepsilon)$ and $(q_1(l) \to r) \notin R'$, insert $(q_1(l) \to r)$ into R'.

It is easy to see that every iteration respects the invariant $L(M') = L(M)$. Moreover, after termination of the algorithm, M' is obviously augmented.

Observe that the maximal number of rules of a pta over the terminal alphabet Σ is in $\mathcal{O}(|Q|^2 \cdot |\Gamma| + |\Sigma| \cdot |Q|^{m+1})$, where m is the maximal rank of a symbol from Σ. As a rule is added in every iteration, the algorithm terminates eventually. Since Σ is fixed, the number of iterations is polynomial in the input.   □

A derivation with no unneccessary turns is called *succinct*. More precisely, $d \in \mathcal{D}_M$ is *succinct* if there are $e_1 \in R_{\downarrow}^*$, $e_2 \in R_{\uparrow}^*$, $\omega \in R_{\Sigma}$, $k \in \mathbb{N}$ and d_1, ..., $d_k \in \mathcal{D}_M$ such that $d = e_1 e_2 \omega d_1 \dots d_k$ and for every $i \in [k]$, d_i is succinct. The set of succinct derivations of $\xi \in T_{\Sigma}$ from $q(\eta) \in \mathcal{C}_M$ in M is denoted by $\mathcal{DS}_M(q(\eta), \xi)$, and the set of all succinct elements of $\mathcal{D}_M$ by $\mathcal{DS}_M$. The following lemma means that in an augmented pta M, we need only consider succinct derivations. We omit its proof, which is based on the observation that in a derivation d of M, it is not necessary to apply a push rule ρ_1 right before a pop rule ρ_2, as we can always replace $\rho_1 \rho_2$ in d by some other rule ρ' of M.

Lemma 3. *Let M be augmented, $q(\eta) \in \mathcal{C}_M$, and $\xi \in T_{\Sigma}$. If $q(\eta) \Rightarrow_M^* \xi$, then there is also a succinct derivation $d \in \mathcal{DS}_M(q(\eta), \xi)$.*

4.2   Compact pts

Besides avoiding unnecessary turns of M, there is still one problem to solve. We might refer to it as M being too verbose in its pushdowns. E.g., in the derivation

$$q(\varepsilon) \Rightarrow_M q'(\gamma) \Rightarrow_M q''(\delta\gamma) \Rightarrow_M \sigma\big(u(\delta\gamma), p(\delta\gamma)\big) \Rightarrow_M^2 \sigma\big(\alpha, p(\gamma)\big) \Rightarrow_M \sigma\big(\alpha, p(\varepsilon)\big)$$

one could save time and space – i.e., derivation steps and pushdown cells – if there was some pushdown symbol $[\delta\gamma]$ such that

$$q(\varepsilon) \Rightarrow_M q''([\delta\gamma]) \Rightarrow_M \sigma\big(u([\delta\gamma]), p([\delta\gamma])\big) \Rightarrow_M^2 \sigma\big(\alpha, p(\varepsilon)\big).$$

We will construct a pts $M^{\sharp}$ with such pushdown symbols, i.e., with all symbols of the form $[\eta]$, where $\eta \in \Gamma^+$. It is said to be a *compact* pts, since, as will be proved later on, for every tree $\xi \in L(M^{\sharp})$, only polynomially many steps in $|\xi|$ are required for a derivation of ξ in $M^{\sharp}$, and the sizes of the pushdowns in the derivation can also bounded in this manner. Evidently, $M^{\sharp}$ can be infinite. However, considering $M^{\sharp}$ makes the following proofs easier, hence we stick with it for now, and deal with the question of a finite representation of $M^{\sharp}$ later.

The pushdown words of $M^{\sharp}$ can be understood as *subdivisions* of those of M. Because we must subdivide $M^{\sharp}$'s pushdown words even further in some proofs, the following definitions are needed. Choose two symbols '[' and ']' not from Γ, and let $\mathcal{S}(\Gamma) = \{[\eta] \mid \eta \in \Gamma^+\}$. Let $\eta = \gamma_1 \dots \gamma_n$ from Γ^+, where $\gamma_i \in \Gamma$ for $i \in [n]$. Given $k_0, \dots, k_m \in \mathbb{N}$ with $0 = k_0 < \dots < k_m = n$ for some $m > 0$, the $(k_0, \dots, k_m)$-*subdivision of* η is the word $[\gamma_{k_0+1} \dots \gamma_{k_1}] \dots [\gamma_{k_{m-1}+1} \dots \gamma_{k_m}] \in \mathcal{S}(\Gamma)^+$. Moreover, the ε-*subdivision of* ε is ε. An $\eta' \in \mathcal{S}(\Gamma)^*$ is a *subdivision* of an $\eta \in \Gamma^*$, denoted by $\eta' \sqsubseteq \eta$, if η' is an $\boldsymbol{E}$-subdivision of η for some $\boldsymbol{E} \in \mathbb{N}^*$. This $\boldsymbol{E}$ is unique; we denote it by $\boldsymbol{E}(\eta')$. If $\boldsymbol{E}(\eta') = (k_1, \dots, k_m)$, then let $E(\eta') = \{k_1, \dots, k_m\}$. Define $\iota \colon \Gamma^* \to \mathcal{S}(\Gamma)^*$ by $\iota(\varepsilon) = \varepsilon$ and $\iota(\eta) = [\eta]$ for $\eta \in \Gamma^+$. Let now $\eta \in \Gamma^*$ and $\eta', \eta'' \in \mathcal{S}(\Gamma)^*$ with $\eta', \eta'' \sqsubseteq \eta$. We write $\eta' \sqsubseteq \eta''$ if $E(\eta') \supseteq E(\eta'')$. We denote the unique $\kappa \sqsubseteq \eta$ with $E(\kappa) = E(\eta') \cup E(\eta'')$ by $\eta' \sqcap \eta''$. Note that $\eta' \sqcap \eta'' \sqsubseteq \eta'$ and $\eta' \sqcap \eta'' \sqsubseteq \eta''$. Regarding the length of $\eta' \sqcap \eta''$ as an element of $\mathcal{S}(\Gamma)^*$,

$$|\eta' \sqcap \eta''| = |E(\eta' \sqcap \eta'')| - 1 \leq |E(\eta')| + |E(\eta'')| - 3 = |\eta'| + |\eta''| - 1 \quad (1)$$

whenever $\eta \in \Gamma^+$, and if $\eta = \varepsilon$, then obviously $|\eta' \sqcap \eta''| = |\eta'| + |\eta''| = 0$. Finally, if $\eta' \in \mathcal{S}(\Gamma)^*$ is the $(k_1, \ldots, k_m)$-subdivision of η, then let $\widetilde{\eta}'$ denote the $(|\eta| - k_m, \ldots, |\eta| - k_1)$-subdivision of $\widetilde{\eta}$.

Now we define the *compact pts* $M^\sharp = (Q, \Sigma, \Gamma_\sharp, q_0, R_\sharp)$ *of* M, where $\Gamma_\sharp = \mathcal{S}(\Gamma)$, and $R_\sharp$ contains the rules *(i)* $q_1(y) \to q_2([\eta]y)$ for every $\eta \in \Gamma^+$ such that $q_1(\varepsilon) \Rightarrow_M^{r_1 \cdots r_k} q_2(\eta)$ with $r_1, \ldots, r_k \in R_\uparrow$, denote the resulting rule by $[r_1 \ldots r_k]$; *(ii)* $q_1([\eta]y) \to q_2(y)$ for every $\eta \in \Gamma^+$ such that $q_1(\eta) \Rightarrow_M^{r_1 \cdots r_k} q_2(\varepsilon)$ with $r_1, \ldots, r_k \in R_\downarrow$, denote the resulting rule by $[r_1 \ldots r_k]$; *(iii)* and for every rule $\omega \in R_\Sigma$, the rule $[\omega]$, which is identical to ω.

Obviously, $L(M^\sharp) = L(M)$. By the notation for the rules of $M^\sharp$, we have $R_\sharp \subseteq \mathcal{S}(R)$. The notion of subdivision, of the relation $\sqsubseteq$, and of the operation $\sqcap$, carries over to derivations of $M^\sharp$ in a straightforward manner. In a derivation d in $M^\sharp$, a subdivision η' of a pushdown η determines a corresponding subdivision d' of d, and vice versa. The following lemma circumstantiates this observation.

Lemma 4. *Let* $q, p \in Q$, $\eta \in \Gamma^*$, *and* $d \in R^*$. *Moreover, let* $d' \sqsubseteq d$ *and* $\eta' \sqsubseteq \eta$.

(i) *If* $d \in R_\downarrow^*$ *with* $q(\eta) \Rightarrow_M^d p(\varepsilon)$, *then* $q(\eta') \Rightarrow_{M^\sharp}^{d'} p(\varepsilon)$ *iff* $\mathbf{E}(\eta') = \mathbf{E}(d')$.

(ii) *If* $d \in R_\uparrow^*$ *with* $q(\varepsilon) \Rightarrow_M^d p(\eta)$, *then* $q(\varepsilon) \Rightarrow_{M^\sharp}^{d'} p(\eta')$ *iff* $\mathbf{E}(\widetilde{\eta}') = \mathbf{E}(d')$.

The following restricted mode of derivation is important as well. Let $\mu \in \mathbb{N}$ and $\zeta \in \mathcal{DF}_M$. We say that ζ *has μ-bounded pushdowns* if there is no infix $\kappa \in \Gamma^*$ of ζ with $|\kappa| > \mu$. Thus the size of every pushdown occurring in ζ is at most μ. Let moreover $\zeta_1, \zeta_2 \in \mathcal{DF}_M$. We write $\zeta_1 \overset{(\mu)}{\underset{M}{\Longrightarrow}}{}^\rho \zeta_2$ if $\zeta_1 \Rightarrow_M^\rho \zeta_2$ and both ζ_1 and ζ_2 have μ-bounded pushdowns. The relations $\overset{(\mu)}{\underset{M}{\Longrightarrow}}$ and $\overset{(\mu)}{\underset{M}{\Longrightarrow}}{}^d$, for some $d \in R^*$, are defined analogously. In the latter case, *all* intermediate derivation forms of d are required to have μ-bounded pushdowns.

In the following lemmas, we establish polynomial bounds for the lengths of successful derivations in $M^\sharp$, and for the sizes of the pushdowns "along the way."

Lemma 5. *Let* M *be augmented, and let* $q(\eta) \in \mathcal{C}_M$, $\eta' \sqsubseteq \eta$, $d \in \mathcal{DS}_M$, $d' \sqsubseteq d$, $\xi \in T_\Sigma$, *and* $\mu \in \mathbb{N}$ *with* $q(\eta) \Rightarrow_M^d \xi$ *and* $q(\eta') \overset{(\mu)}{\underset{M^\sharp}{\Longrightarrow}}{}^{d'} \xi$. *For every* $\eta'' \sqsubseteq \eta'$, *there is a* $d'' \sqsubseteq d'$ *such that* $q(\eta'') \overset{(\mu')}{\underset{M^\sharp}{\Longrightarrow}}{}^{d''} \xi$, *and* $\mu' = \mu + |\eta''| - |\eta'|$.

Proof. Presume η, η', d, d', q, ξ, and μ as above, and let $\eta'' \sqsubseteq \eta'$ and $\mu' = \mu + |\eta''| - |\eta'|$. The proof is by structural induction on ξ, hence suppose $k \in \mathbb{N}$, $\sigma \in \Sigma^{(k)}$ and $\xi_1, \ldots, \xi_k \in T_\Sigma$ such that $\xi = \sigma(\xi_1, \ldots, \xi_k)$. As $d \in \mathcal{DS}_M$, there are $e_1 \in R_\downarrow^*$, $e_2 \in R_\uparrow^*$, $\omega \in R_\Sigma$, $u, p \in Q$, $p_i(\theta) \in \mathcal{C}_M$ and $d_i \in \mathcal{DS}_M(p_i(\theta), \xi_i)$ for every $i \in [k]$ such that $d = e_1 e_2 \omega d_1 \ldots d_k$ and

$$q(\eta_1 \eta_2) \Rightarrow_M^{e_1} u(\eta_2) \Rightarrow_M^{e_2} p(\eta_3 \eta_2) \Rightarrow_M^\omega \sigma(p_1(\theta), \ldots, p_k(\theta)) \Rightarrow_M^{d_1} \cdots \Rightarrow_M^{d_k} \xi,$$

for some $\eta_1, \eta_2, \eta_3 \in \Gamma^*$ with $\eta = \eta_1 \eta_2$ and $\theta = \eta_3 \eta_2$.

By definition of $M^\sharp$, we have $d' = e_1' e_2' [\omega] d_1' \ldots d_k'$ for some $e_1' \sqsubseteq e_1$, $e_2' \sqsubseteq e_2$, and $d_i' \sqsubseteq d_i$, for every $i \in [k]$. Furthermore, $q(\eta') \overset{(\mu)}{\underset{M^\sharp}{\Longrightarrow}}{}^{e_1' e_2'} p(\theta')$, where $\eta' = \eta_1' \eta_2'$,

$\theta' = \eta'_3\eta'_2$, and $\eta'_i \sqsubseteq \eta_i$ for every $i \in [3]$. Observe that $|\theta'| \leq \mu$. As $\eta'' \sqsubseteq \eta'$, there must be $\eta''_1 \sqsubseteq \eta'_1$ and $\eta''_2 \sqsubseteq \eta'_2$ such that $\eta'' = \eta''_1\eta''_2$. Note that $|\eta''_1| \geq |\eta'_1|$. Let e''_1 be the $\boldsymbol{E}(\eta''_1)$-subdivision of η_1, then $q(\eta''_1\eta''_2) \Rightarrow^{e''_1}_{M\sharp} u(\eta''_2)$. In fact,

$$|\eta''| = |\eta'| + |\eta''| - |\eta'| \leq \mu + |\eta''| - |\eta'|\,,$$

hence $q(\eta''_1\eta''_2) \overset{(\mu') e''_1}{\Longrightarrow}_{M\sharp} u(\eta''_2)$. Moreover, $u(\eta''_2) \Rightarrow^{e'_2}_{M\sharp} p(\eta'_3\eta''_2)$. Let $\theta'' = \eta'_3\eta''_2$, then

$$|\theta''| = |\eta'_3| + |\eta''_2| + |\eta''_2| - |\eta'_2| = |\theta'| + |\eta''_1| + |\eta''_2| - (|\eta'_1| + |\eta'_2|)$$
$$\leq \mu + |\eta''_1| + |\eta''_2| - (|\eta'_1| + |\eta'_2|) = \mu + |\eta''| - |\eta'|\,,$$

and thus $u(\eta''_2) \overset{(\mu') e'_2}{\Longrightarrow}_{M\sharp} p(\eta'_3\eta''_2)$. Since $\theta'' \sqsubseteq \theta'$, the induction hypothesis implies that for every $i \in [k]$, there are some $d''_i \sqsubseteq d'_i$ such that $p_i(\theta'') \overset{(\mu') d''_i}{\Longrightarrow}_{M\sharp} \xi_i$ and $\mu'' = \mu + |\theta''| - |\theta'|$. We have

$$\mu'' = \mu + |\theta''| - |\theta'| = \mu + |\eta'_3| + |\eta''_2| - |\eta'_3| - |\eta'_2|$$
$$\leq \mu + |\eta''_1| + |\eta''_2| - |\eta'_1| - |\eta'_2| = \mu + |\eta''| - |\eta'| = \mu'\,.$$

The inequation holds because $|\eta''_1| \geq |\eta'_1|$. Thus for each $i \in [k]$, $p_i(\theta'') \overset{(\mu') d''_i}{\Longrightarrow}_{M\sharp} \xi_i$. We set $d'' = e''_1 e'_2[\omega]d''_1 \ldots d''_k$, yielding $q(\eta'') \overset{(\mu') d''}{\Longrightarrow}_{M\sharp} \xi$. $\square$

In the following, we denote the number $2 \cdot |\xi|$ by $\mu(\xi)$, for every tree $\xi \in T_\Sigma$.

Lemma 6. *Suppose that M is augmented. For every $q(\eta) \in C_M$, $\xi \in T_\Sigma$ and $d \in \mathcal{DS}_M(q(\eta), \xi)$, there are $\eta' \sqsubseteq \eta$ and $d' \sqsubseteq d$ such that $q(\eta') \overset{(\mu(\xi)) d'}{\Longrightarrow}_{M\sharp} \xi$.*

Proof. Assume $q(\eta)$, ξ and d as given above. The proof is by structural induction on ξ, therefore let $\xi = \sigma(\xi_1, \ldots, \xi_k)$ for some $k \in \mathbb{N}$, $\sigma \in \Sigma^{(k)}$ and $\xi_1, \ldots, \xi_k \in T_\Sigma$. Moreover, let $d = e_1 e_2 \omega d_1 \ldots d_k$ such that $e_1 \in R^*_\downarrow$, $e_2 \in R^*_\uparrow$, $\omega \in R_\Sigma$, and d_1, $\ldots$, $d_k \in \mathcal{DS}_M$. Thus there are η_1, η_2, η_3, and $\theta \in \Gamma^*$ with $\eta = \eta_1\eta_2$ and $\theta = \eta_3\eta_2$, as well as $u, p, p_1, \ldots, p_k \in Q$, satisfying

$$q(\eta_1\eta_2) \Rightarrow^{e_1}_M u(\eta_2) \Rightarrow^{e_2}_M p(\eta_3\eta_2) \Rightarrow^\omega_M \sigma(p_1(\theta), \ldots, p_k(\theta)) \Rightarrow^{d_1}_M \cdots \Rightarrow^{d_k}_M \xi.$$

By the induction hypothesis, for every $i \in [k]$, there are a $\theta'_i \sqsubseteq \theta$ and a $d'_i \sqsubseteq d_i$ such that $|\theta'_i| \leq \mu(\xi_i)$ and $p_i(\theta'_i) \overset{(\mu(\xi_i)) d'_i}{\Longrightarrow}_{M\sharp} \xi_i$. Set $\theta' = \theta'_1 \sqcap \cdots \sqcap \theta'_k \sqcap (\iota(\eta_3)\iota(\eta_2))$. Note that if $k = 0$, then $\theta' = \iota(\eta_3)\iota(\eta_2)$. By applying (1) k times,

$$|\theta'| \leq \left(\sum_{i\in[k]} |\theta'_i|\right) + 2 - k \leq \left(\sum_{i\in[k]} \mu(\xi_i)\right) + 2 - k = \mu(\xi) - k \leq \mu(\xi)\,. \qquad (2)$$

Thus $p(\theta') \overset{(\mu(\xi)) [\omega]}{\Longrightarrow}_{M\sharp} \sigma(p_1(\theta'), \ldots, p_k(\theta'))$. Let $j \in [k]$. Because $\theta' \sqsubseteq \theta'_j$, by Lem. 5, there is some $d''_j \sqsubseteq d'_j$ such that $p_j(\theta') \overset{(\mu') d''_j}{\Longrightarrow}_{M\sharp} \xi_j$, and where

$$\mu' = \mu(\xi_j) + |\theta'| - |\theta'_j| \leq \mu(\xi_j) + \left(\sum_{i\in[k]} |\theta'_i|\right) + 2 - |\theta'_j| \leq \left(\sum_{i\in[k]} \mu(\xi_i)\right) + 2 = \mu(\xi)\,.$$

Thus also $p_j(\theta') \overset{(\mu(\xi))\ d_j''}{\Longrightarrow}_{M^\sharp} \xi_j$. By definition of θ', there must be some $\eta_2' \sqsubseteq \eta_2$ and $\eta_3' \sqsubseteq \eta_3$ such that $\theta' = \eta_3'\eta_2'$. Set $\eta' = \iota(\eta_1)\eta_2'$. If $k = 0$, then clearly $|\eta_2'| = 1 < \mu(\xi)$. If $k > 0$, then by (2), $|\eta_2'| \leq |\theta'| < \mu(\xi)$. Thus in both cases $|\eta'| \leq \mu(\xi)$. Hence $q(\eta') \overset{(\mu(\xi))\ \iota(e_1)}{\Longrightarrow}_{M^\sharp} u(\eta_2')$. Moreover, as $\eta_3' \sqsubseteq \iota(\eta_3)$, by Lem. 4, there is some $e_2' \sqsubseteq e_2$ with $u(\eta_2') \overset{(\mu(\xi))\ e_2'}{\Longrightarrow}_{M^\sharp} p(\eta_3'\eta_2')$. Set $d' = \iota(e_1)e_2'[w]d_1'' \ldots d_k''$, then $q(\eta') \overset{(\mu(\xi))\ d'}{\Longrightarrow}_{M^\sharp} \xi$, and the proof is concluded. $\qquad\square$

Lemma 7. *Let M be augmented. For every $\xi \in L(M)$, there is a derivation $d' \in \mathcal{D}_{M^\sharp}(q_0(\varepsilon), \xi)$ with $|d'| \leq \mu(\xi)^2 + \mu(\xi)$.*

Proof. Let $\xi \in L(M)$, let $d \in \mathcal{DS}_M(q_0(\varepsilon), \xi)$, and consider the derivation d' as constructed in Lem. 6. Let $w \in \mathrm{pos}(\xi)$, and let d'' be an infix of d' such that $d'' \in \mathcal{D}_{M^\sharp}(q(\eta'), \xi|_w)$, for some $q(\eta') \in \mathcal{C}_{M^\sharp}$. We prove that $|d''| \leq (\mu(\xi)+1) \cdot \mu(\xi|_w)$ by well-founded induction using the relation "is child node of" on $\mathrm{pos}(\xi)$. For this purpose, let $\xi|_w = \sigma(\xi_1, \ldots, \xi_k)$ for some $k \in \mathbb{N}$, $\sigma \in \Sigma^{(k)}$, and $\xi_1, \ldots, \xi_k \in T_\Sigma$. Observe that d'' is of the form $e_1 e_2[w]d_1' \ldots d_k'$ for some $e_1 \in (R_\sharp)_\downarrow^*$, $e_2 \in (R_\sharp)_\uparrow^*$, $\omega \in R_\Sigma$, $u, p_1, \ldots, p_k \in Q$, $\kappa', \theta' \in \Gamma_\sharp^*$, and $d_i' \in \mathcal{D}_{M^\sharp}(p_i(\theta'), \xi_i)$, for $i \in [k]$, and

$$q(\eta') \overset{(\mu(\xi))\ e_1}{\Longrightarrow}_{M^\sharp} u(\kappa') \overset{(\mu(\xi))\ e_2}{\Longrightarrow}_{M^\sharp} p(\theta') \overset{(\mu(\xi))\ [\omega]}{\Longrightarrow}_{M^\sharp} \sigma(p_1(\theta'), \ldots, p_k(\theta')).$$

As the pushdowns η' and θ' are bounded in their size by $\mu(\xi)$, we must have $|e_1 e_2[w]| \leq 2 \cdot \mu(\xi) + 1$. By the induction hypothesis, $|d_i'| \leq (\mu(\xi)+1) \cdot \mu(\xi_i)$, thus

$$|d''| \leq 2 \cdot (\mu(\xi) + 1) + \sum_{i \in [k]} \left((\mu(\xi) + 1) \cdot \mu(\xi_i) \right) = (\mu(\xi) + 1) \cdot \mu(\xi|_w).$$

The lemma follows with $w = \varepsilon$ and $d'' = d'$. $\qquad\square$

4.3   Representing $M^\sharp$ by a Finite Object

Finally we show how to construct from M a finite representation $M^\dagger$ of $M^\sharp$. Let $\Gamma_\dagger = \mathcal{P}(Q \times Q)$ and define a mapping $h\colon \Gamma \to \Gamma_\dagger$ such that, for every $\gamma \in \Gamma$, $h(\gamma) = \{(q, p) \mid q(\gamma y) \to p(y) \text{ in } R\}$. We set $M^\dagger = (Q, \Sigma, \Gamma_\dagger, q_0, R_\dagger)$, where $R_\dagger$ is the smallest set R' such that (i) $R_\Sigma \subseteq R'$, (ii) for every rule $q(y) \to p(\gamma y)$ in R, the rule $q(y) \to p(h(\gamma)y)$ is in R', (iii) whenever $q(y) \to p(Uy)$ and $p(y) \to u(Vy)$ are in R', then also $q(y) \to u((V \circ U)y)$ is in R', (iv) for every $U \in \Gamma_\dagger$ and $(q, p) \in U$, the rule $q(Uy) \to p(y)$ is in R'. Note that $R_\dagger$ is given effectively by these conditions. The size of $M^\dagger$ is in general exponential in $|M|$.

We show that $M^\dagger$ is indeed a faithful representation of $M^\sharp$. Extend h to $\tilde{h}\colon \Gamma^+ \to \Gamma_\dagger$ by $\tilde{h}(\gamma_1 \ldots \gamma_k) = h(\gamma_1) \circ \cdots \circ h(\gamma_k)$ for $k > 0$ and $\gamma_1, \ldots, \gamma_k \in \Gamma$. Further, extend $\tilde{h}$ to $\hat{h}\colon \Gamma_\sharp^* \to \Gamma_\dagger^*$ by $\hat{h}([\eta_1] \ldots [\eta_k]) = \tilde{h}(\eta_1) \ldots \tilde{h}(\eta_k)$ for every $k \in \mathbb{N}$ and $\eta_1, \ldots, \eta_k \in \Gamma^+$. We identify h, $\tilde{h}$, and $\hat{h}$ in the following. There is the following close relation between $M^\sharp$ and $M^\dagger$; the uncomplicated proof is omitted.

Algorithm 1. Nondeterministic decision procedure for uniform membership

---

Input: pta $M = (Q, \Sigma, \Gamma, q_0, R)$, $\xi \in T_\Sigma$
Output: "Yes" if $\xi \in L(M)$, diverges otherwise
  $\zeta \leftarrow q_0(\varepsilon)$
  loop
    select leftmost $w \in \text{pos}(\zeta)$ such that $\zeta(w) = q(\eta)$ for some $q(\eta) \in \mathcal{C}_{M^\dagger}$
    either
      choose a rule $q(y) \to \sigma(p_1(y), \dots, p_k(y)) \in R$
      $\zeta \leftarrow \zeta[\sigma(p_1(\eta), \dots, p_k(\eta))]_w$
    or
      choose a rule $q(y) \to p(\gamma y) \in R$ and set $u \leftarrow p$, $U \leftarrow h(\gamma)$
      repeat n times for some $n \in \mathbb{N}$
        choose a rule $u(y) \to v(\gamma y) \in R$ and set $u \leftarrow v$, $U \leftarrow h(\gamma) \circ U$
      end repeat
      $\zeta \leftarrow \zeta[u(U\eta)]_w$
    or if $\eta = U\kappa$ for some $U \in \Gamma_\dagger$, $\kappa \in \Gamma_\dagger^*$
      choose some $(u, p) \in U$ such that $u = q$
      $\zeta \leftarrow \zeta[p(\kappa)]$
    end either
    if $\zeta = \xi$ **then** return "Yes" **else if** $\zeta \in T_\Sigma$ **then** diverge **endif**
  end loop

Lemma 8. *For every n, $\mu \in \mathbb{N}$, $q(\eta) \in \mathcal{C}_{M^\sharp}$, and for every $\xi \in T_\Sigma$, we have that*
$$q(\eta) \overset{(\mu)}{\underset{M^\sharp}{\Longrightarrow}}{}^n \xi \text{ iff } q(h(\eta)) \overset{(\mu)}{\underset{M^\dagger}{\Longrightarrow}}{}^n \xi.$$

Suppose that M is augmented. Then the lemma implies together with Lem. 6 and 7 that while $|M^\dagger|$ may be exponential in $|M|$, we may assume nevertheless that for every $\xi \in L(M)$, there is a derivation d of ξ in $M^\dagger$ such that both the length of d, as well as the size of every configuration in d, are bounded by a polynomial in $|\xi|$.

5   The Uniform Membership Problem

Employing $M^\dagger$, we can now investigate the complexity of the uniform membership problem of cftg. We begin with the upper bound.

Theorem 1. *The uniform membership problem of cftg over Σ is in* PSPACE.

Proof. Let $\xi \in T_\Sigma$ and let G be a cftg over Σ. Construct an augmented pta $M = (Q, \Sigma, \Gamma, q_0, R)$ with $L(M) = L(G)$. By Lem. 1 and 2, this takes time (and thus space) polynomial in $|G|$. Recall the mapping $h: \Gamma^+ \to \Gamma_\dagger$ from the definition of $M^\dagger$. Alg. 1 contains a nondeterministic procedure which decides $\xi \in L(M)$ in space restricted to $2 \cdot |\xi|^2 \cdot Q^2$. It works by emulating a derivation d' in the compact pta $M^\dagger$ as constructed above. The construction of d' is "on-the-fly." In each loop, the leftmost configuration $q(\eta)$ in the current derivation form ζ is selected, and a rule ρ is chosen. We may choose ρ to be a stay or pop rule of

$M^\dagger$, it is then applied to $q(\eta)$. Then again, we may choose a nonzero number of push rules of M with compatible states, apply h to the symbols they push, and combine the results by the product of binary relations. Clearly, this procedure can emulate exactly the derivations in $M^\dagger$.

If $\xi \in L(M)$, then there is a succinct derivation $d \in \mathcal{DS}_M(q_0(\eta), \xi)$, and, by Lem. 6 and 8, a derivation $d' \sqsubseteq d$ in $M^\dagger$ that has $(2 \cdot |\xi|)$-bounded pushdowns. Each pushdown symbol that occurs in d' is a subset of $Q \times Q$, and can thus be stored within space $|Q|^2$. As the number of configurations occurring in an intermediate derivation form ζ of d is bounded by $|\xi|$, the space bound of $2 \cdot |\xi|^2 \cdot |Q|^2$ is sufficient to store ζ. By [12, Thm. 1], the procedure is also computable in deterministic space polynomial in $|\xi|$ and $|M|$. □

Theorem 2. *The uniform membership problem of cftg over Σ is PSPACE-hard.*

Proof. Recall the following decision problem. Let Δ be an alphabet. The *intersection problem* is specified as follows.

Input: Deterministic finite-state automata $A_1, \ldots, A_k$ over Δ for some $k \in \mathbb{N}$
Question: Is $\bigcap_{i=1}^k L(A_i) = \emptyset$?

This problem is PSPACE-complete [5]. We give a reduction of its complement to the uniform membership problem of cftg. Then, as PSPACE = coPSPACE, the latter problem is PSPACE-hard. The reduction's idea is to construct a pta M which guesses some $w \in \Delta^*$ on its pushdown, copies it as often as needed (by stay rules with some symbol σ of at least binary rank), and then simulates the automata $A_1, \ldots, A_k$ on the respective copies. If A_i accepts w, M outputs some symbol α on the i-th branch, else it blocks. The search for $w \in \bigcap_{i=1}^k L(A_i)$ is thus reduced to the question $\xi \in L(M)$, for a tree $\xi \in T_\Sigma$ that is independent of w.

Formally, assume deterministic finite-state automata $A_i = (Q_i, \Delta, q_0^i, F_i, \delta_i)$, defined as usual, for some $k \in \mathbb{N}$ and each $i \in [k]$. We require the state sets Q_i to be pairwise disjoint, and $\alpha \notin \Delta$. By assumption, Σ contains some $\alpha^{(0)}$ and $\sigma^{(n)}$ with $n \geq 2$. Construct the pta $M = (Q, \Sigma, \Delta \cup \{\#\}, q_0, R)$ where $Q = \{q_0\} \cup \{u_0, \ldots, u_k\} \cup \bigcup_{i=1}^k Q_i$, with $q_0, u_0, \ldots, u_k$ distinct states, and R contains the rules

$$q_0(y) \rightarrow u_k(\#y), \quad u_k(y) \rightarrow u_k(by), \quad u_i(y) \rightarrow \sigma\big(q_0^i(y), u_{i-1}(y), u_0(y), \ldots, u_0(y)\big)$$

for every $i \in [k]$ and $b \in \Delta$. Moreover, for every $i \in [k]$, $b \in \Delta$, $q, p \in Q_i$ such that $\delta_i(q, b) = p$, and $f \in F_i$, the rule set R contains $q(by) \rightarrow p(y)$ and $f(\#y) \rightarrow \alpha$. Finally, for every $\gamma \in \Delta \cup \{\#\}$, the rule $u_0(\gamma y) \rightarrow \alpha$ is in R. Let ξ be the tree $\sigma(\alpha, \sigma(\alpha, \cdots \sigma(\alpha, \ldots, \alpha) \cdots, \alpha, \ldots, \alpha), \alpha, \ldots, \alpha)$ with exactly k occurrences of σ. Both M and ξ are logspace-computable from the input. It is easy to show that $\xi \in L(M)$ iff there is some $w \in \Delta^*$ such that $w \in \bigcap_{i=1}^k L(A_i)$. □

5.1 Uniform Membership of ε-free Indexed Grammars

Following suit to earlier research, established results on cftg can also give new insight on indexed languages.

Let us recall indexed grammars [1]. In the spirit of [11], an *indexed grammar* is a tuple $G = (N, \Sigma, \Gamma, S, P)$, where N, Σ, and Γ are alphabets, $S \in N$, and P is a finite set of productions of the forms *(i)* $A(y) \to B_1(y) \ldots B_k(y)$, *(ii)* $A(y) \to B(\gamma y)$, *(iii)* $A(\gamma y) \to B(y)$, and *(iv)* $A(y) \to a$, for A, B, B_1, ..., $B_k \in N$, $k \in \mathbb{N}$, $\gamma \in \Gamma$, and $a \in \Sigma$. We call G *ε-free* if all its productions of form *(i)* satisfy $k \geq 1$.

The similarity of indexed grammars to pta is apparent, and $\Rightarrow_G$, as well as $L(G)$, are defined analogously. This similarity is captured in the "yield theorem" [10, p. 115]. Although its original formulation applies to cftg, we restate it for pta, cf. [4, Prop. 8]. The remark on logspace-computability is easily reexamined.

Lemma 9 ([4,10]). *Let $L \subseteq (\Sigma^{(0)})^*$. Then, there is a pta M over Σ such that $L = \mathrm{yield}(L(M))$ iff there is an ε-free indexed grammar G over $\Sigma^{(0)}$ such that $L = L(G)$. Also, G is logspace-computable from M, and vice versa.*

The following corollary is then a direct consequence of Thms. 1 and 2 together with the yield theorem.

Corollary 1. *The uniform membership problem of ε-free indexed grammars is* PSPACE-*complete.*

In contrast, the uniform membership problem of indexed grammars *with* ε-rules is EXP-complete [13].[2]

6   The Non-Uniform Membership Problem

In this section, we intend to show that the pta $M^\dagger$ may also be useful for other means, by presenting an alternative proof of the NP upper bound of non-uniform membership of cftg. Note that this bound is already known: the class of output languages of compositions of macro tree transducers, a proper superclass of the context-free tree languages, is in NP [7, Thm. 8].

If we regard trees from T_Σ as well-parenthesized words over $\Sigma \cup C$, as defined in the preliminaries, then a context-free tree language can be also understood as an indexed string language. Therefore, the following upper bound is as well a consequence of the containment of the indexed languages in NP. Its proof in [11] rests on the correctness of the Turing machine from [1].

Theorem 3. *The membership problem of cftg over Σ is in* NP.

Proof. Let G be the cftg fixed in the membership problem. Construct an equivalent augmented pta M, as well as $M^\dagger$ as defined above. As G is not part of the input, $M^\dagger$ is constructible in constant time. Consider the nondeterministic decision procedure in Alg. 2. By Lem. 8, $L(M^\dagger) = L(M^\sharp)$, and moreover $L(M^\sharp) = L(M)$. So if the procedure returns "Yes", then there is some

---

[2] Still, the emptiness problem of ε-free indexed grammars remains EXP-complete. This follows from a small modification of the indexed grammar witnessing EXP-hardness in [13]. Hence, by Lem. 9, the emptiness problem of cftg is EXP-complete, too.

Algorithm 2. Nondeterministic decision procedure for membership of cftg

---

Input: $\xi \in T_\Sigma$
Output: "Yes" if $\xi \in L(M)$, diverges otherwise
   choose some $d \in R_\dagger^*$ with $|d| \leq \mu(\xi)^2 + \mu(\xi)$ and $\mu(\xi)$-bounded pushdowns
   if $q_0(\eta) \Rightarrow_{M\dagger}^d \xi$ **then** return "Yes" **else** diverge **endif**

$d \in \mathcal{D}_{M\dagger}(q_0(\varepsilon), \xi)$, and hence $\xi \in L(M)$. Conversely, if $\xi \in L(M)$, then there must be some $d' \in \mathcal{D}_{M\sharp}(q_0(\varepsilon), \xi)$, and by Lem. 7, we may assume that $|d'| \leq \mu(\xi)^2 + \mu(\xi)$. By Lem. 8, there is a $d \in \mathcal{D}_{M\dagger}(q_0(\varepsilon), \xi)$ with equal length bound. Thus the procedure returns "Yes".   □

Hardness of the problem can be demonstrated in the same manner as for indexed grammars [11, Prop. 1], by devising a cftg G such that $L(G)$ encodes the set of all satisfiable propositional formulas in 3-conjunctive normal form. For the sake of completeness, we restate the respective theorem.

Theorem 4. *There are a ranked alphabet Σ and a cftg G over Σ such that the membership problem of G is NP-hard.*

7 Conclusion

In this paper, the complexity of the uniform membership problem of cftg was proven to be PSPACE-complete. A corollary for uniform membership of indexed grammars was obtained. As a by-product, we could state an alternative proof for the NP-completeness of the non-uniform membership problem of cftg.

References

1. Aho, A.V.: Indexed grammars–an extension of context-free grammars. J. ACM **15**(4), 647–671 (1968)
2. Asveld, P.: Time and space complexity of inside-out macro languages. Int. J. Comput. Math. **10**(1), 3–14 (1981)
3. Engelfriet, J., Schmidt, E.M.: IO and OI. J. Comput. Syst. Sci. **15**(3), 328–353 (1977); and **16**(1), 67–99 (1978)
4. Guessarian, I.: Pushdown tree automata. Math. Syst. Theory **16**(1), 237–263 (1983)
5. Kozen, D.: Lower bounds for natural proof systems. In: Proc. 18th Symp. Foundations of Computer Science, pp. 254–266 (1977)
6. Lohrey, M.: On the parallel complexity of tree automata. In: Middeldorp, A. (ed.) RTA 2001. LNCS, vol. 2051, p. 201. Springer, Heidelberg (2001)
7. Inaba, K., Maneth, S.: The complexity of tree transducer output languages. In: Proc. of FSTTCS 2008, pp. 244–255 (2008)
8. Papadimitriou, C.H.: Computational Complexity. John Wiley and Sons (2003)
9. Rounds, W.C.: Mappings and grammars on trees. Theor. Comput. Syst. **4**(3), 257–287 (1970)

10. Rounds, W.C.: Tree-oriented proofs of some theorems on context-free and indexed languages. In: Proc. 2nd ACM Symp. Theory of Comput., pp. 109–116 (1970)
11. Rounds, W.C.: Complexity of recognition in intermediate level languages. In: Proc. 14th Symp. Switching and Automata Theory, pp. 145–158 (1973)
12. Savitch, W.J.: Relationships between nondeterministic and deterministic tape complexities. J. Comput. Syst. Sci. **4**(2), 177–192 (1970)
13. Tanaka, S., Kasai, T.: The emptiness problem for indexed language is exponential-time complete. Syst. Comput. Jpn. **17**(9), 29–37 (1986)
14. Yoshinaka, R.: An attempt towards learning semantics: Distributional learning of IO context-free tree grammars. In: Proc. of TAG+11, pp. 90–98 (2012)

Attacking BEAR and LION Schemes in a Realistic Scenario

Matteo Piva, Marco Pizzato[⊠], and Massimiliano Sala

University of Trento, Trento, Italy
{pvimtt,marco.pizzato1,maxsalacodes}@gmail.com

Abstract. BEAR and LION are block ciphers introduced by Biham and Anderson in 1996. Their special feature is that they use very efficiently a hash function and a stream cipher, so that the hardware implementation of BEAR and LION becomes straightforward, assuming that the two other primitives are already present. In this paper we discuss their security starting from the strength of their building blocks.

Keywords: Block-ciphers · Stream-ciphers · Hash functions · BEAR · LION · Key-recovery

1 Introduction

Traditional stream ciphers and hash functions enjoy cheaper hardware implementation compared to traditional block ciphers (such as AES or Kasumi). Although this is not always the case (see e.g. [8]), it is not uncommon to have a device already endowed with an implementation of a stream cipher and a hash function, but still missing a block cipher. In this situation it would be very convenient to construct a block cipher simply by reusing the other two primitives. The extreme case happens with a family of block ciphers proposed in [1] by Anderson and Biham in 1996, inspired by the famous Luby-Rackoff construction presented in [2], whose most famous members are called BEAR and LION and which need only a couple of extra XOR's to get a secure block cipher, provided either the hash function or the stream cipher is robust. In particular, under some conditions of ideality for the two primitives, it is possible to show that no single-pair key-recovery attack exists for BEAR/LION. Unfortunately, their security was questioned in the paper [5] of same year, and so these systems do not enjoy the popularity they probably deserve. A deeper analysis of their security has been performed much later in [6], where alternative assumptions on the primitives ensure even the non-existence of multi-pair key-recovery attacks.

Other similar constructions can be found in [2–4].

In this paper we present two types of results. While in Section 2 we revisit the system definition and known results, in Section 3 we present three attacks, two for BEAR and one for LION and we provide an estimate of the attack cost in the (more realistic) situation offered by non-ideal primitives. Finally, in Section 4 we improve and generalize some results of [6].

© Springer International Publishing Switzerland 2015
A. Maletti (Ed.): CAI 2015, LNCS 9270, pp. 189–195, 2015.
DOI: 10.1007/978-3-319-23021-4_17

2   Preliminaries and Known Results

In this paper we adopt the following notation. By $\mathbb{F}$ we denote the field $\mathbb{F}_2$, with a capital R or a capital L we mean elements of $\mathbb{F}^r$ and $\mathbb{F}^l$ respectively, with $r > l$. The set of plaintexts is composed of messages of the form $(L_i, R_i) \in \mathbb{F}^{l+r}$. The key space is denoted by $\mathcal{K}$. Its elements are of the form $K = (K_1, K_2) \in \mathbb{F}^k \times \mathbb{F}^k$, with $k \geq l$.

2.1   BEAR

The structure of BEAR is based on a stream cipher S and a keyed hash function H_K. They have to satisfy the following properties.
The keyed hash function $H_K : \mathbb{F}^r \to \mathbb{F}^l$:

- is based on a unkeyed hash function $H' : \mathbb{F}^{r+k} \to \mathbb{F}^l$, in which the key is appended or prepended to the message,
- is one-way and collision free, in the sense that it is hard given Y to find X such that $H'(X) = Y$, and to find unequal X and Y such that $H'(X) = H'(Y)$,
- is pseudo-random, in that even given $H'(X_i)$ for any set of inputs, it is hard to predict any bit of $H'(Y)$ for a new input Y.

The stream cipher $S : \mathbb{F}^l \to \mathbb{F}^r$:

- resists key recovery attacks, i.e. it is hard to find the seed X given $S(X)$,
- resists expansion attacks, i.e. it is hard to expand any partial stream of Y.

We recall the BEAR encryption/decryption scheme, where "+" denotes XOR, the standard sum in the vector space $\mathbb{F}^n$.

ENCRYPTION	DECRYPTION
$\overline{L} = L + H_{K_1}(R)$	$\overline{L} = L' + H_{K_2}(R')$
$R' = R + S(\overline{L})$	$R = R' + S(\overline{L})$
$L' = \overline{L} + H_{K_2}(R')$	$L = \overline{L} + H_{K_1}(R)$

In their article, Anderson and Biham proved the following results.

Theorem 1. *An oracle which finds the key of BEAR, given one plaintext/ciphertext pair, can efficiently and with high probability find the seed M of the stream cipher S for any output $Y = S(M)$.*

Theorem 2. *An oracle which finds the key of BEAR, given one plaintext/ciphertext pair, can efficiently and with high probability find preimages and collisions of the hash function H'.*

As a consequence we have

Corollary 1. *If it is impossible to find efficiently the seed of S or it is impossible to find efficiently preimages and collisions for H', then no efficient key-recovery single-pair attack exists for BEAR.*

In [6], the authors introduced the following criterion of security for a hash function, which is related to the definition of MAC algorithm in [7].

Definition 1. *Given a keyed hash function $\mathcal{H} = \{H_K\}_{K \in \mathbb{F}^k}$, $H_K : \mathbb{F}^r \to \mathbb{F}^l$ for any $K \in \mathbb{F}^k$, we say that $\mathcal{H}$ is key-resistant if, given a pair (Z, R) such that $Z = H_K(R)$ for a random K and a random R, it is hard to find K.*

With this definition they claim the following result.

Theorem 3. *Let $n \geq 1$. Let $\mathcal{A}_n$ be an oracle able to find the key of BEAR given any set of n plaintext-ciphertext pairs $\{((L_i, R_i), (L'_i, R'_i))\}_{1 \leq i \leq n}$. Then $\mathcal{A}_n$ is able to solve efficiently any equation $Z = H_{K_1}(R)$, knowing Z and R, for any random $R \in \mathbb{F}^r$ and any random $K_1 \in \mathbb{F}^k$.*

2.2 LION

The structure of LION is based on a stream cipher S and a hash function H. They have to satisfy the following properties.
The hash function $H : \mathbb{F}^r \to \mathbb{F}^l$:

- is one-way and collision free, i.e. it is hard given Y to find X such that $H(X) = Y$, and to find unequal X and Y such that $H(X) = H(Y)$.

The stream cipher $S : \mathbb{F}^l \to \mathbb{F}^r$:

- is pseudo-random,
- resists key recovery attacks, i.e. it is hard to find the seed X given $S(X)$,
- resists expansion attacks, i.e. it is hard to expand any partial stream of Y.

We recall the LION encryption/decryption scheme.

ENCRYPTION	DECRYPTION
$\overline{R} = R + S(L + K_1)$	$\overline{R} = R' + S(L' + K_2)$
$L' = L + H(\overline{R})$	$L = L' + H(\overline{R})$
$R' = \overline{R} + S(L' + K_2)$	$R = \overline{R} + S(L + K_1)$

Note that the key space for LION is $\mathcal{K} = \mathbb{F}^{2l}$.

In their article, Anderson and Biham claimed that results similar to Theorem 1 and Theorem 2 hold also for LION. In [6] the authors proved the following result for the stream cipher.

Theorem 4. *Let $H : \mathbb{F}^r \to \mathbb{F}^l$ and $S : \mathbb{F}^l \to \mathbb{F}^r$. An oracle $\mathcal{A}_1$ which finds the key of LION, given one plaintext/ciphertext pair, can efficiently and with high probability find the seed M of the stream cipher S for any particular output $Y = S(M)$.*

They also proved a similar result for the hash function, although assuming a restrictive condition as follows.

Definition 2. *Let H and S be functions, $H : \mathbb{F}^r \to \mathbb{F}^l$ and $S : \mathbb{F}^l \to \mathbb{F}^r$, with $r \geq l$. We say that (S, H) is a good pairing if for a random $Y \in \mathbb{F}^l$ we have $H^{-1}(Y) \cap \mathrm{Im}(S) \neq \emptyset$.*

Theorem 5. *Assume that (S, H) is a good pairing. An oracle $\mathcal{A}_1$ which finds the key of LION, given one plaintext/ciphertext pair, can efficiently and with high probability find preimages and collisions of the hash function H.*

As in the BEAR case, the authors introduced another concept of security for the stream cipher S.

Definition 3. *Let $\mathcal{K} = \mathbb{F}^{2l}$. Given a stream cipher $S : \mathbb{F}^l \to \mathbb{F}^r$, we say that S is key-resistant if, given a pair (Z, L) such that $Z = S(L + K_1)$ for a random $(K_1, K_2) \in \mathcal{K}$ and a random $L \in \mathbb{F}^l$, it is hard to find K_1.*

Theorem 6. *Let $n \geq 1$. Let $\mathcal{A}_n$ be an oracle able to find the key of LION given any set of n plaintext-ciphertext pairs $\{((L_i, R_i), (L_i', R_i'))\}_{1 \leq i \leq n}$. Then $\mathcal{A}_n$ is able to solve any equation $Z = S(L + K_1)$, knowing Z and L, for any random $L \in \mathbb{F}^l$ and any random $K_1 \in \mathbb{F}^k$.*

3   Our Attacks

The results of [1] and [6] ensure that no practical key-recovery attack is possible for BEAR and LION if the hash function or the stream cipher are ideal. Nevertheless, no consequences are explored if both the primitives are not perfectly secure. In this paper we deal with more realistic situations where stream ciphers and hash functions are not ideal. We will explain the meaning of "not ideal" in Definitions 4, 5, 6 and 7.

In this section we present three attacks, two on BEAR and one on LION.

3.1   Attacks on BEAR

The following definitions are useful to measure the non-ideality of our primitives.

Definition 4. *Let $H_K : \mathbb{F}^r \mapsto \mathbb{F}^l$ be a keyed hash function. We say that H_K is $t-$key-resistant if, for every K and R, given $Z = H_K(R)$ and R, there is an algorithm recovering the key K with an expected cost of 2^t H-evaluations.*

Definition 5. *Let $S : \mathbb{F}^l \mapsto \mathbb{F}^r$ be an injective stream cipher. We say that S is u-resistant if, given $S(L)$, there is an algorithm recovering the seed L with an expected cost of 2^u S-evaluations.*

Remark 1. Since, by an exhaustive search involving, on average, half of all keys we will be able to solve the equation $Z = H_K(R)$, it is obvious that an ideal $\{H_K\}_{K \in \mathbb{F}^k}$ is $(k-1)-$key-resistant. Similarly, an ideal stream cipher S is $(l-1)-$resistant. Therefore, $t \leq k-1$ and $u \leq l-1$.

We can now describe two attacks, attack **A** and attack **B**. The first is a chosen-plaintext attack which assumes that the keyed hash function is not ideal, that is, it is t−key-resistant (but the stream cipher could be ideal). In the second attack we assume that also the stream cipher is not ideal, that is, it is u-resistant, but we let the attacker mount a known-plaintext attack. The description of the two attacks is given below, followed by Theorem 7, where their complexity is shown. Essentially, Theorem 7 states that Attack **B** is more effective than Attack **A**. We find this of interest because normally known-plaintext attacks need more input data and computations than the corresponding chosen-plaintext attacks, but in this case the known-plaintext version is using the big advantage of an additional weaker primitive (i.e., the stream cipher).

ATTACK A

Let $Y = S(0)$ and choose a random $\tilde{R}$. An $\tilde{L} \in \mathbb{F}^l$ exists s.t. $\tilde{L} = H_{K_1}(\tilde{R})$. We do not know $\tilde{L}$ but we claim that we can recover it, as follows. Observe that if someone encrypts $(\tilde{L}, \tilde{R})$ he will get $\bar{L} = 0$ and so $R' = \tilde{R} + S(0) = \tilde{R} + Y$. We encrypt some plaintexts $\{(L, \tilde{R}) \mid L \in \mathbb{F}^l\}$, until we get a ciphertext (L', R') with

$$\tilde{R} + R' = Y \tag{1}$$

which requires, on average, 2^{l-1} attempts. We note that equation (1) ensures that $\tilde{L} = H_{K_1}(\tilde{R})$, since the stream S is injective. We can now recover the keys. In fact, from $L' = H_{K_2}(R')$ we obtain K_2 with an expected cost of 2^t H-evaluations and, from $\tilde{L} = H_{K_1}(\tilde{R})$, we obtain K_1 with an expected cost of 2^t H-evaluations.

ATTACK B

Given any random plaintext/ciphertext pair, (L, R) and (L', R'), we compute $R + R' = S(\bar{L})$. We recover $\bar{L}$ with an expected cost of 2^u S-evaluations. Then we compute $\bar{L} + L = H_{K_1}(R)$. With an expected cost of 2^t H-evaluations we recover K_1. Finally, we obtain $L' + \bar{L} = H_{K_2}(R')$. Again, with 2^t H-evaluations we can recover K_2.

Theorem 7
*The expected cost of Attack **A** is 2^{t+1} H-evaluations plus 2^{l-1} encryptions, which is $(2^{t+1} + 2^l)$ H-evaluations plus 2^{l-1} S-evaluations.*
*The expected cost of Attack **B** is 2^{t+1} H-evaluations plus 2^u S-evaluations.*

Proof. As for Attack **A**, we have to consider 2^{l-1} encryptions in order to find $\tilde{L}$, which amount to 2^l H-evaluations plus 2^{l-1} S-evaluations. We need then to consider the 2^{t+1} H-evaluations required to recover the key $K = (K_1, K_2)$.

Now consider Attack **B**. We need 2^u S-evaluations in order to obtain $\bar{L}$ and 2^{t+1} H-evaluations to recover the key $K = (K_1, K_2)$.

3.2   Attack on LION

The following definitions are useful to measure the non-ideality of our primitives.

For any T subset of an additive group G and any $g \in G$, we write

$$T + g = \{t + g \mid t \in T\}.$$

Definition 6. *Let $H : \mathbb{F}^r \mapsto \mathbb{F}^l$ be a hash function and $S : \mathbb{F}^l \mapsto \mathbb{F}^r$ a stream cipher. We say that H is t−resistant if, for random $R, Y \in \mathbb{F}^r$, given $Z = H(R)$, there is an algorithm recovering a preimage of Z lying in $\mathrm{Im}(S) + Y$ with an expected cost of 2^t H-evaluations.*

Remark 2. The expected number of preimages $H^{-1}(Z)$ in $\mathbb{F}^r$ is 2^{r-l}, so, on average, we have only one preimage of Z in a random translate of $\mathrm{Im}(S)$. Clearly, the best the attacker can hope for in an inversion attack on H is to be able to list this preimage using only 1 H-evaluation, which means that H is at least 0−resistant.

An ideal H will force the attacker to perform nearly all possible H-evaluations (namely 2^l) in order to find the right preimage R, therefore H will be l−resistant. In other words, $0 \le t \le l$.

Definition 7. *Let $S : \mathbb{F}^l \mapsto \mathbb{F}^r$ be a stream cipher. We say that S is u−key-resistant if, for random K and L, given $S(L + K)$ and L, there is an algorithm recovering K with an expected cost of 2^u S-evaluations.*

Remark 3. We note that the two definitions of u−resistant (keyed and unkeyed) for S are equivalent, since translations act regularly. Therefore, $0 \le u \le l - 1$, with $l - 1$ being the ideal value.

We can now describe a known-plaintext attack (attack **C**), assuming the hash function is t−resistant and the stream cipher is u−key-resistant.

ATTACK C

Consider any random plaintext/ciphertext pair, (L, R) and (L', R'). We compute $Y = L + L'$. For an unknown $\bar{R}$, we have $Y = H(\bar{R})$. With 2^t operations we can find the preimage $H^{-1}(Y)$ lying in $\mathrm{Im}(S) + R$. It follows that this preimage must be $\bar{R}$ (Remark 2). Let us call this preimage $\bar{R}$. Now we perform the following operations. First, we compute $R + \bar{R}$ and we solve $R + \bar{R} = S(L + K_1)$, recovering K_1 with expected cost of 2^u S-evaluations. Then, we compute $R' + \bar{R}$ and we solve $R' + \bar{R} = S(L' + K_2)$, recovering K_2 with expected cost of 2^u S-evaluations.

We note that we might have been unlucky in the sense that $\bar{R}$ could not have been the only preimage of Y in $\mathrm{Im}(S) + R$. This is unlikely, since the plaintext/ciphertext pair was random. Even in this case, it is enough to try a few encryptions and check if the candidate key pair (K_1, K_2) is the correct one. The cost of these tries is negligible (in real life situations) compared to the cost of the equation solving.

Theorem 8

*The expected cost of attack **C** is 2^t H-evaluations plus 2^{u+1} S-evaluations.*

Proof. We need 2^t H-evaluations to invert Y and 2^u S-evaluations for each half of the key.

4   Good-Pairing Is Not Necessary

In this section we generalize the proof of Theorem 1.8 of [6], removing the useless hypothesis on the good pairing and thus proving the original result as claimed in [1].

Remark 4. Essentially, the idea in the following generalization is that we do not need to assume the hypothesis on the good pairing, i.e. $H^{-1}(Y) \cap \text{Im}(S) \neq \emptyset$ for a random Y, since it is almost always true that $H^{-1}(Y) \cap (\text{Im}(S) + R) \neq \emptyset$, for a suitable R.

Theorem 9. *An oracle $\mathcal{A}_1$ which finds the key of LION, given one plaintext-ciphertext pair, can efficiently and with high probability find preimages and collisions of the hash function H.*

Proof. Consider some $\tilde{R}$ and $Y = H(\tilde{R})$. With high probability we find R such that $H^{-1}(Y) \cap (\text{Im}(S) + R) \neq \emptyset$. Fix also some element $L \in \mathbb{F}^l$. There exists K_1 such that $\bar{R} \in H^{-1}(Y)$, with $\bar{R} = S(L+K_1)+R$. We can also suppose (otherwise we choose another R) that $\bar{R} \neq \tilde{R}$. We obtain $L' = L + H(\bar{R}) = L + Y$. Now with $K_2 = K_1 + Y$ we have $R' = S(L + K_1) + R + S(L + Y + K_1 + Y) = R$. We give to $\mathcal{A}_1$ as input the pair $((L, R), (L + Y, R))$ and it returns (K_1, K_2). We can then compute $\bar{R} = S(L + K_1) + R$ and we have $H(\bar{R}) = Y$, finding a collision $H(\tilde{R}) = H(\bar{R}) = Y$.

To find a preimage, we can argument as above with a given Y and supposing we do not know $\tilde{R}$.

Acknowledgements. The authors would like to thank the anonymous referees for their valuable comments.

References

1. Anderson, R., Biham, E.: Two practical and provably secure block ciphers. In: Gollmann, D. (ed.) FSE 1996. LNCS, vol. 1039, pp. 113–120. Springer, Heidelberg (1996)
2. Luby, M., Rackoff, C.: How to construct pseudorandom permutations from pseudorandom functions. SIAM J. Comput., 373–386 (1988)
3. Lucks, S.: Faster luby-rackoff ciphers. In: Gollmann, D. (ed.) FSE 1996. LNCS, vol. 1039, pp. 189–203. Springer, Heidelberg (1996)
4. Lucks, S.: BEAST: A fast block cipher for arbitrary blocksizes. In: Proc. of IFIP, pp. 144–153 (1996)
5. Morin, P.: Provably secure and efficient block ciphers. In: Proc. of Selected Areas in Cryptography, pp. 30–37 (1996)
6. Maines, L., Piva, M., Rimoldi, A., Sala, M.: On the provable security of BEAR and LION schemes. Applicable Algebra in Engineering, Communication and Computing **22**(5–6), 413–423 (2011)
7. Preneel, B.: The state of cryptographic hash functions. In: Damgård, I.B. (ed.) LDS. LNCS, vol. 1561, pp. 158–182. Springer, Heidelberg (1999)
8. Preneel, B., Rijmen, V., Knudsen, L.R.: Evaluation of ZUC. ABT Crypto. Tech. Report **7** (2010)

Weighted Restarting Automata
and Pushdown Relations

Qichao Wang, Norbert Hundeshagen, and Friedrich Otto[⊠]

Fachbereich Elektrotechnik/Informatik, Universität Kassel, 34109 Kassel, Germany
{wang,hundeshagen,otto}@theory.informatik.uni-kassel.de

Abstract. Weighted restarting automata have been introduced to study quantitative aspects of computations of restarting automata. Here we study the special case of assigning words as weights from the semiring of formal languages over a given (output) alphabet, in this way generalizing the restarting transducers introduced by Hundeshagen (2013). We obtain several new classes of word relations in terms of restarting automata, which we relate to various types of pushdown relations.

Keywords: Weighted restarting automaton · Restarting transducer · Pushdown relation · Quasi-realtime pushdown relation

1 Introduction

Analysis by reduction is a linguistic technique that is used to check the correctness of sentences of natural languages through sequences of local simplifications. The *restarting automaton* was invented as a formal model for the analysis by reduction [7]. In order to study quantitative aspects of computations of restarting automata, *weighted restarting automata* were introduced in [10]. These automata are obtained by assigning an element of a given semiring S as a weight to each transition of a restarting automaton. Then the product (in S) of the weights of all transitions that are used in a computation yields a weight for that computation, and by forming the sum over the weights of all accepting computations for a given input $w \in \Sigma^*$, a value from S is assigned to w. Thus, a partial function $f : \Sigma^* \dashrightarrow S$ is obtained. Here we consider the special case that S is the semiring of formal languages over some finite (output) alphabet Δ. Then f is a transformation from Σ^* into the languages over Δ. Thus, we obtain a generalization of the notion of a *restarting transducer* as introduced in [6].

It is well known (see, e.g., [8]) that the class of languages that are accepted by monotone RWW- and RRWW-automata (see Section 2 for the definitions) coincides with the class of context-free languages. Accordingly, we are interested in the classes of transformations that are computed by various types of weighted restarting automata that are *monotone*. In this paper we compare some of these classes to each other and we relate them to the class of *pushdown relations* and some of its subclasses. In particular, we prove that monotone weighted RRWW-automata compute strictly more transformations than

A. Maletti (Ed.): CAI 2015, LNCS 9270, pp. 196–207, 2015.
DOI: 10.1007/978-3-319-23021-4_18

monotone weighted RWW-automata. The latter in turn compute a class that properly includes the quasi-realtime pushdown relations, which we will show to coincide with the transformations that are computed by monotone RWW- and RRWW-transducers.

This paper is structured as follows. In Section 2 we recall some basic notions concerning weighted restarting automata, and in Section 3 we look at the pushdown relations and some of their subclasses. Then, in Section 4 we study the classes of transformations that are computed by (monotone) restarting transducers, and in Section 5 we investigate the computational power of weighted RWW- and RRWW-automata that are monotone. The paper closes with a short summary and some problems for future work.

2   Weighted Restarting Automata

We assume that the reader is familiar with the standard notions and concepts of theoretical computer science, such as monoids, finite automata, and semirings. Throughout the paper we will use $|w|$ to denote the length of a word w and λ to denote the empty word. Further, $\mathbb{P}(X)$ denotes the power set of a set X, and $\mathbb{P}_{\text{fin}}(X)$ denotes the set of all finite subsets of X.

A *restarting automaton* (or RRWW-automaton for short) is a nondeterministic machine with a finite-state control, a flexible tape with endmarkers, and a read/write window of a fixed finite size. Formally, it is described by an 8-tuple $M = (Q, \Sigma, \Gamma, \text{¢}, \$, q_0, k, \delta)$, where Q is a finite set of states, Σ is a finite input alphabet, Γ is a finite tape alphabet containing Σ, the symbols $\text{¢}, \$ \notin \Gamma$ are used as markers for the left and right border of the work space, respectively, $q_0 \in Q$ is the initial state, $k \geq 1$ is the size of the *read/write window*, and δ is the (partial) *transition relation* that associates finite sets of transition steps to pairs of the form (q, w), where q is a state and w is a possible content of the read/write window. There are four types of transition steps. A *move-right step* (MVR) causes M to shift its read/write window one position to the right and to change the state. A *rewrite step* causes M to replace the content w of the read/write window by a shorter string v, thereby reducing the length of the tape, and to change the state. Further, the read/write window is placed immediately to the right of the string v. However, occurrences of the delimiters ¢ and $\$$ can neither be deleted nor newly created by a rewrite step. A *restart step* causes M to place its read/write window over the left end of the tape, so that the first symbol it sees is the left sentinel ¢, and to reenter the initial state q_0, and, finally, an *accept step* causes M to halt and accept.

If $\delta(q, w)$ is undefined for some pair (q, w), then M necessarily halts in a corresponding situation, and we say that M *rejects*. Finally, if each rewrite step is combined with a restart step into a joint rewrite/restart operation, then M is called an *RWW-automaton*.

A *configuration* of M is a string $\alpha q \beta$, where $q \in Q$, and either $\alpha = \lambda$ and $\beta \in \{\text{¢}\} \cdot \Gamma^* \cdot \{\$\}$ or $\alpha \in \{\text{¢}\} \cdot \Gamma^*$ and $\beta \in \Gamma^* \cdot \{\$\}$; here q is the current state, and $\alpha\beta$ is the current content of the tape, where it is understood that the window contains the first k symbols of β or all of β when $|\beta| \leq k$. A *restarting configuration* is of the form $q_0 \text{¢} w\$$. If $w \in \Sigma^*$, then $q_0 \text{¢} w\$$ is an *initial configuration*.

We observe that any computation of M consists of certain phases. A phase, called a *cycle*, starts in a restarting configuration, the head moves along the tape performing move-right operations and a single rewrite operation until a restart operation is performed and thus a new restarting configuration is reached. If no further restart operation is performed, the computation necessarily finishes in a halting configuration – such a phase is called a *tail*. It is required that in each cycle M performs *exactly one* rewrite step. A word $w \in \Sigma^*$ is accepted by M, if there is an accepting computation which starts from the initial configuration $q_0 \text{¢} w\$$. By $L(M)$ we denote the language consisting of all (input) words that are accepted by M.

Next we come to the notion of *monotonicity*. Let $C := \alpha q \beta$ be a *rewrite configuration* of an RRWW-automaton M, that is, a configuration in which a rewrite step is to be applied. Then $|\beta|$ is called the *right distance* of C, which is denoted by $D_r(C)$. A *sequence of rewrite configurations* $S = (C_1, C_2, \ldots, C_n)$ is called *monotone* if $D_r(C_1) \geq D_r(C_2) \geq \cdots \geq D_r(C_n)$, that is, if the distance of the place of rewriting to the right end of the tape does not increase from one rewrite step to the next. A *computation* of an RRWW-automaton M is called *monotone* if the sequence of rewrite configurations that is obtained from the cycles of that computation is monotone. Observe that here the rewrite configuration is not taken into account that corresponds to the possible rewrite step that is executed in the tail of the computation considered. Finally, an RRWW-automaton M is called *monotone* if all its computations that start with an initial configuration are monotone. We use the prefix mon- to denote monotone types of restarting automata.

For studying quantitative aspects of computations of restarting automata, the weighted restarting automaton has been introduced in [10]. A *weighted restarting automaton* of type X, a wX-automaton for short, is a pair (M, ω), where M is a restarting automaton of type X, and ω is a *weight function* from the transitions of M into a semiring S. This *weight function* assigns an element $\omega(t) \in S$ as a weight to each transition t of M. Here we only consider the case that S is the semiring $S = (\mathbb{P}(\Delta^*), \cup, \cdot, \emptyset, \{\lambda\})$ of languages over Δ with the operations of union and product.

Let $\mathcal{M} = (M, \omega)$ be a weighted restarting automaton, where $M = (Q, \Sigma, \Gamma, \text{¢}, \$, q_0, k, \delta)$. Let $\text{AC}_M(w) = \{A_1, A_2, \cdots, A_m\}$ be the set of all accepting computations of M on input w. We assume that the computation $A_i \in \text{AC}_M(w)$ $(1 \leq i \leq m)$ uses the transitions $t_{i,1}, t_{i,2}, \cdots, t_{i,n_i}$ of M. Then the weight of a transition $t_{i,j}$ $(1 \leq j \leq n_i)$ is a language $\omega(t_{i,j}) = L_{i,j}$ over Δ, and the weight of the computation A_i is $\omega(A_i) = L_{i,1} \cdot L_{i,2} \cdot \ldots \cdot L_{i,n_i} = \hat{L}_i \in \mathbb{P}(\Delta^*)$. Finally, $f_\omega^M(w) = \hat{L}_1 \cup \hat{L}_2 \cup \cdots \cup \hat{L}_m \in \mathbb{P}(\Delta^*)$ is the language over Δ that is associated by $\mathcal{M}$ to w, that is, f_ω^M is a transformation from Σ^* into $\mathbb{P}(\Delta^*)$. If $w \notin L(M)$, then $\text{AC}_M(w) = \emptyset$, and accordingly, $f_\omega^M(w) = \emptyset$. In this way, the weighted restarting automaton $\mathcal{M} = (M, \omega)$ on Σ yields the relation

$Rel(\mathcal{M}) = \{\, (u, v) \mid u \in L(M), v \in f_{\omega}^{M}(u) \,\} \subseteq \Sigma^* \times \Delta^*$. By $\mathcal{R}(\mathsf{wX})$ we denote the class of relations that are computed by weighted restarting automata of type wX.

As the weight of a transition of M can be any language over Δ, the general model of weighted restarting automata is quite powerful. Therefore, we introduce some more restricted types of weighted restarting automata.

Definition 1. *A weighted restarting automaton* $\mathcal{M} = (M, \omega)$ *of type* wX *is called a* finitely weighted restarting automaton *(a* $\mathsf{w_{FIN}X}$*-automaton for short), if the weight function* ω *maps the transitions of* M *into a semiring of the form* $S = (\mathbb{P}_{\mathrm{fin}}(\Delta^*), \cup, \cdot, \emptyset, \{\lambda\})$. *It is called a* word-weighted restarting automaton *(a* $\mathsf{w_{word}X}$*-automaton for short), if the weight of each transition* t *of* M *is of the form* $\omega(t) = \{v\}$ *for some* $v \in \Delta^*$.

It is rather obvious that $\mathcal{R}(\mathsf{w_{word}X}) = \mathcal{R}(\mathsf{w_{FIN}X}) \subsetneq \mathcal{R}(\mathsf{wX})$ for each type $\mathsf{X} \in \{\mathsf{mon\text{-}RWW}, \mathsf{mon\text{-}RRWW}, \mathsf{RWW}, \mathsf{RRWW}\}$. We close this section with a simple example of a relation that is computed by a weighted restarting automaton.

Example 2. Let $M_1 = (Q, \Sigma, \Gamma, \mathvisiblespace, \$, q_0, k, \delta)$ be the mon-RWW-automaton that is defined by taking $Q := \{q_0\}$, $\Gamma := \Sigma := \{a\}$, and $k := 2$, where δ is defined as follows:

$$t_1 : (q_0, \mathvisiblespace a) \to (q_0, \mathsf{MVR}), \quad t_3 : (q_0, aa) \to (q_0, \mathsf{MVR}),$$
$$t_2 : (q_0, \mathvisiblespace\$) \to \mathsf{Accept}, \qquad t_4 : (q_0, a\$) \to \$.$$

Here t_4 is the only (combined) rewrite/restart operation of M_1. It is easily seen that $L(M_1) = \{\, a^n \mid n \geq 0 \,\}$.

Let $(\mathbb{P}_{\mathrm{fin}}(\Delta^*), \cup, \cdot, \emptyset, \{\lambda\})$ be the semiring of finite languages over $\Delta = \{c\}$, let ω_1 be the weight function that assigns the set $\{c\}$ to the MVR transitions t_1 and t_3, and that assigns the set $\{\lambda\}$ to all other transitions, and let $\mathcal{M}_1 = (M_1, \omega_1)$. It follows easily that

$$f_{\omega_1}^{M_1}(w) = \begin{cases} \{c^{\frac{1}{2}(n+1)n}\}, & \text{for } w = a^n, \ n \geq 0, \\ \emptyset, & \text{for } w \notin L(M_1), \end{cases}$$

and hence, $Rel(\mathcal{M}_1) = \{\, (a^n, c^{\frac{1}{2}(n+1)n}) \mid n \geq 0 \,\}$.

Finally, we recall the notion of *restarting transducer* from [3]. In analogy to finite transducers and pushdown transducers, a restarting transducer is a restarting automaton that is equipped with an additional output function which gives an output word for each restart and each accept transition. Hence, restarting transducers are a special type of word-weighted restarting automata. By $\mathcal{R}(\mathsf{X\text{-}Td})$ we denote the class of relations that are computed by restarting transducers of type X.

3 Pushdown Relations

A *pushdown transducer* (PDT for short) is defined by an 8-tuple $T = (Q, \Sigma, \Delta, X, q_0, Z_0, F, E)$, where Q is a finite set of states, Σ is an input alphabet, Δ is an output alphabet, X is a pushdown alphabet, $q_0 \in Q$ is the initial state, $Z_0 \in X$ is the bottom marker of the pushdown, $F \subseteq Q$ is the set of final states, and $E \subset Q \times (\Sigma \cup \{\lambda\}) \times X \times Q \times X^* \times \Delta^*$ is a finite transition relation that produces a (possible empty) output word in each step (see, e.g., [2]). The output produced during a computation is then simply the concatenation of all outputs produced during that computation.

A configuration of T is written as a 4-tuple (q, u, α, v), where $q \in Q$ is the current state, $u \in \Sigma^*$ is the still unread part of the input, $\alpha \in X^*$ is the current content of the pushdown, and $v \in \Delta^*$ is the output produced so far. The relation $Rel(T)$ computed by T is defined as

$$Rel(T) = \{\, (u, v) \in \Sigma^* \times \Delta^* \mid \exists q \in F, \alpha \in X^* : (q_0, u, Z_0, \lambda) \vdash^* (q, \lambda, \alpha, v) \,\}.$$

A relation $R \subseteq \Sigma^* \times \Delta^*$ is called a *pushdown relation* if $R = Rel(T)$ holds for some PDT T. By PDR we denote the class of all pushdown relations.

A pushdown relation R is called *linearly bounded* if there exists a constant $c \in \mathbb{N}$ such that $|v| \leq c \cdot |u|$ holds for all pairs $(u, v) \in R$. By lbPDR we denote the class of all linearly bounded pushdown relations.

A pushdown relation R is called *realtime* if it is computed by a PDT $T = (Q, \Sigma, \Delta, X, q_0, Z_0, F, E)$ that does not perform any λ-steps, that is, its set of transitions E satisfies the condition $E \subset Q \times \Sigma \times X \times Q \times X^* \times \Delta^*$. By rtPDR we denote the class of all realtime pushdown relations.

Finally, a pushdown relation R is called *quasi-realtime* if it is computed by a PDT $T = (Q, \Sigma, \Delta, X, q_0, Z_0, F, E)$ for which each λ-step pops a symbol from the pushdown, that is, if $(q, \lambda, x, q', x', v) \in E$, then $x' = \lambda$. By qrtPDR we denote the class of all quasi-realtime pushdown relations.

Proposition 3. rtPDR $\subsetneq$ qrtPDR $\subsetneq$ lbPDR $\subsetneq$ PDR.

Proof. The first and the third inclusions are obvious. Concerning the second inclusion, assume that R is computed by a quasi-realtime PDT T. Let $(u, v) \in R$. On reading an input symbol, T can push a string of length c (for some constant $c \geq 1$) onto its pushdown, and so altogether at most $c \cdot |u|$ symbols are pushed. Hence, T can execute at most $c \cdot |u|$ λ-transitions, which means that, on input u, T executes at most $(c+1) \cdot |u|$ steps. Thus, the output v produced during this computation satisfies the inequality $|v| \leq d \cdot (c+1) \cdot |u|$, where d is the maximal length of any output string produced by T in a single step.

It remains to prove that all the inclusions above are proper. The transduction $R_{uu^R} = \{\, (u, uu^R) \mid u \in \{a, b\}^* \,\}$ is quasi-realtime: a PDT T can output its input u letter by letter, also pushing each letter onto the pushdown. At the end of the input, which T can guess, it empties its pushdown letter by letter, producing the output u^R. On the other hand, this transduction is not realtime, as in a realtime pushdown relation the final output syllable is produced when the last input symbol is being read, which is not possible for the relation R_{uu^R}.

The relation $R_{a^m b^m c^n} = \{\, (a^m b^m c^n, c^n a^m b^m) \mid m, n \geq 1 \,\}$ is a linearly bounded pushdown relation. A PDT T can first guess c^n, outputting this factor and pushing it onto the pushdown. Then it compares the syllables a^m and b^m, producing the output $a^m b^m$. Finally, it checks the syllable c^n against the c-syllable on its pushdown. However, $R_{a^m b^m c^n}$ is not quasi-realtime. The output syllable c^n must be produced first, but the pushdown must be used for comparing a^m to b^m, which are the first two syllables of the input. In addition, when this comparison is made, then the output $a^m b^m$ must be produced. Hence, the output syllable c^n must already be produced before the input syllable c^n is being read, that is, the output c^n is produced through λ-transitions that do not pop from the pushdown.

Finally, the relation $R_+ = \{ (a^m, b^n a^m b^n) \mid m, n \geq 1 \}$ is a pushdown relation. Obviously, it is not linearly bounded.   □

The class of pushdown relations can be characterized in terms of context-free languages and morphisms. For that we recall the following concept from [1].

Definition 4. *A language $L \subseteq \Gamma^*$ characterizes a relation $R \subseteq \Sigma^* \times \Delta^*$ if there exist two morphisms $h_1 : \Gamma^* \to \Sigma^*$ and $h_2 : \Gamma^* \to \Delta^*$ such that $R = \{ (h_1(w), h_2(w)) \mid w \in L \}$.*

In [1] it was shown that the pushdown relations are characterized by the context-free languages. For the case that Σ and Δ are disjoint, an even stronger result was shown that assumes that $\Gamma = \Sigma \cup \Delta$ and that h_1 (h_2) is the projection from Γ^* onto Σ^* (Δ^*). In terms of [1] this is expressed by saying that the pushdown relations are *strongly characterized* by the context-free languages. In the following we extend this result to lbPDR.

Lemma 5. *Every linearly bounded pushdown relation is strongly characterized by a context-free language.*

Proof. Let $R \subseteq \Sigma^* \times \Delta^*$ be an lbPDR, and let c be a constant such that $|v| \leq c \cdot |u|$ for all $(u, v) \in R$. From Definition 4 it follows that R is characterized by a context-free language $L \subseteq \Gamma^*$ and two morphisms $h_1 : \Gamma^* \to \Sigma^*$ and $h_2 : \Gamma^* \to \Delta^*$. Thus, for each pair $(u, v) \in R$, there is a word $w \in L$ such that $h_1(w) = u$ and $h_2(w) = v$. Now a strong characterization would put the additional restriction $|w| \leq |u| + |v| \leq (c + 1) \cdot |u|$ on the length of w, which is not necessarily the case for the above characterization in terms of L.

To simplify the discussion, we assume that Γ, Σ, and Δ are pairwise disjoint. We introduce an additional alphabet $\Gamma' = \{ x' \mid x \in \Gamma, h_2(x) \neq \lambda \}$ and take $\Gamma_0 = \Gamma \cup \Gamma'$. Further, we define a morphism $h : \Gamma^* \to \Gamma_0^*$, where $x \in \Gamma$:

$$h(x) = \begin{cases} xx', & \text{if } h_1(x) \neq \lambda \text{ and } h_2(x) \neq \lambda, \\ x', & \text{if } h_1(x) = \lambda \text{ and } h_2(x) \neq \lambda, \\ x, & \text{otherwise,} \end{cases}$$

and we extend h_1 and h_2 to morphisms $h_1' : \Gamma_0^* \to (\Gamma' \cup \Sigma)^*$ and $h_2' : (\Gamma' \cup \Sigma)^* \to (\Sigma \cup \Delta)^*$ through $h_1'(x) = \begin{cases} h_1(x), x \in \Gamma \\ x, \quad x \in \Gamma' \end{cases}$ and $h_2'(x') = \begin{cases} h_2(x), x' \in \Gamma' \\ x', \quad x' \in \Sigma \end{cases}$.

Clearly, the language $L' = h_2'(h_1'(h(L))) \subseteq (\Sigma \cup \Delta)^*$ is context-free. Let π^Σ and π^Δ be the projections from $(\Sigma \cup \Delta)^*$ onto Σ^* and Δ^*. Then R is strongly characterized by L' and the two projections π^Σ and π^Δ.   □

4 Pushdown Relations and Restarting Transducers

Every relation that is computed by a restarting transducer is *linearly bounded* in the sense of the class lbPDR, as a restarting transducer outputs symbols only during restart and accept steps, and any computation on an input of length n contains at most $n + 1$ such steps. It follows that restarting transducers cannot compute all pushdown relations. Naturally, the question arises of whether they can at least compute all linearly bounded pushdown relations. In [3] it was claimed that monotone RWW- and RRWW-transducers do exactly compute these relations, but actually, only a weaker result was proven there. Here we show that these transducers actually characterize the class qrtPDR. By Proposition 3 this means that they cannot realize all relations from the class lbPDR.

Theorem 6. $\mathcal{R}(\text{mon-RWW-Td}) = \mathcal{R}(\text{mon-RRWW-Td}) = \text{qrtPDR}$.

To prove this result we present two lemmas.

Lemma 7. $\text{qrtPDR} \subseteq \mathcal{R}(\text{mon-RWW-Td})$.

Proof. Let $R \subseteq \Sigma^* \times \Delta^*$ be the relation that is computed by the quasi-realtime PDT $T = (Q, \Sigma, \Gamma, \Delta, \delta, q_0, Z_0, F)$. We now simulate T by a mon-RWW-Td using a construction from [9].

Let $l := \max\{ |\gamma| \mid \exists (q, a, A, p, \gamma, v) \in \delta \}$, and let $\Gamma' := \Gamma'_1 \cup \Gamma'_2$, where $\Gamma'_1 := \{ (x) \mid x \in \Gamma^+, |x| \leq 2l \}$ and $\Gamma'_2 := \{ (y) \mid y \in \Gamma^{2l} \}$. Thus, a symbol $(x) \in \Gamma'_1$ encodes a word $x \in \Gamma^*$ of length at most $2l$, while a symbol $(y) \in \Gamma'_2$ encodes a word $y \in \Gamma^*$ of length $2l$. Finally, let M be the RWW-Td $M = (Q_M, \Sigma, \Gamma', \Delta, \mathfrak{c}, \$, 4, \delta')$ that simulates T as follows.

In each cycle M simulates two steps of T. Assume that an accepting computation of T on input $w = a_0 a_1 \cdots a_n$ begins by first applying the transition $(q_1, B_1 \cdots B_{m_1} C_1, v_1) \in \delta(q_0, a_0, Z_0)$ and then the transition $(q_2, B_{m_1+1} \cdots B_{m_1+m_2} C_2, v_2) \in \delta(q_1, a_1, C_1)$. As $m_1 < l$ and $m_2 < l$, $|B_1 \cdots B_{m_1+m_2} C_2| < 2l$ holds. Accordingly, starting with the input configuration corresponding to input w, M can execute the rewrite step $\mathfrak{c} a_0 a_1 a_2 \rightarrow \mathfrak{c}(x C_2) a_2$, where $x := B_1 \cdots B_{m_1+m_2}$, producing the output $v_1 v_2$.

Assume that by executing the next two steps, the PDT T reaches the configuration $(q_4, a_4 \cdots a_n, B_1 \cdots B_{m_1+m_2-1} x_1)$, that is, the factor $a_2 a_3$ is read from the input tape, the internal state changes to q_4, the two topmost symbols $B_{m_1+m_2} C_2$ on the pushdown are rewritten into the string $x_1 \in \Gamma^*$, and the output $v_3 v_4$ is produced. If $m_1 + m_2 - 1 + |x_1| \leq 2l$, then M rewrites $(x C_2) a_2 a_3 a_4$ into $(x') a_4$, where $x' = B_1 \cdots B_{m_1+m_2-1} x_1$, and if $m_1 + m_2 - 1 + |x_1| > 2l$, then M rewrites $(x C_2) a_2 a_3 a_4$ into $(x')(x'') a_4$, where $x' x'' = B_1 \cdots B_{m_1+m_2-1} x_1$ and $|x'| = 2l$.

In addition, if T executes a λ-step, then it changes its state, pops a symbol from the pushdown, and produces an output syllable. In order for M to simulate this in a length-reducing fashion, we must combine up to $2l$ λ-steps of T (or several λ-steps together with the next non-λ-step) into a single simulation step of M. This is rather technical, but nevertheless fairly standard.

Continuing in this way it follows that the tape content of M is always of the form $\alpha(u)a_j \cdots a_n$, where $(u) \in \Gamma_1'$, and $\alpha \in \Gamma_2'^*$. Here αu encodes the current content of the pushdown of T, and $a_j \cdots a_n$ is the suffix of the input that T still has to read. As long as $j < n-1$, M can simulate the next two steps of T by rewriting the four symbols $(x_i)(x_{i+1})a_j a_{j+1}$ either into $(x_i)(x_{i+1})(x_{i+2})$, into $(x_i)(x_{i+2})$, or into (x_{i+2}), depending on the way in which the contents of the pushdown of T is modified by these steps. This simulation continues until either T rejects (and then M rejects as well), or until $j = n-1$ is reached. At that point M can detect whether T will accept or reject, and it will then act likewise. It follows that M is monotone, and that $Rel(M) = R$ holds.   □

Lemma 8. $\mathcal{R}(\text{mon-RRWW-Td}) \subseteq \text{qrtPDR}$.

Proof. Let M be a mon-RRWW-Td. Using the simulation technique from [8] it can be shown that M can be simulated by a PDT T. Let $\mathbb{c}uqvw\$$ be a rewrite configuration within an accepting computation of M, and assume that M now executes the rewrite step $(q', v') \in \delta(q, v)$. Then the next cycle starts from the restarting configuration $q_0\mathbb{c}uv'w\$$, and as M is monotone, the next rewrite operation is performed within a suffix of $uv'w$ of length at most $|vw|$. Thus, the prefix uv' can be stored on the pushdown of T, while the input contains the suffix w still unread. As an RRWW-transducer, M moves to the right after performing the above rewrite step, and (without loss of generality) it only restarts and produces its output at the right end of the tape, provided the state reached leads to a restart operation. As T cannot scan its input completely each time it simulates a rewrite step, it guesses the output z produced by M at the end of the current cycle, and it keeps the state q' reached by the above rewrite step and the output z guessed in its finite-state control. When it processes further letters from w, it updates this state information. Finally, when w has been processed completely, then T checks whether all the states of M stored in its finite-state control correspond to restart steps and the corresponding output strings.

In fact, as M is monotone, it can be checked quite easily that T is quasi-realtime, that is, whenever T executes a λ-transition, then it pops a symbol from its pushdown. In addition, whenever T simulates a rewrite step of M, then it must remember the state q' that M enters through this rewrite step and the output z that M will produce in the current cycle. Luckily, there are only finitely many pairs of the form (q', z) of M, and hence, T can actually store all the pairs occurring in the computation being simulated in its finite-state control.   □

As $\mathcal{R}(\text{mon-RWW-Td}) \subseteq \mathcal{R}(\text{mon-RRWW-Td})$, Lemmas 7 and 8 imply the characterization in Theorem 6. Next it can be shown that all linearly bounded pushdown relations are accepted by (non-monotone) RRWW-transducers.

Theorem 9. $\text{lbPDR} \subseteq \mathcal{R}(\text{RRWW-Td})$.

Proof. Let $R \subseteq \Sigma^* \times \Delta^*$ be a linearly bounded pushdown relation. W.l.o.g. we assume that Σ and Δ are disjoint. By Lemma 5, R is strongly characterized by a context-free language $L \subseteq (\Sigma \cup \Delta)^*$ and the two projections $h_i : (\Sigma \cup \Delta)^* \to \Sigma^*$

and $h_o : (\Sigma \cup \Delta)^* \to \Delta^*$. Furthermore, there is a constant k such that, for all $(u, v) \in R$, there exists a word $w \in L$ such that $|w| \leq k \cdot |u|$ and $(u, v) = (h_i(w), h_o(w))$. Let M be a PDA for L. Now an RRWW-Td T for R can be constructed that proceeds in two steps. For a given pair $(u, v) \in R$,

1. T guesses a characterizing word w of (u, v) and produces the output $h_o(w)$,
2. T verifies that $w \in L$ by simulating the PDA M on w.

The main problem in constructing T is the fact that we have to ensure that these steps are realized in a length-reducing manner. ☐

Actually, the inclusion in Theorem 9 has already been stated in [4] and its journal version [5] by relating restarting transducers to transducing observer systems. The proof above can easily be converted to the latter, in this way correcting the proof given in these papers, which only proves a weaker result.

5   Relations Computed by Monotone Weighted RWW- and RRWW-Automata

In the previous section we have shown that monotone RWW- and RRWW-transducers compute the relations in qrtPDR. Are (word-weighted) RWW- and RRWW-automata that are monotone more expressive?

We begin this investigation by studying the relation between the classes $\mathcal{R}(\text{mon-wRWW})$ and $\mathcal{R}(\text{mon-wRRWW})$. Let $\tau_1 \subseteq \{a, b, c\}^* \times \{d, e\}^*$ be the relation

$$\tau_1 = \{ (a^k b^k c^m, d^m e^k) \mid k, m \geq 1 \}.$$

Lemma 10. $\tau_1 \notin \mathcal{R}(\text{mon-wRWW})$.

Proof. Assume that $\tau_1 \in \mathcal{R}(\text{mon-wRWW})$, that is, there exists a weighted mon-RWW-automaton M and a weight function ω' that maps the transitions of M into subsets of $\{d, e\}^*$ such that $\tau_1 = Rel((M, \omega'))$. As τ_1 is actually a (partial) function, we see that ω' can be replaced by a weight function ω that maps each transition of M into a singleton, which means that $\mathcal{M} = (M, \omega)$ is a word-weighted mon-RWW-automaton. Interpreting the weight $\omega(t)$ of a transition as output, we see that, for an input of the form $a^k b^k c^m$, $\mathcal{M}$ first outputs the symbol d m-times, which is the number of c-symbols in the input, and then it outputs k e-symbols, which is the number of a- and b-symbols in the input.

As the language $L = \{ a^k b^k c^m \mid k, m \geq 1 \}$ is not regular, M needs to execute rewrite steps in all its accepting computations on input $a^k b^k c^m$, if k is sufficiently large. At what position can the first of these rewrite steps be applied?

(1) Assume that the first rewrite step is applied within the suffix c^m. While processing this suffix, M can easily produce the output d^m. Then M must compare the prefix a^k to the infix b^k, and while doing so it should produce the output e^k. However, M is monotone, which means that the position of a rewrite step in a cycle cannot have a larger right distance than the rewrite step in the previous

cycle. Accordingly, the infix b^k must be reduced by rewrites to a word that fits into the window of M, which means that M cannot distinguish between b^k and b^{k+r} for some positive integer r. Thus, together with $a^k b^k c^m$, M would also accept the word $a^k b^{k+r} c^m$, contradicting our assumption on M.

(2) From the arguments above, it follows that the first rewrite step must be executed within the prefix a^k or at the border between the prefix a^k and the infix b^k. This means that M must first compare the syllables a^k and b^k, and since by this process the information on the exponent k is being destroyed, it must produce the output e^k during this process. However, as the output syllable e^k is preceded by the prefix d^m, M must already output the syllable d^m before it starts to output e-symbols. As shown in [10], the length of any computation of M on an input of length n is at most $\frac{1}{2}(n+2)(n+3)-1$. This means that during the processing of the prefix $a^k b^k$, M can perform at most $\frac{1}{2}(2k+2)(2k+3)-1$ steps. Choose $l \geq 1$ to be a constant such that $|\omega(t)| \leq l$ for all transitions t of M, and choose m such that $m > (\frac{1}{2}(2k+2)(2k+3)-1) \cdot l$. Then M is not able to produce m d-symbols, while it is processing the prefix $a^k b^k$. Thus, it either stops producing d-symbols before it has erased all information on the number k, which means that not enough d-symbols are produced, or it keeps on producing d-symbols while erasing all information on k. In the latter case it will then not be able to produce the correct number of e-symbols.   □

Obviously, $\mathcal{R}(\text{mon-wRWW})$ is contained in $\mathcal{R}(\text{mon-wRRWW})$. We now prove that this inclusion is proper.

Theorem 11. $\mathcal{R}(\text{mon-xRWW}) \subsetneq \mathcal{R}(\text{mon-xRRWW})$ *for all* $\mathsf{x} \in \{\mathsf{w}, \mathsf{w_{FIN}}, \mathsf{w_{word}}\}$.

Proof. By Lemma 10, $\tau_1 \notin \mathcal{R}(\text{mon-wRWW})$. On the other hand, it is easy to construct a monotone word-weighted RRWW-automaton $\mathcal{M} = (M, \omega)$ such that $Rel(\mathcal{M}) = \tau_1$. This automaton $\mathcal{M}$ proceeds as follows. Let $w = a^k b^k c^m$ be given as input. In the first cycle, $\mathcal{M}$ places a marking on the prefix of w by encoding the first two symbols into a combined (new) symbol, and then it moves to the suffix c^m of w. While scanning this suffix, it outputs a d-symbol for each c-symbol that it encounters, and at the right delimiter, it restarts. In the subsequent cycles, on seeing the marking at the left end of the tape, $\mathcal{M}$ realizes that it has already produced the d-symbols. Hence, it now moves to the boundary between the prefix a^k and the infix b^k to compare them. In each subsequent rewrite step, it removes a single a-symbol and a single b-symbol, producing a single e-symbol as output (via ω). It follows that $Rel(\mathcal{M}) = \tau_1$, which completes the proof.   □

We remark that Theorem 11 is the first result that establishes a difference in the computational power between a model of the monotone RWW-automaton and the corresponding model of the monotone RRWW-automaton.

The relation $\tau_1 = \{ (a^k b^k c^m, d^m e^k) \mid k, m \geq 1 \}$ considered above is a linearly bounded pushdown relation that is not computed by any monotone weighted RWW-automaton. On the other hand, the relation considered in Example 2 is computed by a monotone word-weighted RWW-automaton. Its domain a^* is context-free, while its range $\{ c^{\frac{1}{2}(n+1)n} \mid n \geq 0 \}$ is not. Hence, this relation is not a pushdown relation. Thus, we have the following incomparability result.

Theorem 12

For each prefix $x \in \{w, w_{FIN}, w_{word}\}$, *the class of relations* $\mathcal{R}(\text{mon-xRWW})$ *is incomparable to the classes* lbPDR *and* PDR *with respect to inclusion.*

Finally, we turn to the class of relations that are computed by monotone wRRWW-automata. Let $\tau_2 \subseteq \{a, b, c\}^* \times \{d, e\}^*$ be the relation

$$\tau_2 = \{\, (a^k b^k c^{m+l} a^l, d^m e^k d^m e^l) \mid k, l, m \geq 1 \,\}.$$

Lemma 13. $\tau_2 \notin \mathcal{R}(\text{mon-wRRWW})$.

Proof. The relation τ_2 is a partial function. Thus, if τ_2 is computed by a monotone wRRWW-automaton, then it is also computed by a monotone word-weighted RRWW-automaton $\mathcal{M} = (M, \omega)$. Interpreting the weight $\omega(t) \in \{d, e\}^*$ of a transition t as output, $\mathcal{M}$ first outputs the syllable d^m, then e^k, then d^m again, and finally e^l given the word $a^k b^k c^{m+l} a^l$ as input.

As $\mathcal{M}$ is monotone, we see that $\mathcal{M}$ must first compare the prefix a^k to the infix b^k (see the proof of Lemma 10). Since the information about the exponent k is lost during this process, $\mathcal{M}$ must produce the output syllable e^k during this process. Hence, the prefix d^m of the output must be produced before this process starts, which means that $\mathcal{M}$ can only perform rewrites on the prefix a^k of the input while it produces the output d^m.

The exact value of m is unknown, that is, while moving right across the input syllable c^{m+l}, $\mathcal{M}$ must guess it. After comparing the numbers of a- and b-symbols and outputting correspondingly many e-symbols, $\mathcal{M}$ must again produce m d-symbols, that is, it must somehow remember this number. However, as $\mathcal{M}$ must not perform any rewrite steps on the suffix $c^{m+l} a^l$ before a^k has been compared to b^k, it must encode the number m within the prefix a^k. However, if m is sufficiently large, then this is not possible. Hence, it follows that τ_2 cannot be computed by any weighted RRWW-automaton that is monotone. ☐

Clearly τ_2 is a linearly bounded pushdown relation, too. Hence, from Example 2 and Lemma 13 the following incomparability result follows.

Theorem 14

For each prefix $x \in \{w, w_{FIN}, w_{word}\}$, *the class of relations* $\mathcal{R}(\text{mon-xRRWW})$ *is incomparable to the classes* lbPDR *and* PDR *with respect to inclusion.*

6   Conclusion

We have studied the classes of (binary) relations that are computed by weighted RWW- and RRWW-automata that are monotone, relating them to the classes of relations that are computed by monotone RWW- and RRWW-transducers and to some classes of pushdown relations. The inclusion results obtained are summarized in the diagram in Figure 1. In particular, we have shown that the monotone RWW- and RRWW-transducers characterize the class of quasi-realtime pushdown relations, and we have seen that monotone (word-) weighted RWW-automata are strictly weaker in computational power than monotone (word-) weighted RRWW-automata. The latter is the first known case where it has been shown that a version of the (nondeterministic) monotone RWW-automaton differs in expressive power from the corresponding version of the (nondeterministic) monotone RRWW-automaton. Of course, it remains to derive a characterization of the classes of relations computed by these automata in terms of other types of devices.

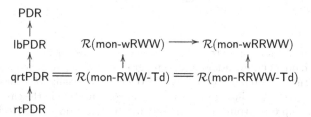

Fig. 1. Hierarchy of classes of (binary) relations that are computed by monotone R(R)WW-transducers and (word-)weighted R(R)WW-automata. An arrow denotes a proper inclusion, and classes that are not connected are incomparable with respect to inclusion.

References

1. Aho, A.V., Ullman, J.D.: The Theory of Parsing, Translation, and Compiling. Prentice-Hall Inc., Upper Saddle River (1972)
2. Choffrut, C., Culik II, K.: Properties of Finite and Pushdown Transducers. SIAM J. Comput. **12**(2), 300–315 (1983)
3. Hundeshagen, N.: Relations and Transductions Realized by Restarting Automata. Ph.D. thesis, Fachbereich Elektrotechnik/Informatik, Universität Kassel (2013)
4. Hundeshagen, N., Leupold, P.: Transducing by observing and restarting transducers. In: Freund, R., Holzer, M., Truthe, B., Ultes-Nitsche, U. (eds.) NCMA 2012. books@ocg.at, vol. 290, pp. 93–106. Österreichische Computer Gesellschaft, Vienna (2012)
5. Hundeshagen, N., Leupold, P.: Transducing by Observing Length-Reducing and Painter Rules. RAIRO - Theor. Inform. Appl. **48**(1), 85–105 (2014)
6. Hundeshagen, N., Otto, F.: Characterizing the rational functions by restarting transducers. In: Dediu, A.-H., Martín-Vide, C. (eds.) LATA 2012. LNCS, vol. 7183, pp. 325–336. Springer, Heidelberg (2012)
7. Jančar, P., Mráz, F., Plátek, M., Vogel, J.: Restarting automata. In: Reichel, H. (ed.) FCT 1995. LNCS, vol. 965, pp. 283–292. Springer, Heidelberg (1995)
8. Jančar, P., Mráz, F., Plátek, M., Vogel, J.: On Monotonic Automata with a Restart Operation. J. Auto. Lang. Comb. **4**(4), 287–312 (1999)
9. Kutrib, M., Messerschmidt, H., Otto, F.: On Stateless Two-Pushdown Automata and Restarting Automata. Int. J. Found. Comp. Sci. **21**, 781–798 (2010)
10. Otto, F., Wang, Q.: Weighted Restarting Automata. The results of this paper have been announced at WATA 2014 in Leipzig, May 2014 (submitted)

Equivalence Checking Problem for Finite State Transducers over Semigroups

Vladimir A. Zakharov[✉]

Institute for System Programming RAS,
National Research University Higher School of Economics, Moscow, Russia
zakh@cs.msu.su

Abstract. Finite state transducers over semigroups can be regarded as a formal model of sequential reactive programs. In this paper we introduce a uniform technique for checking effectively functionality, k-valuedness, equivalence and inclusion for this model of computation in the case when a semigroup these transducers operate over is embeddable in a decidable group.

1   Introduction

Finite state transducers extend the finite state automata to model functions on strings or lists, that is why they are used in fields as diverse as computational linguistics and model-based testing. In software engineering transducers provide a suitable formal model for various device drivers for manipulating with strings, transforming images, filtering dataflows, inserting fingerprints, etc. (see [1,10]). Algorithms for building compositions of transducers, checking equivalence, reducing their state space considerably enhance the effectiveness of designing, testing, verification and maintenance of such software routines.

Transducers may be used also as simple models of sequential reactive programs. These programs operate in the interaction with the environment permanently receiving data (requests) from it. At receiving a piece of data such program performs a sequence of actions. When certain control points are achieved a program outputs the current results of computation as a response. It is significant that different sequences of actions may yield the same result. Therefore, the basic actions of a program may be viewed as generating elements of some appropriate semigroup, and the result of computation may be regarded as the composition of actions performed by the program.

Imagine, for example, that a radio-controlled robot moves on the earth surface. It can make one step moves in any of 4 directions N, E, S, W. When such robot receives a control signal syg in a state q it must choose and carry out a sequence of steps (say, N, N, W, S), and enter to the next state q'. At some distinguished states q_{fin} robot reports its current location. The most simple model of computation which is suitable for designing such a robot and analyzing its behaviour is non-deterministic finite state transducer operating on free Abelian group of rank 2.

A. Maletti (Ed.): CAI 2015, LNCS 9270, pp. 208–221, 2015.
DOI: 10.1007/978-3-319-23021-4_19

These considerations give rise to the concept of a transducer which has some finitely generated semigroup S for the set of outputs. In this paper we study the equivalence checking problem and some related problems for finite state transducers over semigroups. The study of these problems for classical transducers that operate on words began in the early 60s. First, it was shown that the equivalence checking problem is undecidable for non-deterministic transducers [6] even over 1-letter input alphabet [8]. But the undecidability displays itself only in the case of unbounded transduction when an input word may have arbitrary many images. At the next stage bound-valued transducers were studied. It was proved that it is decidable in polynomial time whether the cardinality of the image of every word by a given transducer is bounded [15] and whether it is bounded by a given integer k [7]. The equivalence checking problem was shown also to be decidable for deterministic [3], functional (single-valued) transducers [2,13], and k-valued transducers [5,16]. In a series of papers [4,11,12,14] a construction to decompose k-valued transducers into a sum of functional and unambiguous ones was developed and used for checking bounded valuedness, k-valuedness and equivalence of finite state transducers over words.

This paper offers an alternative technique for the analysis of finite state transducers over semigroups. To check the equivalence of transducers π_1 and π_2 we associate with them a Labeled Transition System Γ_{π_1,π_2}. Each path in this LTS represents all possible runs of π_1 and π_2 on the same input word. Every node u of Γ_{π_1,π_2} keeps track of the states of π_1 and π_2 achieved at reading some input word and the *deficiency* of the output words computed so far. If both transducers reach their final states and the deficiency of their outputs is nonzero then this indicates that π_1 and π_2 produce different images for the same word, and, hence, they are not equivalent. The nodes of Γ_{π_1,π_2} that capture this effect are called *rejecting* nodes. Thus, the equivalence checking of π_1 and π_2 is reduced to checking the reachability of rejecting nodes in LTS Γ_{π_1,π_2}. We show that one needs to analyze only a bounded fragment of Γ_{π_1,π_2} to certify (un)reachability of rejecting nodes. The size of this fragment is polynomial of the size of π_1 and π_2 if both transducers are deterministic, and single-exponential if they are k-bounded. The same approach is applicable for checking k-valuedness of transducers over semigroups.

Initially, this LTS-based approach was introduced and developed in [17] for equivalence checking sequential programs in polynomial time. The concept of deficiency and a similar way of its application to the analysis of classical transducers was independently introduced in [4] and used in [12,14] under the names "Advance or Delay Action" (ADA), or "Lead or Delay Action" (LDA). The main advantage of our approach (apart from the fact that it is applicable to a more general type of transducers) is twofold. First, unlike one used in [11,16], it does not require a pre-processing (decomposition) of transducers to be analyzed and can be applied to any given transducers at once. Second, the checking procedure does not rely on the specific features of internal structures (like the analysis of strongly connected components used in [12,14]) of transducers under consideration and makes a plain depth-first search of rejecting nodes in the corresponding LTS. Complexity issues of our technique are shortly discussed in the conclusion.

2   Preliminaries

Given a finite *alphabet* A, denote by A^* the set of all finite *words* over A. A *finite state automaton* over A is 5-tuple $\mathcal{A} = \langle A, Q, init, F, \varphi \rangle$, where Q is a finite set of *states*, *init* is an *initial state*, F is a subset of *final states*, and $\varphi, \varphi \subseteq Q \times A \times Q$, is a *transition relation*. An automaton $\mathcal{A}$ *accepts* a word $w = a_1 a_2 \ldots a_n$ if there exists a sequence of states $q_0, q_1, \ldots, q_n$ such that $q_0 = init$, $q_n \in F$, and $(q_{i-1}, a_i, q_i) \in \varphi$ holds for every $i, 1 \le i \le n$. A *language* $L(\mathcal{A})$ of $\mathcal{A}$ is the set of all words accepted by $\mathcal{A}$. We write $\mathcal{A}[q]$ for the automaton $\langle A, Q, q, F, \varphi \rangle$ which has q for the initial state.

Let $S = (B, \cdot, e)$ be a semigroup generated by a set of elements B and having the identity element e. A *finite state transducer* over S is 6-tuple $\pi = \langle A, S, Q, q_0, F, T \rangle$, where Q is a finite set of states, q_0 is an initial state, F is a subset of final states, and $T, T \subseteq Q \times A \times S \times Q$, is a finite transition relation. Quadruples (q, a, s, q') in T are called *transitions* and denoted by $q \xrightarrow{a/s} q'$. We will denote by $\mathcal{A}_\pi$ the underlying finite state automaton $\langle A, Q, q_0, F, \varphi_\pi \rangle$, where

$$\varphi_\pi = \{(q, a, q') : q \xrightarrow{a/s} q' \text{ for some } s \text{ in } S\}.$$

A *run* of π on a word $w = a_1 a_2 \ldots a_n$ is a sequence of transitions of the form

$$q_i \xrightarrow{a_1/s_1} q_{i+1} \xrightarrow{a_2/s_2} \cdots \xrightarrow{a_{n-1}/s_{n-1}} q_{i+n-1} \xrightarrow{a_n/s_n} q_{i+n} . \tag{1}$$

The element $s = s_1 \cdot s_2 \cdots s_n$ of the semigroup S is called an *image* of w, and the pair (w, s) is called the *label* of a run (1). We will use notation $q_i \xrightarrow{w/s} q_{i+n}$ for a run of a transducer. If $q_i = q_0$ then (1) is an *initial* run. If $q_{i+n} \in F$ then (1) is a *final* run. A run which is both initial and final is called *complete*. By $Lab(\pi)$ we denote the *transduction* relation realized by π which is the set of labels (w, s) of all complete runs of π. A state q is *useful* if at least one complete run passes via q. In what follows we will assume that all states of the transducers under consideration are useful; in [4] such transducers are called *trim*. A transducer π is *deterministic* if for every letter a and a state q the set T contains at most one transition of the form $q \xrightarrow{a/s} q'$. A transducer π is *k-valued*, where k is a positive integer, if for every input word w the transduction relation $Lab(\pi)$ contains at most k labels of the form (w, s). A 1-valued transducer π is also called *functional*. Transducers π' and π'' are *equivalent* ($\pi' \sim \pi''$ in symbols) if $Lab(\pi') = Lab(\pi'')$.

In the rest of the paper we define and study procedures for checking equivalence and k-valuedness of finite state transducers over a semigroup S which can be embedded in a group. A semigroup S is *embeddable* in a group G if this group includes a semigroup S' isomorphic to S. The set of necessary and sufficient conditions for the embeddability of a semigroup in a group were given in [9]. The conditions are countably infinite in number and no finite subset will suffice. In fact, a free semigroup is embeddable in a free group, and any commutative semigroup can be embedded in a group iff it is cancellative. Without loss of generality in what follows it will be assumed that the transducers under consideration operate over a finitely generated decidable group G (i.e. there exists an algorithm for checking whether two words in the generators of G represent the same element), and, given an element s, we write s^- for the element of G which is inverse of s.

3   Equivalence Checking Deterministic Transducers

Let $\pi = \langle A, G, Q, q_0, F, T \rangle$ and $\pi' = \langle A, G, Q', q_0', F', T' \rangle$ be deterministic transducers over a finitely generated decidable group G. To check their equivalence we define the Labeled Transition System (LTS) $\Gamma^0_{\pi,\pi'} = \langle Q \times Q' \times G, \Rightarrow \rangle$. The nodes of $\Gamma^0_{\pi,\pi'}$ are triples of the form (q, q', g), where $q \in Q$, $q' \in Q'$, and $g \in G$. The third component g in this triple is called a *deficiency* (of initial runs arriving at the states q and q').

The transition relation $\Rightarrow$ is defined as follows: for every pair of nodes $v_1 = (q_1, q_1', g_1)$ and $v_2 = (q_2, q_2', g_2)$, and for every letter a a relation $v_1 \overset{a}{\Rightarrow} v_2$ holds iff $q_1 \xrightarrow{a/s} q_2$ and $q_1' \xrightarrow{a/s'} q_2'$ are transitions in π and π' respectively, and $g_2 = s^- g_1 s'$.

Given a word $w = a_1 a_2 \ldots a_n$ and a pair of nodes $v = (q_1, q_1', g_1)$ and $u = (q_2, q_2', g_2)$ we write $v \overset{w}{\Rightarrow} u$ as shorthand notation of a sequence $v \overset{a_1}{\Rightarrow} v_1 \overset{a_2}{\Rightarrow} \ldots \overset{a_{n-1}}{\Rightarrow} v_{n-1} \overset{a_n}{\Rightarrow} u$ which is called a *path* in $\Gamma^0_{\pi,\pi'}$. In this case we say that a node u is *reachable* from a node v. It is easy to see that $v \overset{w}{\Rightarrow} u$ holds iff $q_1 \xrightarrow{w/s} q_2$, $q_1' \xrightarrow{w/s'} q_2'$, and $s^- g_1 s' = g_2$.

The node $v_{src} = (q_0, q_0', e)$, where e is the identity element of G, is called the *source node* of $\Gamma^0_{\pi,\pi'}$. Denote by $V^0_{\pi,\pi'}$ the set of nodes reachable in LTS $\Gamma^0_{\pi,\pi'}$ from v_{src}. A node (q, q', g) is called *rejecting* if it satisfies at least one of the following requirements:

1. both q and q' are final states of π and π', and $g \neq e$;
2. exactly one of the states q or q' is final;
3. for some letter a only one of the states q or q' has a a-transition, whereas the other state does not.

The set of all rejecting nodes of LTS $\Gamma^0_{\pi,\pi'}$ is denoted by $R^0_{\pi,\pi'}$.

Lemma 1. *Deterministic transducers π and π' are equivalent iff $V^0_{\pi,\pi'} \cap R^0_{\pi,\pi'} = \emptyset$.*

Proof. Follows immediately from the definitions of LTS $\Gamma^0_{\pi,\pi'}$, $V^0_{\pi,\pi'}$, $R^0_{\pi,\pi'}$ and the equivalence $\sim$ in view of the fact that π and π' are both deterministic and trim. □

Thus, the equivalence checking of deterministic trim transducers is reduced to the searching of rejecting nodes in the set of reachable nodes of LTS $\Gamma^0_{\pi,\pi'}$. Next we show how to cut down the search space.

Lemma 2. *If the set $V^0_{\pi,\pi'}$ contains a pair of nodes $v_1 = (q, q', g_1)$ and $v_2 = (q, q', g_2)$ such that $g_1 \neq g_2$ then $V^0_{\pi,\pi'} \cap R^0_{\pi,\pi'} \neq \emptyset$.*

Proof. Suppose, to the contrary, that $V^0_{\pi,\pi'} \cap R^0_{\pi,\pi'} = \emptyset$, and, hence, $\pi \sim \pi'$. By the definition of $\Gamma^0_{\pi,\pi'}$ there is such a word w_0 that $q_0 \xrightarrow{w_0/s_0} q$, $q'_0 \xrightarrow{w_0/s'_0} q'$, and $g_1 = s_0^- s'_0$. Since the state q is useful, there is such a word w that $q_0 \xrightarrow{w_0/s_0} q \xrightarrow{w/s} p$ is a complete run of π. As far as $\pi \sim \pi'$ and π' is deterministic, the run $q'_0 \xrightarrow{w_0/s'_0} q' \xrightarrow{w/s'} p'$ of π' is complete and $s_0 s = s'_0 s'$. Hence, $g_1 = s(s')^-$. Inasmuch as $q \xrightarrow{w/s} p$ and $q' \xrightarrow{w/s'} p'$, there is a path $v_2 \xRightarrow{w} (p, p', g)$ in $\Gamma^0_{\pi,\pi'}$, where $g = s^- g_2 s'$. Having in mind that (p, p', g) is in $V^0_{\pi,\pi'}$, both states p and p' are final, and assuming that $V^0_{\pi,\pi'} \cap R^0_{\pi,\pi'} = \emptyset$, we arrive at the equality $g = e$. Therefore, $g_2 = s(s')^- = q_1$ which contradicts the premise of Lemma. $\qquad\square$

By Lemmata 1 and 2, to check the equivalence of deterministic trim transducers π and π' it is sufficient to analyze at most $|Q||Q'| + 1$ nodes reachable from the source node of LTS $\Gamma^0_{\pi,\pi'}$. This consideration brings us to

Theorem 1. *The equivalence problem for deterministic transducers over finitely generated decidable group G is decidable. Moreover, if the word problem for G is decidable in polynomial time then the equivalence problem for deterministic transducers over G is decidable in polynomial time as well.*

4   Checking Functional Transducers

To check the functionality of a transducer $\pi = \langle A, G, Q, q_0, F, T \rangle$ we also take advantage of LTSs. Let $\pi = \langle A, G, Q, q_0, F, T \rangle$ and $\pi' = \langle A, G, Q', q'_0, F', T' \rangle$ be a pair of transducers. Define a LTS $\Gamma^1_{\pi,\pi'} = \langle Q \times Q' \times G, \Rightarrow \rangle$ as follows: for every pair of nodes $v_1 = (q_1, q'_1, g_1)$ and $v_2 = (q_2, q'_2, g_2)$, and for every letter a, a relation $v_1 \xRightarrow{a} v_2$ holds iff there exist transitions $q_1 \xrightarrow{a/s} q_2$ and $q'_1 \xrightarrow{a/s'} q'_2$ such that $g_2 = s^- g_1 s'$, and $L(\mathcal{A}_\pi[q_2]) \cap L(\mathcal{A}_{\pi'}[q'_2]) \neq \emptyset$. The set of all nodes of LTS $\Gamma^1_{\pi,\pi'}$ reachable from the source node (q_0, q_0, e) is denoted by $V^1_{\pi,\pi'}$. We say that (q_1, q_2, g) is a *rejecting* node if q_1 and q_2 are final states, and $g \neq e$. The set of all rejecting nodes of LTS $\Gamma^1_{\pi,\pi'}$ is denoted by $R^1_{\pi,\pi'}$. The lemmata below can be proved using the same reasoning as in previous section.

Lemma 3. *A transducer π is functional iff $V^1_{\pi,\pi} \cap R^1_{\pi,\pi} = \emptyset$.*

Lemma 4. *If the set $V^1_{\pi,\pi}$ includes a pair of nodes $v_1 = (q, p, g_1)$ and $v_2 = (q, p, g_2)$ such that $g_1 \neq g_2$ then $V^1_{\pi,\pi} \cap R^1_{\pi,\pi} \neq \emptyset$.*

As it follows from Lemmata 3 and 4, to check functionality of a transducer π one needs only to analyze at most $|Q|^2 + 1$ nodes reachable from the source node of $\Gamma^1_{\pi,\pi}$.

Theorem 2. *The functionality of transducers over finitely generated decidable group G can be checked effectively. Moreover, if the word problem for G is decidable in polynomial time then the functionality checking can be performed in polynomial time as well.*

The equivalence of functional transducers π and π' can be checked in the same way by means of LTS $\Gamma^1_{\pi,\pi'}$. But now we need to check in advance that $L(\mathcal{A}_\pi[q_0]) = L(\mathcal{A}_{\pi'}[q'_0])$ since, unlike the case of deterministic transducers, the nodes (q_1, q_2, g) in $\Gamma^1_{\pi,\pi'}$ such that exactly one of the states q_1 and q_2 is final can not be regarded as rejecting ones.

Lemma 5. *If $L(\mathcal{A}_\pi[q_0]) = L(\mathcal{A}_{\pi'}[q'_0])$ then functional transducers π and π' are equivalent iff $V^1_{\pi,\pi'} \cap R^1_{\pi,\pi'} = \emptyset$.*

Lemma 6. *If the set $V^1_{\pi,\pi'}$ includes a pair of nodes $v_1 = (q, q', g_1)$ and $v_2 = (q, q', g_2)$ such that $g_1 \neq g_2$ then $V^1_{\pi,\pi'} \cap R^1_{\pi,\pi'} \neq \emptyset$.*

Theorem 3. *The equivalence problem for functional transducers over finitely generated decidable group G is decidable. Moreover, if the word problem for G is decidable in polynomial time then the equivalence problem for functional transducers is PSPACE-complete.*

5   Checking 2-Valuedness of Transducers

The LTS-based techniques put forward in Sections 3 and 4 for checking the equivalence of deterministic and functional transducers can be developed further to cope with the analysis of k-valued transducers. For the sake of clarity we consider in details only the case of $k = 2$; the same arguments supplied with a bit more cumbersome combinatorics gives a general solution to the checking problems for k-valued finite state transducers.

We begin with checking 2-valuedness of transducers over a decidable group. Given a transducer $\pi = \langle A, G, Q, q_0, F, T \rangle$ define a LTS $\Gamma^2_\pi = \langle Q \times (Q \times G)^2, \Rightarrow \rangle$ as follows: for every pair of nodes $v_1 = (q_1, (q_2, g_{12}), (q_3, g_{13}))$ and $v_2 = (q'_1, (q'_2, g'_{12}), (q'_3, g'_{13}))$, and a letter a, a transition $v_1 \overset{a}{\Rightarrow} v_2$ takes place if there exist transitions $q_1 \overset{a/s_1}{\longrightarrow} q'_1$, $q_2 \overset{a/s_2}{\longrightarrow} q'_2$, and $q_3 \overset{a/s_3}{\longrightarrow} q'_3$ such that the equalities $g'_{12} = s_1^- g_{12} s_2$ and $g'_{13} = s_1^- g_{13} s_3$ hold, and $L(\mathcal{A}_\pi[q'_1]) \cap L(\mathcal{A}_\pi[q'_2]) \cap L(\mathcal{A}_\pi[q'_3]) \neq \emptyset$.

A triple of states (q_1, q_2, q_3) will be called a *type* of a node $(q_1, (q_2, g_{12}), (q_3, g_{13}))$. As in the case of 1-valuedness, we define the set V^2_π of all nodes reachable in LTS Γ^2_π from the source node $(q_0, (q_0, e), (q_0, e))$. From the definitions of Γ^2_π and V^2_π it follows that a node $v = (q_1, (q_2, g_{12}), (q_3, g_{13}))$ is in V^2_π iff there exists such a word w that $q_0 \overset{w/s_1}{\longrightarrow} q_1$, $q_0 \overset{w/s_2}{\longrightarrow} q_2$, $q_0 \overset{w/s_3}{\longrightarrow} q_3$, and $g_{12} = s_1^- s_2$, $g_{13} = s_1^- s_3$.

The set R^2_π of rejecting nodes includes all such nodes $(q_1, (q_2, g), (q_3, h))$ that q_1, q_2, q_3 are final states, and $g \neq e$, $h \neq e$, $g \neq h$ hold.

Lemma 7. *A transducer π is 2-valued iff $V^2_\pi \cap R^2_\pi = \emptyset$.*

Proof. Follows from the definitions of V^2_π, R^2_π, and 2-valuedness property.   □

Now we need to cut off the space of V^2_π for searching the rejecting nodes. This is achieved by means of the following two lemmata. Their proofs are based on the pigeonhole principle and basic group-theoretic properties.

Lemma 8. *Suppose that V_π^2 includes four nodes $v_i = (q, (q', g_i'), (q'', g_i''))$, $1 \leq i \leq 4$, of the same type such that the inequalities $g_i' \neq g_j'$, $g_i'' \neq g_j''$, and $g_i'(g_i'')^- \neq g_j'(g_j'')^-$ hold for every pair of indices i, j, $1 \leq i < j \leq 4$. Then $V_\pi^2 \cap R_\pi^2 \neq \emptyset$.*

Proof. Since all nodes v_i, $1 \leq i \leq 4$, are in V_π^2 then $L(\mathcal{A}_\pi[q]) \cap L(\mathcal{A}_\pi[q']) \cap L(\mathcal{A}_\pi[q'']) \neq \emptyset$. Hence, there exists such a word w that $q \xrightarrow{w/s} p$, $q' \xrightarrow{w/s'} p'$, and $q'' \xrightarrow{w/s''} p''$ are final runs of the transducer π. Then, by definition of Γ_π^2, the set of reachable nodes V_π^2 includes four nodes $u_i = (p, (p', s^- g_i' s'), (p'', s^- g_i'' s''))$, $1 \leq i \leq 4$. If u_1 is not a rejecting node then at least one of the equalities hold: $s^- g_1' s' = e$, $s^- g_1'' s'' = e$, or $s^- g_1' s' = s^- g_1'' s''$. Without loss of generality consider the case of $s^- g_1' s' = e$ (two other possibilities are treated in the similar way). Since $g_1' \neq g_2'$, this case implies $s^- g_2' s' \neq e$. Therefore, if u_2 is not a rejecting node then this is due to one of the equalities $s^- g_2'' s'' = e$, or $s^- g_2' s' = s^- g_2'' s''$. Consider the case of $s^- g_2'' s'' = e$ (the other possibility is treated similarly). Since $g_1' \neq g_3'$ and $g_2'' \neq g_3''$, the above equalities $s^- g_1' s' = e$ and $s^- g_2'' s'' = e$ imply $s^- g_3' s' \neq e$ and $s^- g_3'' s'' \neq e$. Therefore, if u_3 is not a rejecting node then $s^- g_3' s' = s^- g_3'' s''$. But, taking into account that $g_1' \neq g_4'$, $g_2'' \neq g_4''$, and $g_3'(g_3'')^- \neq g_4'(g_4'')^-$, the equalities $s^- g_1' s' = e$, $s^- g_2'' s'' = e$, and $s^- g_3' s' = s^- g_3'' s''$ bring us to the conclusion that $s^- g_4' s' \neq e$, $s^- g_4'' s'' \neq e$, and $s^- g_4' s' \neq s^- g_4'' s''$, which means that $v_4 \in R_\pi^2$. $\quad\square$

Lemma 9. *Let $v_i = (q, (q', g_i'), (q'', g_i''))$, $1 \leq i \leq 4$, be four pairwise different nodes in LTS Γ_π^2 that satisfy one of the following requirements:*

  a) $g_i' = g_j'$ holds for every pair i, j, $1 \leq i < j \leq 4$;
  b) $g_i'' = g_j''$ holds for every pair i, j, $1 \leq i < j \leq 4$;
  c) $(g_i')^- g_i'' = (g_j')^- g_j''$ holds for every pair i, j, $1 \leq i < j \leq 4$.
If a rejecting node is reachable from v_4 then some rejecting node is reachable from one of the nodes v_1, v_2, v_3.

Proof. We consider only the case when all nodes satisfy the first requirement $g_i' = g'$ for every $i, 1 \leq i \leq 4$. The similar reasoning is adequate for the other alternatives.

Suppose that a rejecting node $u_4 = (p, (p', h'), (p'', h_4''))$ is reachable from v_4 through some word w. Then there are three final runs $q \xrightarrow{w/s} p$, $q' \xrightarrow{w/s'} p'$, and $q'' \xrightarrow{w/s''} p''$ of π such that $h' = s^- g' s'$ and $h_4'' = s^- g_4'' s''$. Since u_4 is a rejecting node, we have $h' \neq e$.

The definition of Γ_π^2 guarantees that for every i, $1 \leq i \leq 3$, there is a path from the node v_i to the node $u_i = (p, (p', h'), (p'', h_i''))$, where $h_i'' = s^- g_i'' s''$. If $u_1 \notin R_\pi^2$ then either $h_1'' = e$ or $(h')^- h'' = e$. Consider only the case $h_1'' = e$ (the other possibility is treated in the same way). Since $g_2'' \neq g_1''$ and $g_1'' \neq g_3''$, we have $h_2'' \neq e$ and $h_3'' \neq e$. Therefore, if $u_2 \notin R_\pi^2$ then $(h')^- h_2'' = e$. But, as far as $g_2'' \neq g_3''$, it is true that $(h')^- h_3'' \neq e$. Thus, we conclude that u_3 is a rejecting node. $\quad\square$

With Lemmata 8 and 9 in hand we are able to prove

Theorem 4. *If G is a finitely generated decidable group then 2-valuedness is a decidable property of transducers over G.*

Proof. By Lemma 7 we can check 2-valuedness of a transducer π through the reachability analysis of rejecting nodes in LTS Γ_π^2. To this end we introduce a depth-first search of rejecting nodes. It begins with the source node $(q_0, (q_0, e), (q_0, e))$ and keeps track of useful nodes only. Suppose that at some step the traversal reaches a node $v = (q, (q', g'), (q'', g''))$ in Γ_π^2 which has not been visited yet. Then the following 4 cases are possible.

1) If v is a rejecting node then the search stops and announces that π is not 2-valued.

2) Otherwise, check if there exist 3 previously visited useful nodes $v_i = (q, (q', g_i'), (q'', g_i''))$, $1 \le i \le 3$, of the same type as v that satisfy one of the following requirements:

    a) $g' = g_i'$ for every i, $1 \le i \le 3$;

    b) $g'' = g_i''$ for every i, $1 \le i \le 3$;

    c) $(g')^- g'' = (g_i')^- g_i''$ for every i, $1 \le i \le 3$.

If so then v is regarded as useless and a backtracking step is made from this node.

3) Otherwise, if 27 useful nodes $v_i = (q, (q', g_i'), (q'', g_i''))$, $1 \le i \le 27$, of the same type as v has been already visited then the search stops and announces that π is not 2-valued.

4) Otherwise, the node v is regarded as *useful*, and the search procedure continues its depth-first traversal of LTS Γ_π^2.

If the search backtracks finally to the source node then π is recognized 2-valued.

As it can be seen from the definition of the search procedure, it always terminates at visiting at most $27|Q|^3$ useful nodes of Γ_π^2. Lemma 9 guarantees that by skipping useless nodes we do not miss possible paths to some rejecting nodes. This certifies the completeness of our search. To prove its correctness we need to show that case 3) of the search is correct. Indeed, simple combinatorial considerations disclose that if we have 28 nodes (v and v_i, $1 \le i \le 27$) such that neither 4 nodes of them fall under the premise of Lemma 9 (i.e., the nodes are useful) then this set of nodes includes a quadruple of nodes that satisfy the assumptions of Lemma 8. $\qquad\square$

Corollary 1. *If the word problem for a group G is decidable in polynomial time then 2-valuedness property of transducers over G can be checked in polynomial time.*

Both Lemmata 8 and 9, as well as the decision procedure defined in Theorem 4 can be readily extended to the case of an arbitrary k: the nodes of LTS Γ_π^2 are $(k+1)$-tuples $(q_0, (q_1, h_1), \ldots, (q_k, h_k))$, and to certify the reachability of a rejecting node in Γ_π^2 it suffices to visit at most $\binom{k+1}{2}^{\binom{k+1}{2}} |Q|^{k+1} + 1$ useful nodes.

6   Checking the Equivalence of 2-Valued Transducers

Instead of solving the equivalence checking problem for finite state transducers we study a more general inclusion checking problem: given a pair of transducers π and π' check whether $Lab(\pi') \subseteq Lab(\pi)$. The LTS-based approach is invoked once again.

Let $\pi = \langle A, G, Q, q_0, F, T \rangle$ and $\pi' = \langle A, G, Q', q_0', F', T' \rangle$ be a pair of trim 2-valued transducers. Clearly, if $Lab(\pi') \subseteq Lab(\pi)$ then $L(\mathcal{A}_{\pi'}) \subseteq L(\mathcal{A}_{\pi})$. Therefore, in this section we deal only with the case of π and π' such that $L(\mathcal{A}_{\pi'}) \subseteq L(\mathcal{A}_{\pi})$.

To define an LTS $\Gamma_{\pi,\pi'}^3$ corresponding to the inclusion checking problem for transducers π and π' we introduce a concept of block of states. Let $\widehat{Q}$ be some multiset of states of transducer π. Then a *block of states in* $\widehat{Q}$ is any maximal (i.e., inextensible) subset B of $\widehat{Q}$ such that $\bigcap_{q \in B} L(\mathcal{A}_{\pi}[q]) \neq \emptyset$, i.e. some word is accepted by every automaton $\mathcal{A}_{\pi}[q]$, $q \in B$, but no such words are accepted by an automaton $\mathcal{A}_{\pi}[q']$ for any $q', q' \in \widehat{Q} \setminus B$.

LTS $\Gamma_{\pi,\pi'}^3 = \langle V, \Rightarrow \rangle$ is defined as follows. The set of nodes V consists of all such pairs $u = (q', X)$, where $q' \in Q'$, and $X = \{(q_1, g_1), \ldots, (q_m, g_m)\} \subseteq Q \times G$, that satisfy the requirement $L(\mathcal{A}_{\pi'}[q']) \cap \bigcap_{i=1}^{m} L(\mathcal{A}_{\pi}[q_i]) \neq \emptyset$. The pair $(q', \{q_1, \ldots, q_m\})$ will be referred to as a *type* of the node u. For every letter a and a pair of nodes $u = (q', X)$ and $v = (p', Y)$ of types (q', B_u) and (p', B_v) respectively a transition $u \overset{a}{\Rightarrow} v$ takes place iff

1. there is transition $q' \xrightarrow{a/s'} p'$ in the transducer π',
2. B_v is a block of states in the multiset $\widehat{Q} = \{\widehat{q} : \exists q \ (q \in B_u \text{ and } q \xrightarrow{a/s} \widehat{q}\ )\ \}$, and
3. a pair (p, h) is in Y if and only if $p \in B_v$ and there exists such a pair (q, g) in X that $q \xrightarrow{a/s} p$ is a transition of transducer π and $h = (s')^{-}gs$.

As usual, given a word w we write $u \overset{w}{\Rightarrow} v$ for the composition of corresponding 1-letter transitions of LTS. The node $v_{src} = (q_0', \{(q_0, e)\})$ is the *source* node of LTS $\Gamma_{\pi,\pi'}^3$. By $V_{\pi,\pi'}^3$ we denote the set of all nodes reachable from v_{src}. A node (q', X) such that $q' \in F'$, and for every pair (q, g) in X either $q \notin F$, or $g \neq e$, is called a *rejecting* node. The set of rejecting nodes of $\Gamma_{\pi,\pi'}^3$ is denoted by $R_{\pi,\pi'}^3$. The intended meaning of LTS $\Gamma_{\pi,\pi'}^3$ with regard to the inclusion checking of π and π' is clarified in the propositions below.

Proposition 1. *Let w_0 and w_1 be arbitrary words, and $q_0' \xrightarrow{w_0/s_0'} q_1' \xrightarrow{w_1/s_1'} q_2'$ be a complete run of transducer π'. Then there exists such a node $v = (q_1', X)$ that $v_{src} \overset{w_0}{\Rightarrow} v$ and for every complete run $q_0 \xrightarrow{w_0/s_0} q_1 \xrightarrow{w_1/s_1} q_2$ of transducer π the multiset X includes a pair $(q_1, (s_0')^{-}s_0)$.*

Proposition 2. *Suppose that* $v_{src} \overset{w_0}{\Rightarrow} (q', X)$. *Then there exist such a word* w_1 *and a complete run* $q_0' \overset{w_0/s_0'}{\longrightarrow} q_1' \overset{w_1/s_1'}{\longrightarrow} q_2'$ *of transducer* π' *that for every complete run* $q_0 \overset{w_0/s_0}{\longrightarrow} q_1 \overset{w_1/s_1}{\longrightarrow} q_2$ *of transducer* π *the multiset* X *includes a pair* $(q_1, (s_0')^- s_0)$.

Both propositions can be proved by induction on the length of w_0 relying on the definition of transition relation $\Rightarrow$ only. The correctness of these propositions is due to the fact that the type of every reachable node is specified as block of states.

Lemma 10. $Lab(\pi') \subseteq Lab(\pi) \iff V^3_{\pi,\pi'} \cap R^3_{\pi,\pi'} = \emptyset$.

Proof. Follows from Propositions 1,2 above and the definition of rejecting node. □

We show that, even though the set $V^3_{\pi,\pi'}$ may be infinite, only finitely many nodes must be checked to verify (un)reachability of rejecting nodes.

Consider an arbitrary reachable node v of type (q', B). Since the transducer π is 2-valued, for every state q of π at most two copies of q may occur in the multiset B. Therefore, $|B| \leq 2|Q|$, and the total number of types of reachable nodes in $\Gamma^3_{\pi,\pi'}$ does not exceed $|Q'|3^{|Q|}$.

Consider the language $L = L(\mathcal{A}_{\pi'}[q']) \cap \bigcap_{q \in B} L(\mathcal{A}_\pi[q])$; it will be called a *language of type* (q', B). By definition of $\Gamma^3_{\pi,\pi'}$, this language is non-empty. The set of types of all reachable nodes can be divided into three classes depending on the properties of L. A type (q', B) will be called *A-type* iff there exists such a word w in L which has two different images s_1' and s_2' of w in two final runs $q' \overset{w/s_1'}{\longrightarrow} p_1'$ and $q' \overset{w/s_2'}{\longrightarrow} p_2'$ of transducer π'. A type (q', B) will be called *B-type* iff it does not belong to the class A and there exist a state q in the multiset B and a word w in L which has two different images s_1 and s_2 in two final runs $q \overset{w/s_1}{\longrightarrow} p_1$ and $q \overset{w/s_2}{\longrightarrow} p_2$ of transducer π. All other types will be called *C-types*. Lemmata below elucidate some properties of these classes that are crucial for the solution of the inclusion checking problem.

Lemma 11. *Suppose that* $Lab(\pi') \subseteq Lab(\pi)$, *and* (q', B) *be a A-type. Then at most* $2^{|B|}$ *nodes of this type are reachable from the source node.*

Proof. Let L be the language of type (q', B). Consider an arbitrary node $v = (q', X)$ of type (q', B) such that $v_{src} \overset{w_0}{\Rightarrow} v$, and an arbitrary pair (q, g) from X. Since (q', B) is A-type, there exists such a word w in L which has different images s_1' and s_2' in two final runs $q' \overset{w/s_1'}{\longrightarrow} p_1'$ and $q' \overset{w/s_2'}{\longrightarrow} p_2'$ of transducer π'. By definition of L, the transducer π has a final run $q \overset{w/s}{\longrightarrow} q_1$. Notice, that the elements s_1', s_2' and s depend on the type (q', B) and the state q only. By Proposition 2, transducers π and π' have initial runs $q_0 \overset{w_0/s_0}{\longrightarrow} q$ and $q_0' \overset{w_0/s_0'}{\longrightarrow} q'$ such that $g = (s_0')^- s_0$. Then the transducer π' has two complete runs $q_0' \overset{w_0/s_0'}{\longrightarrow} q' \overset{w/s_1'}{\longrightarrow} p_1'$ and $q_0' \overset{w_0/s_0'}{\longrightarrow} q' \overset{w/s_2'}{\longrightarrow} p_2'$, and the transducer π has a complete run $q_0 \overset{w_0/s_0}{\longrightarrow} q \overset{w/s}{\longrightarrow} q_1$.

Since π is a 2-valued transducer, $s_0' s_1' \neq s_0' s_2'$, and $Lab(\pi') \subseteq Lab(\pi)$, we may be sure that at least one of the equalities $s_0 s = s_0' s_1'$ or $s_0 s = s_0' s_2'$ holds. Hence, either $g = s_1' s^-$, or $g = s_2' s^-$. The assertion of the Lemma follows from the fact that both possible values of g depend on the type (q', B) and the state q only.   □

Lemma 12. *Suppose that $Lab(\pi') \subseteq Lab(\pi)$, and (q', B) is a B-type. Then at most $3^{|B|}$ nodes of this type are reachable from the source node.*

Proof. Let L be the language of type (q', B). Consider an arbitrary node $v = (q', X)$ of type (q', B) such that $v_{src} \overset{w_0}{\Rightarrow} v$. Let a pair (q, g) in X be such that for some word w in L final runs $q \overset{w/s_1}{\longrightarrow} p_1$ and $q \overset{w/s_2}{\longrightarrow} p_2$ of transducer π yield different images of w. Consider an arbitrary pair (p, h) in X. Since $w \in L$, there exist final runs $p \overset{w/s}{\longrightarrow} p_3$ and $q' \overset{w/s'}{\longrightarrow} p'$ of π and π'. By referring to Proposition 2 we conclude the following. Since $Lab(\pi') \subseteq Lab(\pi)$, exactly one of the equalities $s' = gs_1$ or $s' = gs_2$ holds. Since π is a 2-valued transducer, exactly one of the equalities $gs_1 = hs$ or $gs_2 = hs$ is valid. Hence, either $h = s's^-$, or $h = s'(s_1)^- s_2 s^-$, or $h = s'(s_1)^- s_2 s^-$. The assertion of Lemma follows from the fact that these possible values of h depend on the type (q', B) and the states q and p only.   □

Let (q', B) be a C-type, where $B = \{q_1, \ldots, q_m\}$, and L be the language of this type. Associate with (q', B) any word w_0 from L and consider a final run $q' \overset{w_0/s'}{\longrightarrow} p'$ of transducer π' and final runs $q_i \overset{w_0/s_i}{\longrightarrow} p_i$ for every i, $1 \leq i \leq m$. The tuple $(s', s_1, \ldots, s_m)$ of elements in S will be called a w_0-*characteristics* of the type (q', B). This characteristics will help us to narrow the search space. Suppose that $u = (q', \{(q_1, g_1), \ldots, (q_m, g_m)\})$ is a reachable node of the C-type (q', B). If $s' \neq g_i s_i$ holds for every i, $1 \leq i \leq m$, then, by definition of LTS $\Gamma^3_{\pi,\pi'}$, a rejecting node is reachable from u. We will say that such a node u is *pre-rejecting* node of the type (q', B). Otherwise, the set X can be split into two subsets $X_0 = \{(q_i, g_i) : s' = g_i s_i, 1 \leq i \leq m\}$ and $X_1 = \{(q_j, g_j) : s' \neq g_j s_j, 1 \leq j \leq m\}$ such that $X_0 \neq \emptyset$. We will use a notation $(q', X_0 \oplus X_1)$ for such a node u. Note that since π is a 2-valued transducer, $g_i s_i = g_j s_j$ holds for every two pairs (q_i, g_i), (q_j, g_j) from X_1.

Lemma 13. *Let (q', B) be a C-type, $B = \{q_1, \ldots, q_m\}$, and $k = 2m$. Suppose that $k + 1$ nodes $u_1 = (q', X_0 \oplus X_1), \ldots, u_{k+1} = (q', X_0 \oplus X_{k+1})$ of type (q', B) are reachable from the source node. Then a rejecting node is reachable from one of the nodes $u_1, \ldots, u_{k+1}$ iff a rejecting node is reachable from one of the nodes $u_1, \ldots, u_k$.*

Proof. Let $(s', s_1, \ldots, s_m)$ be a characteristics of the type (q', B). Assume that $X_0 = \{(q_1, g_1), \ldots, (q_\ell, g_\ell)\}$ and $X_j = \{(q_{\ell+1}, g_{\ell+1j}), (q_m, g_{mj})\}$ for every j, $1 \leq j \leq k + 1$.

Suppose that $u_{k+1} \overset{w}{\Rightarrow} v$ holds for some rejecting node v and a word w. Then, by definition of $\Gamma^3_{\pi,\pi'}$, the transducer π' has a final run $q' \overset{w/s'}{\longrightarrow} p'$ and for every i, $1 \leq i \leq m$, the transducer π either has no final runs on the word w from the state q_i, or every final run $q_i \xrightarrow{w/t_i} p_i$ yields an image t_i of w such that $s' \neq g_{ik+1}t_i$ (actually, at most two such images t_{i1} and t_{i2} are possible due to the fact that π is a 2-valued transducer). We analyze the worst case when the second alternative is achieved for every state q_i, $1 \leq i \leq m$. Thus, we have at most $2(m-1)$ elements $t_{i\sigma}$, $\sigma \in \{1,2\}$ from G that are images of w on final runs from the states $q_{\ell+1}, \ldots, q_m$.

If a rejecting node is not reachable from, say, a node u_1 then for some (q_i, g_{i1}) from X_1 and for some image t of the word w the equality $s' = g_{i1}t$ holds, i.e. $g_{i1} = s't^-$. Recall that for any other pair (q_j, g_{j1}) we have $g_{i1}s_i = g_{j1}s_j$, i.e. $g_{j1} = s't^- s_i s_j^-$. This means that the image t completely defines all elements g_{j1}, $\ell + 1 \leq j \leq m$, in X_1. Clearly, different images of the word w define the elements in the different sets X_i. Since the amount of images of w does not exceed $2(m-1) < k$, there exists such node u_i, $1 \leq i < k$, that $s' \neq g_{ji}t_{j\sigma}$ holds for every component (q_j, g_{ji}) of X_i and image(s) $t_{j\sigma}$ of the word w. The latter means that a rejecting node is reachable from u_i.   □

Theorem 5. *If G is a finitely generated decidable group G then inclusion problem $Lab(\pi') \subseteq Lab(\pi)$ for 2-valued transducers over G is decidable.*

Proof. The search of rejecting nodes in $\Gamma^3_{\pi,\pi'}$ begins with the source node v_{src}. Suppose that at some step the traversal reaches a node $u = (q', X)$ of a type (q', B), and u has not been visited yet. Then the following 6 cases are possible.
1) If $u \in R^3_{\pi,\pi'}$ then the search stops and announces that π does not include π'.
2) Otherwise, if (q', B) is a A-type and $2^{|B|}$ nodes of the same type have been already visited then the search stops and announces that π does not include π'.
3) Otherwise, if (q', B) is a B-type and $3^{|B|}$ nodes of the same type have been already visited then the search stops and announces that π does not include π'.
4) Otherwise, if (q', B) is a C-type, and u is a pre-rejecting node of this type then the search stops and announces that π does not include π'.
5) Otherwise, if (q', B) is a C-type, $u = (q', X_0 \oplus X_1)$, and $2|B|$ nodes of the form $u_i = (q', X_0 \oplus X_{1i})$ have been already visited then the search backtracks from u.
6) Otherwise, the search procedure continues its depth-first traversal of LTS Γ^2_π. If the backtracking ends in the source node then the inclusion $Lab(\pi') \subseteq Lab(\pi)$ holds.

Termination, correctness and completeness of this search procedure follow from Lemmata 10-13. As it can be seen from the description of the search procedure, to check the inclusion $Lab(\pi') \subseteq Lab(\pi)$ less than $|Q'|8^{|Q|}$ nodes of LTS $\Gamma^3_{\pi,\pi'}$ have to be analyzed.   □

Corollary 2. *The equivalence checking problem for 2-valued transducers over finitely generated decidable group G is decidable. Moreover, if the word problem for G is decidable in polynomial time then the equivalence checking problem for 2-valued transducers over G is decidable in single exponential time.*

The same approach is applicable to equivalence checking of k-valued transducers for an arbitrary k. But till now the author did not find adequate means for presenting the general solution of this problem in short terms; this remains the topic for further research.

7   Conclusion

The complexity of checking procedures defined in Sections 3-6 depends on the complexity of the word problem for a group G. The time complexity of our algorithms for the cases when G is the free group is estimated below on the following parameters: n (number of states), m (number of transitions), and ℓ (maximal length of the outputs of transitions).

- deterministic equivalence checking: $O(\ell n^3)$,
- functionality checking: $O(\ell m^2 n^2)$,
- k-valuedness checking: $O((k+1)^{2(k+1)^2} \ell m^{k+1} n^{k+1})$,
- functional equivalence checking: $2^{O(n)}$;
- 2-valued equivalence checking: $2^{O(n \log m)}$.

One can compare these complexity estimates with previously known upper bounds for the complexity of k-valuedness checking $O(2^{(k+1)^4} \ell m^{k+1} n^{k+1})$ obtained in [12] and equivalence checking of k-valued transducers $2^{O(\ell k^5 n^{k+4})}$ presented in [14]. As is easy to see, even the best known algorithms for the analysis of k-valued transducers have the complexity which is exponential of k. So, an open question is if it is possible to check k-valuedness and equivalence of nondeterministic transducers in time polynomial of k.

The author would like to thank the anonymous referees whose keen and valuable comments helped him to improve the original version of the paper.

References

1. Alur, R., Cerny, P.: Streaming transducers for algorithmic verification of single-pass list-processing programs. In: Proc. of 38th ACM SIGACT-SIGPLAN Symposium on Principles of Programming Languages, pp. 599–610 (2011)
2. Blattner, M., Head, T.: Single-valued a-transducers. Journal of Computer and System Sciences **15**, 310–327 (1977)
3. Blattner, M., Head, T.: The decidability of equivalence for deterministic finite transducers. Journal of Computer and System Sciences **19**, 45–49 (1979)
4. Beal, M.-P., Carton, O., Prieur, C., Sakarovitch, J.: Squaring transducers: an efficient procedure for deciding functionality and sequentiality. Theoretical Computer Science **292** (2003)
5. Culik, K., Karhumaki, J.: The equivalence of finite-valued transducers (on HDTOL languages) is decidable. Theoretical Computer Science **47**, 71–84 (1986)
6. Griffiths, T.: The unsolvability of the equivalence problem for ε-free nondeterministic generalized machines. Journal of the ACM **15**, 409–413 (1968)
7. Gurari, E., Ibarra, O.: A note on finite-valued and finitely ambiguous transducers. Mathematical Systems Theory **16**, 61–66 (1983)

8. Ibarra, O.: The unsolvability of the equivalence problem for Efree NGSM's with unary input (output) alphabet and applications. SIAM Journal on Computing **4** (1978)
9. Malcev, A.I.: Uber die Einbettung von assoziativen Systemen. Gruppen, Rec. Math. (Mat. Sbornik) N.S. **6**, 331–336 (1939)
10. Veanes, M., Hooimeijer, P., Livshits, B., et al.: Symbolic finite state transducers: algorithms and applications. In: Proc. of the 39th ACM SIGACT-SIGPLAN Symposium on Principles of Programming Languages (2012)
11. Sakarovitch, J., de Souza, R.: On the decomposition of k-valued rational relations. In: Proc. of 25th International Symposium on Theoretical Aspects of Computer Science, pp. 621–632 (2008)
12. Sakarovitch, J., de Souza, R.: On the decidability of bounded valuedness for transducers. In: Ochmański, E., Tyszkiewicz, J. (eds.) MFCS 2008. LNCS, vol. 5162, pp. 588–600. Springer, Heidelberg (2008)
13. Schutzenberger, M.P.: Sur les relations rationnelles. In: Brakhage, H. (ed.) GI-Fachtagung 1975. LNCS, vol. 33, pp. 209–213. Springer, Heidelberg (1975)
14. de Souza, R.: On the decidability of the equivalence for k-valued transducers. In: Ito, M., Toyama, M. (eds.) DLT 2008. LNCS, vol. 5257, pp. 252–263. Springer, Heidelberg (2008)
15. Weber, A.: On the valuedness of finite transducers. Acta Informatica **27**, 749–780 (1989)
16. Weber, A.: Decomposing finite-valued transducers and deciding their equivalence. SIAM Journal on Computing **22**, 175–202 (1993)
17. Zakharov, V.A.: An efficient and unified approach to the decidability of equivalence of propositional program schemes. In: Proc. of the 25th International Colloquium "Automata, Languages and Programming", pp. 247–258 (1998)

Author Index

Printed in the United States
By Bookmasters